Solutions Manual for Students

?SU

Calculus
and Analytic Geometry
Second Edition

Douglas F. Riddle

D1525748

Wadsworth Publishing Company, Inc., Belmont, California

ISBN 0-534-00501-2

Printed in the United States of America

1 2 3 4 5 6 7 8 9 10---80 79 78 77 76

Contents

Chapter One
Plane Analytic Geometry

1. $d = \sqrt{(2-1)^2 + (5+3)^2} = \sqrt{65}$

5. $d = \sqrt{(-\frac{5}{2} - \frac{1}{2})^2 + (2 - \frac{3}{2})^2} = \frac{\sqrt{37}}{2}$

9. $5 = \sqrt{(x-1)^2 + (2-5)^2}$, $\quad 25 = (x-1)^2 + 9$, $\quad x = 5, -3$

13. $A = (2, 1)$, $B = (4, 3)$, $C = (-1, -2)$;

$\overline{AB} = \sqrt{2^2 + 2^2} = 2\sqrt{2}$, $\quad \overline{BC} = \sqrt{5^2 + 5^2} = 5\sqrt{2}$,

$\overline{AC} = \sqrt{3^2 + 3^2} = 3\sqrt{2}$; $\quad \overline{AB} + \overline{AC} = \overline{BC}$, collinear

17. $A = (-1/2, 2/3)$, $B = (1/4, 3/5)$, $C = (7/4, 7/15)$;

$\overline{AB} = \sqrt{(\frac{3}{4})^2 + (\frac{1}{15})^2} = \frac{\sqrt{2041}}{60}$,

$\overline{BC} = \sqrt{(\frac{3}{2})^2 + (\frac{2}{15})^2} = \frac{2\sqrt{2041}}{60}$,

$\overline{AC} = \sqrt{(\frac{9}{4})^2 + (\frac{3}{15})^2} = \frac{3\sqrt{2041}}{60}$; $\quad \overline{AB} + \overline{BC} = \overline{AC}$, collinear

21. $A = (\sqrt{3} - 2, 2\sqrt{3} + 1)$, $B = (\sqrt{3} + 2, -\sqrt{3} + 1)$,

$C = (2\sqrt{3} - 2, 2\sqrt{3} + 2)$; $\quad \overline{AB} = \sqrt{4^2 + (3\sqrt{3})^2} = \sqrt{43}$,

$\overline{BC} = \sqrt{(\sqrt{3} - 4)^2 + (3\sqrt{3} + 1)^2} = \sqrt{47 - 2\sqrt{3}}$,

$AC = \sqrt{\sqrt{3}^2 + 1^2} = \sqrt{4}$; $\quad \overline{AB}^2 + \overline{AC}^2 \neq \overline{BC}^2$,

not a right triangle

25. $A = (1, 1)$, $B = (4, 1)$, $C = (3, -2)$, $D = (0, -2)$;

$\overline{AB} = \sqrt{3^2} = 3$, $\quad \overline{CD} = 3$; $\quad \overline{BC} = \sqrt{1^2 + 3^2} = \sqrt{10}$,

$\overline{DA} = \sqrt{1^2 + 3^2} = \sqrt{10}$

29. $A = (x_3 - x_2)(y_1 - y_3) - \frac{1}{2}(x_1 - x_2)(y_1 - y_2)$

$\quad - \frac{1}{2}(x_3 - x_2)(y_2 - y_3) - \frac{1}{2}(x_3 - x_1)(y_1 - y_3)$

$= \frac{1}{2}(2x_3 y_1 - 2x_3 y_3 - 2x_2 y_1 + 2x_2 y_3 - x_1 y_1 + x_1 y_2 + x_2 y_1$

$\quad - x_2 y_2 - x_3 y_2 + x_3 y_3 + x_2 y_2 - x_2 y_3 - x_3 y_1 + x_3 y_3$

$$+ x_1y_1 - x_1y_3)$$

$$= \frac{1}{2}(x_1y_2 + x_2y_3 + x_3y_1 - x_1y_3 - x_2y_1 - x_3y_2)$$

Other positions
of the vertices
give the same
results with
the subscripts
permuted. The
terms can then
be rearranged
to give either
the above result
or its negative.
Thus

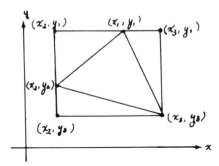

$$A = \frac{1}{2}\left| x_1y_2 + x_2y_3 + x_3y_1 - x_1y_3 - x_2y_1 - x_3y_2 \right|$$

Section 1.3, pp. 11-12

1. $x = 3 + \frac{1}{4}(7 - 3) = 4$, $y = 4 + \frac{1}{4}(0 - 4) = 3$

5. $x = -4 + 3(3 + 4) = 17$, $y = 1 + 3(8 - 1) = 22$

9. $x = \frac{4 + 3}{2} = \frac{7}{2}$, $y = \frac{-1 + 3}{2} = 1$

13. $\frac{x_1 + 4}{2} = 2$, $x_1 = 0$, $\frac{y_1 - 3}{2} = -5$, $y_1 = -7$

17. $r = \frac{2}{5}$; $x = 2 + \frac{2}{5}(4 - 2) = \frac{14}{5}$, $y = -1 + \frac{2}{5}(5 + 1) = \frac{7}{5}$

21. Midpoint $= (\frac{3 + x}{2}, 1)$

$5 = \sqrt{(\frac{1 + x}{2})^2 + 9}$, $25 = \frac{x^2 + 2x + 1}{4} + 9$, $x = 7, -9$

25. $A = (1, 1)$, $B = (4, 1)$, $C = (3, -2)$, $D = (0, -2)$

$X = $ midpt of $AC = (2, -\frac{1}{2})$

29. $M = (a/2, \ b/2)$

$\overline{OM} = \sqrt{a^2/4 + b^2/4}$

$\qquad = \left(\sqrt{a^2 + b^2}\right)/2$

$\overline{AM} = \overline{BM} = \sqrt{a^2/4 + b^2/4}$

$\qquad = \left(\sqrt{a^2 + b^2}\right)/2$

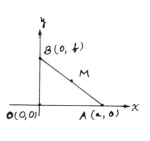

<u>Section 1.5, p. 16</u>

1. $m = \dfrac{8 - 3}{5 - 2} = \dfrac{5}{3}$, $\quad \tan \theta = 1.6667$, $\quad \theta = 59^{\circ}$

5. $m = \dfrac{5 - 2}{-4 + 4}$, no slope, $\quad \theta = 90^{\circ}$

9. $m_1 = \dfrac{-11 + 2}{-2 - 1} = 3$, $\quad m_2 = \dfrac{8 - 2}{2 - 0} = 3$

Taking a point from each line, $(1, -2)$, $(2, 8)$ gives

$m_3 = \dfrac{8 + 2}{2 - 1} = 10 \neq 3$, $\quad$ parallel

13. $m_1 = \dfrac{1 + 1}{1 - 4} = -\dfrac{2}{3}$, $\quad m_2 = \dfrac{3 + 3}{-2 - 7} = -\dfrac{2}{3}$

Taking a point from each line $(1, 1)$, $(-2, 3)$ gives

$m = \dfrac{1 - 3}{1 + 2} = -\dfrac{2}{3}$, $\quad$ coincident

17. $3 = \dfrac{5 - 3}{x - 4}$, $\quad 3x - 12 = 2$, $\quad x = \dfrac{14}{3}$

21. $\dfrac{4 - 7}{x - 3} = \dfrac{-1 - 1}{x - 5}$, $\quad \dfrac{-3}{x - 3} = \dfrac{-2}{x - 5}$, $\quad 2x - 6 = 3x - 15$

$x = 9$

25. $b = \sqrt{x^2 + a^2}$

$b^2 = x^2 + a^2$

$x^2 = b^2 - a^2$

$x = -\sqrt{b^2 - a^2}$

$m_{AC} = \dfrac{a}{b + \sqrt{b^2 - a^2}}$

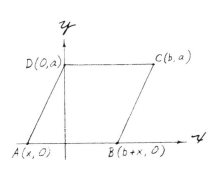

$$m_{BD} = \frac{-a}{b - \sqrt{b^2 - a^2}}$$

$$= \frac{-a(b + \sqrt{b^2 - a^2})}{b^2 - (b^2 - a^2)} = -\frac{b + \sqrt{b^2 - a^2}}{a}$$

Section 1.6, pp. 18-19

1.

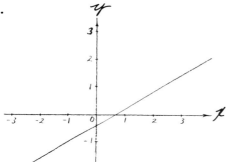

5.

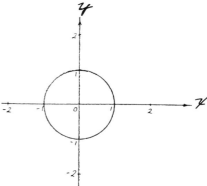

9.

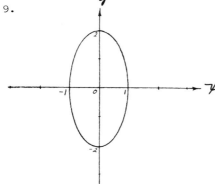

13.

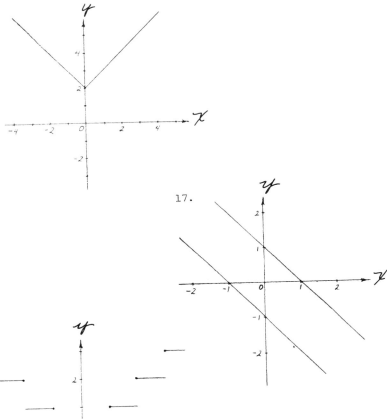

17.

21.

25.

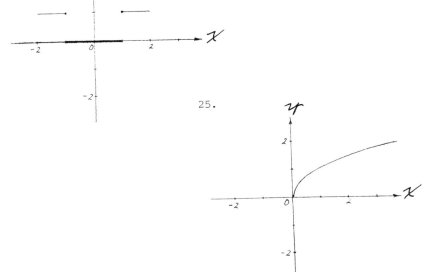

29. $\begin{cases} 3x - 5y = 2 \\ 4x + 2y = 1 \end{cases}$

$6x - 10y = 4$

$20x + 10y = 5$

$26x = 9$

$x = \dfrac{9}{26}$;

$\dfrac{27}{26} - 5y = \dfrac{52}{26}$

$y = -\dfrac{5}{26}$

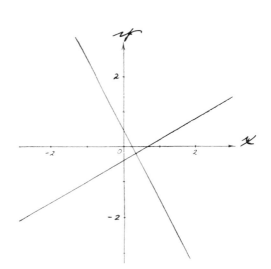

33. $\begin{cases} 2x + y = 2 \\ x^2 - y^2 = 1 \end{cases}$

$y = 2 - 2x$

$x^2 - 4 + 8x - 4x^2 = 1$

$3x^2 - 8x + 5 = 0$

$(3x - 5)(x - 1) = 0$

$x = \dfrac{5}{3}$, $\quad x = 1$,

$\dfrac{10}{3} + y = \dfrac{6}{3}$, $\quad 2 + y = 2$

$y = -\dfrac{4}{3}$, $\quad y = 0$

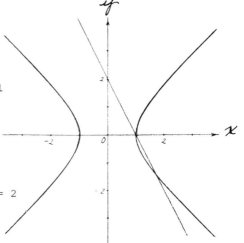

37. $\begin{cases} x^2 - y^2 = 1 \\ x^2 + y^2 = 7 \end{cases}$

$2x^2 = 8$

$x^2 = 4$

$x = \pm 2$,

$y^2 = 3$, $\quad y = \pm\sqrt{3}$

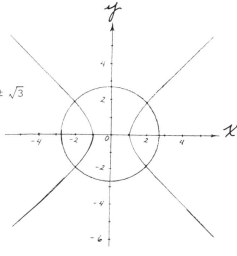

41.

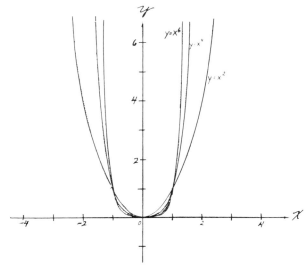

1. $A = (1, 5)$, $B = (-2, -1)$, $C = (4, 10)$

   $\overline{AB} = \sqrt{9 + 36} = \sqrt{45} = 3\sqrt{5}$, $\overline{BC} = \sqrt{36 + 121} = \sqrt{157}$,

   $\overline{AC} = \sqrt{9 + 25} = \sqrt{34}$. Since $\overline{BC} \neq \overline{AB} + \overline{AC}$, they are

   noncollinear. $m_{AB} = \dfrac{5 + 1}{1 + 2} = \dfrac{6}{3} = 2$, $m_{BC} = \dfrac{10 + 1}{4 + 2} = \dfrac{11}{6}$

   Since $m_{AB} \neq m_{BC}$, they are noncollinear.

5. $\begin{cases} 2x + y = 5 \\ x - 3y = 7 \end{cases}$

   $6x + 3y = 15$

   $x - 3y = 7$

   $7x = 22$

   $x = 22/7$

   $y = 5 - 2x$

   $\quad = \dfrac{35}{7} - \dfrac{44}{7}$

   $\quad = -\dfrac{9}{7}$

$(22/7, -9/7)$

9. $E = (\dfrac{a + b}{2}, \ 0)$

   $F = (\dfrac{b + d}{2}, \ \dfrac{e}{2})$

   $G = (\dfrac{d}{2}, \ \dfrac{c + e}{2})$

   $H = (\dfrac{a}{2}, \ \dfrac{c}{2})$

   $m_{EF} = \dfrac{e/2 - 0}{\dfrac{b + d}{2} - \dfrac{a + b}{2}}$

   $\quad = \dfrac{e}{d - a}$

   $m_{GH} = \dfrac{\dfrac{c + e}{2} - \dfrac{c}{2}}{\dfrac{d}{2} - \dfrac{a}{2}} = \dfrac{e}{d - a}$

$D(d, e)$

G

$C (0, c)$

F

H

$A(a, 0)$ E $B(b, 0)$

Since $m_{EF} = m_{GH}$, $\quad EF \parallel GH$

$$m_{EH} = \frac{c/2 - 0}{\dfrac{a}{2} - \dfrac{a+b}{2}} = -\frac{c}{b} \quad , \quad m_{FG} = \frac{\dfrac{c+e}{2} - \dfrac{e}{2}}{\dfrac{d}{2} - \dfrac{b+d}{2}} = -\frac{c}{b}$$

Since $m_{EH} = m_{FG}$, $\quad EH \parallel FG$

13. $D = $ Midpoint of $BC = (2, 1)$

$P = $ Point 2/3 of the way from A to D

$x = 3 + \dfrac{2}{3}(2 - 3) = \dfrac{7}{3}$

$y = 8 + \dfrac{2}{3}(1 - 8) = \dfrac{10}{3}$

$P = (7/3, \ 10/3)$

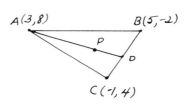

17. $m_{AB} = \dfrac{3 - 1}{6 - 3} = \dfrac{2}{3}$

$\overline{AB} = \sqrt{(6 - 3)^2 + (3 - 1)^2}$

$\quad = \sqrt{13}$

$m_{AC} = \dfrac{y_1 - 1}{x_1 - 3} = -\dfrac{3}{2}$

$y_1 - 1 = -\dfrac{3(x_1 - 3)}{2}$

$\overline{AC} = \sqrt{(x_1 - 3)^2 + (y_1 - 1)^2}$

$\quad = \sqrt{13}$

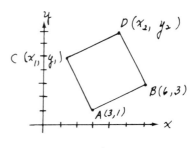

$(x_1 - 3)^2 + (y_1 - 1)^2 = 13$, $\quad (x_1 - 3)^2 + \dfrac{9(x_1 - 3)^2}{4} = 13$

$(x_1 - 3)^2 = 4$, $\quad x_1 - 3 = \pm 2$, $\quad x_1 = 3 \pm 2 = 5$ or 1

If $x_1 = 5$, then $y_1 = -2$ and $C = (5, -2)$. Since this is not in the first quadrant, x_1 must be 1. Thus $y_1 = 4$ and $C = (1, 4)$.

$m_{BD} = \dfrac{y_2 - 3}{x_2 - 6} = -\dfrac{3}{2}$ $\qquad\qquad m_{CD} = \dfrac{y_2 - 4}{x_2 - 1} = \dfrac{2}{3}$

$$2y_2 - 6 = -3x_2 + 18 \qquad\qquad 2x_2 - 2 = 3y_2 - 12$$

$$3x_2 + 2y_2 = 24 \qquad\qquad\qquad 2x_2 - 3y_2 = -10$$

Solving simultaneously, we have $x_2 = 4$ and $y_2 = 6$. Thus $D = (4, 6)$.

Chapter Two
Equations of Lines and Circles

Section 2.1, pp. 22-24

1. $y + 4 = -2(x - 2)$

   $2x + y = 0$

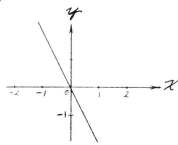

5.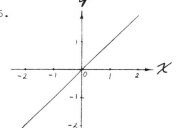

   $y = x$

   $x - y = 0$

9. $m = \dfrac{4 - 5}{1 - 3} = \dfrac{1}{2}$

   $y - 4 = \dfrac{1}{2}(x - 1)$

   $x - 2y + 7 = 0$

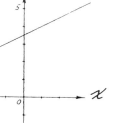

13. $m = \dfrac{5}{1} = 5$

$y = 5x$

$5x - y = 0$

17. (1, 4), (3, 0)   (3, 0), (-1, -2)   (-1, -2), (1, 4)

$m = \dfrac{4 - 0}{1 - 3} = -2$ $m = \dfrac{0 + 2}{3 + 1} = \dfrac{1}{2}$ $m = \dfrac{-2 - 4}{-1 - 1} = 3$

$y - 4 = -2(x - 1)$ $y = \dfrac{1}{2}(x - 3)$ $y + 2 = 3(x + 1)$

$2x + y - 6 = 0$ $x - 2y - 3 = 0$ $3x - y + 1 = 0$

21. Vertices (from Prob. 20): (2, 2), (1, -2), (-3, 1)
Midpoints: (-1, -1/2), (-1/2, 3/2), (3/2, 0)

(2, 2), (-1, -1/2)   (1, -2), (-1/2, 3/2)   (-3, 1), (3/2, 0)

$m = \dfrac{2 + 1/2}{2 + 1} = \dfrac{5}{6}$ $m = \dfrac{3/2 + 2}{-1/2 - 1} = \dfrac{-7}{3}$ $m = \dfrac{1 - 0}{-3 - 3/2} = -\dfrac{2}{9}$

$y - 2 = \dfrac{5}{6}(x - 2)$ $y + 2 = -\dfrac{7}{3}(x - 1)$ $y - 1 = -\dfrac{2}{9}(x + 3)$

$5x - 6y + 2 = 0$ $7x + 3y - 1 = 0$ $2x + 9y - 3 = 0$

25. $m_1 = \dfrac{2 - 6}{4 + 2} = -\dfrac{2}{3}$, $m_2 = \dfrac{3}{2}$ Midpoint: (1, 4)

$y - 4 = \dfrac{3}{2}(x - 1)$, $3x - 2y + 5 = 0$

29. $\sqrt{(x - 2)^2 + (y - 5)^2} = \sqrt{(x - 4)^2 + (y + 1)^2}$

$x^2 - 4x + 4 + y^2 - 10y + 25 = x^2 - 8x + 16 + y^2 + 2y + 1$

$4x - 12y + 12 = 0$, $x - 3y + 3 = 0$

33. $(0, 32)$, $(100, 212)$, $m = \dfrac{32 - 212}{0 - 100} = \dfrac{9}{5}$

   $F - 32 = \dfrac{9}{5}(C - 0)$ $9C - 5F + 160 = 0$

37. Since no two of the three lines are parallel, $\dfrac{A_1}{B_1} \neq \dfrac{A_2}{B_2}$,

   $\dfrac{A_2}{B_2} \neq \dfrac{A_3}{B_3}$, $\dfrac{A_3}{B_3} \neq \dfrac{A_1}{B_1}$. Thus $A_1 B_2 \neq B_1 A_2$ and

   $A_1 B_2 - B_1 A_2 \neq 0$. Solving the first two equations

   simultaneously, we have

$$x = -\frac{\begin{vmatrix} C_1 & B_1 \\ C_2 & B_2 \end{vmatrix}}{\begin{vmatrix} A_1 & B_1 \\ A_2 & B_2 \end{vmatrix}}$$

$$y = -\frac{\begin{vmatrix} A_1 & C_1 \\ A_2 & C_2 \end{vmatrix}}{\begin{vmatrix} A_1 & B_1 \\ A_2 & B_2 \end{vmatrix}}$$

   If the lines are concurrent, the above point of
   intersection is on the third line. Thus

$$-A_3 \frac{\begin{vmatrix} C_1 & B_1 \\ C_2 & B_2 \end{vmatrix}}{\begin{vmatrix} A_1 & B_1 \\ A_2 & B_2 \end{vmatrix}} \quad -B_3 \frac{\begin{vmatrix} A_1 & C_1 \\ A_2 & C_2 \end{vmatrix}}{\begin{vmatrix} A_1 & B_1 \\ A_2 & B_2 \end{vmatrix}} \quad + C_3 = 0$$

$$A_3 \begin{vmatrix} B_1 & C_1 \\ B_2 & C_2 \end{vmatrix} - B_3 \begin{vmatrix} A_1 & C_1 \\ A_2 & C_2 \end{vmatrix} + C_3 \begin{vmatrix} A_1 & B_1 \\ A_2 & B_2 \end{vmatrix} = 0$$

$$\begin{vmatrix} A_1 & B_1 & C_1 \\ A_2 & B_2 & C_2 \\ A_3 & B_3 & C_3 \end{vmatrix} = 0$$

The above steps are reversible.

Section 2.2, pp. 27-28

1. $y = 4x + 2$

   $4x - y + 2 = 0$

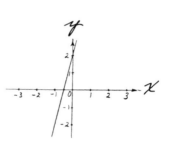

5. $y = \frac{3}{4}x + \frac{2}{3}$

   $9x - 12y + 8 = 0$

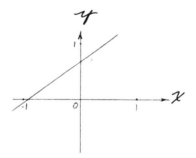

9. $\frac{x}{4} + \frac{y}{2} = 1$

   $x + 2y - 4 = 0$

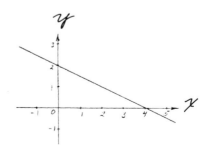

13. $\dfrac{x}{2/3} + \dfrac{y}{-2/5} = 1$

   $3x - 5y - 2 = 0$

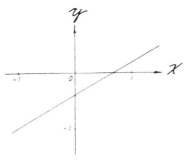

17. $x = 4$

   $x - 4 = 0$

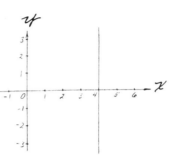

21. $m_1 = -1/2$, $m_2 = 2$, $y - 1 = 2(x - 4)$, $2x - y - 7 = 0$

25. Altitude 1   Altitude 2

   $(1, 4)$, $(7, 3)$ $(7, 3)$, $(2, -3)$

   $m_1 = \dfrac{4-3}{1-7} = -\dfrac{1}{6}$, $m_2 = 6$ $m_1 = \dfrac{3+3}{7-2} = \dfrac{6}{5}$, $m_2 = -\dfrac{5}{6}$

   $y + 3 = 6(x - 2)$ $y - 4 = -\dfrac{5}{6}(x - 1)$

   $6x - y - 15 = 0$ $5x + 6y - 29 = 0$

   Solving simultaneously, we have

$$36x - 6y = 90$$
$$5x + 6y = 29$$
$$x = \frac{119}{41}, \quad y = \frac{99}{41}$$

33. $(x - 2)(x - 3) = 0$

$x = 2, \quad x = 3$

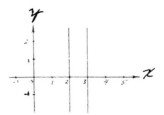

37. $\begin{vmatrix} x & y & 1 \\ x_1 & y_1 & 1 \\ x_2 & y_2 & 1 \end{vmatrix} = 0$ is linear in x and y. By Theorem

2.5 it represents a line. The points (x_1, y_1) and (x_2, y_2) satisfy the equation. (If two rows of a determinant are equal, the value of the determinant is 0). Thus the equation represents the line through (x_1, y_1) and (x_2, y_2).

Section 2.3, pp. 32-34

1. $\dfrac{|2 + 5 - 5|}{\sqrt{1^2 + 1^2}} = \dfrac{2}{\sqrt{2}} = \sqrt{2}$

5. $\dfrac{|3 \cdot 1 + 4 \cdot 1 - 5|}{\sqrt{9 + 16}} = \dfrac{2}{5}$

9. $\dfrac{|3 \cdot 2 + 4|}{3} = \dfrac{10}{3}$

13. $\dfrac{3x - 4y - 2}{\sqrt{9 + 16}} = -\dfrac{4x - 3y + 4}{\sqrt{9 + 16}}, \quad 7x - 7y + 2 = 0$

17. $\dfrac{x + y - 2}{\sqrt{1 + 1}} = -\dfrac{2x - 3}{2}, \quad 2(x + y - 2) = -\sqrt{2}(2x - 3)$

$(2\sqrt{2} + 2)x + 2y - (4 + 3\sqrt{2}) = 0$

21. Dividing the second equation through by 2, we have a constant term of $-3/2$. Thus

$$d = \frac{2 - (-3/2)}{\sqrt{4 + 1}} = \frac{7}{2\sqrt{5}}$$

25. $A = \frac{1}{2} bh = \frac{1}{2} \sqrt{4^2 + 3^2} \, \frac{26}{5} = 13$

29. $mx - y + 1 = 0$, $\quad 3 = \frac{|4m - 1 + 1|}{\sqrt{m^2 + 1}}$

$9(m^2 + 1) = 16m^2$, $\quad m^2 = \frac{9}{7}$, $\quad m = \pm \frac{3}{\sqrt{7}}$

33. Prove that if $P(x_1, \ y_1)$ is a point not on the line $Ax + By + C = 0$ ($C \neq 0$), then

(a) C and $Ax_1 + By_1 + C$ agree in sign if P and the origin are on the same side of the line;

(b) C and $Ax_1 + By_1 + C$ have opposite signs if P and the origin are on opposite sides of the line.

Either $A \neq 0$ or $B \neq 0$. Suppose $B \neq 0$. If P is on the same side as the origin O, then either P and O are both above or both below the line. Thus $Ax_1 + By_1 + C$ and $A \cdot 0 + B \cdot 0 + C = C$ have the same sign. If P and O are on opposite sides of the line, then one is above and the other below the line. Thus $Ax_1 + By_1 + C$ and $A \cdot 0 + B \cdot 0 + C = C$ have opposite signs.

If $B = 0$, then $A \neq 0$ and the same argument can be used, using right and left of the line instead of above and below (see Problem 32).

1. $(x - 1)^2 + (y - 3)^2 = 25$

    $x^2 + y^2 - 2x - 6y - 15 = 0$

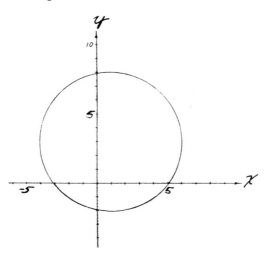

5. $(x - \frac{1}{2})^2 + (y + \frac{3}{2})^2 = 4$

    $x^2 + y^2 - x + 3y - \frac{3}{2} = 0$

    $2x^2 + 2y^2 - 2x + 6y - 3 = 0$

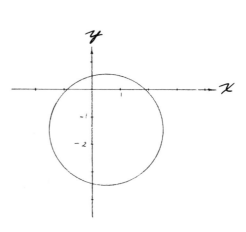

9. $C = (0, -\frac{3}{2})$

$r^2 = 2^2 + (\frac{3}{2})^2 = \frac{25}{4}$

$x^2 + (y + \frac{3}{2})^2 = \frac{25}{4}$

$x^2 + y^2 + 3y - 4 = 0$

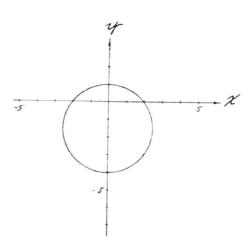

13. $(x - 4)^2 + (y - 1)^2 = 4$

$x^2 + y^2 - 8x - 2y + 13 = 0$

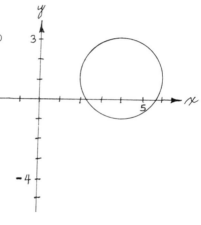

17. $x^2 + y^2 - 2x - 4y + 1 = 0$

$x^2 - 2x + 1 + y^2 - 4y + 4$

$= -1 + 1 + 4$

$(x - 1)^2 + (y - 2)^2 = 4$

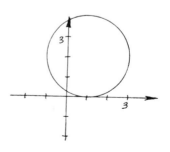

21. $4x^2 + 4y^2 - 4x - 12y + 1 = 0$

$x^2 + y^2 - x - 3y + \frac{1}{4} = 0$

$x^2 - x + \frac{1}{4} + y^2 - 3y + \frac{9}{4}$

$\qquad = -\frac{1}{4} + \frac{1}{4} + \frac{9}{4}$

$(x - \frac{1}{2})^2 + (y - \frac{3}{2})^2 = \frac{9}{4}$

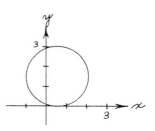

25. $9x^2 + 9y^2 - 6x + 18y + 11 = 0.$

$x^2 + y^2 - \frac{2}{3}x + 2y + \frac{11}{9} = 0$

$x^2 - \frac{2}{3}x + \frac{1}{9} + y^2 + 2y + 1 = -\frac{11}{9} + \frac{1}{9} + \frac{9}{9}$

$(x - \frac{1}{3})^2 + (y + 1)^2 = -\frac{1}{9}$

29. $x^2 + y^2 - x - 3y - 6 = 0$, $\quad 4x - y - 9 = 0$

$y = 4x - 9$, $\quad x^2 + 16x^2 - 72x + 81 - x - 12x + 27 - 6 = 0$

$17x^2 - 85x + 102 = 0$, $\quad x^2 - 5x + 6 = 0$

$(x - 3)(x - 2) = 0$

$x = 3 \qquad\quad x = 2$

$y = 3 \qquad\quad y = -1$

33. $x = 2y - 2$, $\quad 4y^2 - 8y + 4 + y^2 - 4y + 4 + 4y + 1 = 0$

$5y^2 - 8y + 9 = 0$, $\quad y = \dfrac{8 \pm \sqrt{64 - 180}}{2} = 4 \pm \sqrt{-29}$

The circle and line do not intersect.

37. $x^2 + y^2 = a^2$

$$m_1 = \frac{y}{x + a}$$

$$m_2 = \frac{y}{x - a}$$

$$m_1 m_2 = \frac{y^2}{x^2 - a^2} = \frac{y^2}{-y^2} = -1$$

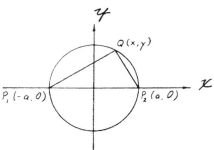

Review 2, pp. 38-39

1. (a) $m = \frac{5 - 3}{1 + 2} = \frac{2}{3}$ (b) $y - 0 = 2(x - 3)$

$y - 5 = \frac{2}{3}(x - 1)$ $2x - y - 6 = 0$

$2x - 3y + 13 = 0$

(c) $m = -1, \quad b = 1/3$ (d) $x = 2$

$y = -x + 1/3$ $x - 2 = 0$

$3x + 3y - 1 = 0$

5. $D = (1, 1), \ E = (5/2, 2), \ F = (-1/2, 4)$

AD: $(1,5), (1,1)$ BE: $(-2,3), (5/2,2)$ CF: $(4,-1), (-1/2,4)$

$x = 1$ $m = \frac{3 - 2}{-2 - 5/2} = -\frac{2}{9}$ $m = \frac{-1 - 4}{4 + 1/2} = -\frac{10}{9}$

$x - 1 = 0$ $y - 3 = -\frac{2}{9}(x + 2)$ $y + 1 = -\frac{10}{9}(x - 4)$

$2x + 9y - 23 = 0$ $10x + 9y - 31 = 0$

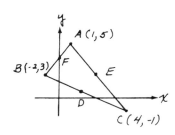

9. $A = (1, 5)$, $B = (-2, 3)$ $\qquad$ $B = (-2, 3)$, $C = (4, -1)$

$m_{AB} = \dfrac{5 - 3}{1 + 2} = \dfrac{2}{3}$ $\qquad\qquad$ $m_{BC} = \dfrac{3 + 1}{-2 - 4} = -\dfrac{2}{3}$

$y - 5 = \dfrac{2}{3}(x - 1)$ $\qquad\qquad$ $y - 3 = -\dfrac{2}{3}(x + 2)$

$2x - 3y + 13 = 0$ $\qquad\qquad$ $2x + 3y - 5 = 0$

$A = (1, 5)$, $C = (4, -1)$ $\qquad$ $A = (1, 5)$, BC: $2x + 3y - 5 = 0$

$m_{AC} = \dfrac{5 + 1}{1 - 4} = -2$ $\qquad\qquad$ $h_1 = \dfrac{|2 \cdot 1 + 3 \cdot 5 - 5|}{\sqrt{4 + 9}} = \dfrac{12}{\sqrt{13}}$

$y - 5 = -2(x - 1)$

$2x + y - 7 = 0$

$B = (-2, 3)$, AC: $2x + y - 7 = 0$

$h_2 = \dfrac{|2(-2) + 3 - 7|}{\sqrt{4 + 1}} = \dfrac{8}{\sqrt{5}}$

$C = (4, -1)$, AB: $2x - 3y + 13 = 0$

$h_3 = \dfrac{|2 \cdot 4 - 3(-1) + 13|}{\sqrt{4 + 9}} = \dfrac{24}{\sqrt{13}}$

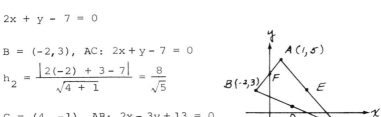

13. $\dfrac{x + y - 4}{\sqrt{2}} = \dfrac{x - 7y + 2}{\sqrt{50}}$

$5(x + y - 4) = x - 7y + 2$

$2x + 6y - 11 = 0$

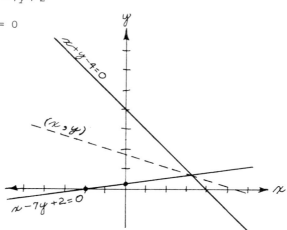

17. $M = \left(\dfrac{a + b}{2} \ , \ \dfrac{\sqrt{r^2 - a^2} + \sqrt{r^2 - b^2}}{2} \right)$

$m_{AB} = \dfrac{\sqrt{r^2 - b^2} - \sqrt{r^2 - a^2}}{b - a}$

$m_{MC} = \dfrac{a - b}{\sqrt{r^2 - b^2} - \sqrt{r^2 - a^2}}$

Chapter Three
Functions

1. Function   5. Function

9. No function   13. $D = \{x \mid x \neq \pm 1\}$

17.

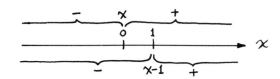

$D = \{x \mid x \geqq 1 \text{ or } x < 0\}$

21. $D = \{x \mid x \geqq 2\}$, $R = \{y \mid y \geqq 0\}$

25. $D = \{x \mid x > 0\}$, $R = \{y \mid y > 0\}$

29. $D = \{x \mid x \neq 0\}$, $R = \{y \mid y \geqq 1 \text{ or } y < 0\}$

33. $f(x) = x - 3$, $D = \{x \mid 0 < x \leqq 6\}$, $R = \{y \mid -3 < y \leqq 3\}$

37. $f(x) = 2x$, $D = \{x \mid x \text{ a positive integer}\}$

$R = \{y \mid y \text{ an even positive integer}\}$

41. $f(-1) = -1$, $f(0)$ does not exist

$f(2) = 1/2$, $f(x + 1) = 1/(x + 1)$

45. $f(y) = y^2 + 1$, $f(x + h) = (x + h)^2 + 1$

49. $A = \pi r^2$, $C = 2\pi r$

53. $h = 2r$

$$V = \frac{1}{3} \pi r^2 h$$
$$= \frac{1}{3} \pi r^2 \cdot 2r$$
$$= \frac{2}{3} \pi r^3$$

24   *Section 3.1*

57. $f(2t + 5) = f(t^2 - 3)$, $a(2t + 5) + b = a(t^2 - 3) + b$

$2t + 5 = t^2 - 3$, $t^2 - 2t - 8 = 0$

$(t - 4)(t + 2) = 0$, $t = 4,$ $t = -2$

Section 3.2, pp. 49-50

1. $\dfrac{f(x + h) - f(x)}{h} = \dfrac{(x + h)^2 + 1 - (x^2 + 1)}{h}$

$\qquad = \dfrac{x^2 + 2hx + h^2 + 1 - x^2 - 1}{h} = \dfrac{2hx + h^2}{h} = 2x + h$

5. $\dfrac{f(2 + h) - f(2)}{h} = \dfrac{2(2 + h) + 3 - 7}{h} = \dfrac{2h}{h} = 2$

9. $\dfrac{f(x) - f(1)}{x - 1} = \dfrac{1/x - 1}{x - 1} = \dfrac{1 - x}{x(x - 1)} = -\dfrac{1}{x}$

13. $\dfrac{f(h) - f(0)}{h} = \dfrac{h \sin 1/h - 0}{h} = \sin 1/h$

17. $\dfrac{1}{f(x) + f(y)} = \dfrac{1}{1/x + 1/y} = \dfrac{xy}{y + x}$

21. $(f + g)(x) = \sqrt{x} + \sqrt{1 - x}$, $(f - g)(x) = \sqrt{x} - \sqrt{1 - x}$

$(fg)(x) = \sqrt{x}\,\sqrt{1 - x} = \sqrt{x - x^2}$

$(f/g)(x) = \dfrac{\sqrt{x}}{\sqrt{1 - x}} = \sqrt{\dfrac{x}{1 - x}}$

$(g \circ f)(x) = g(f(x)) = g(\sqrt{x}) = \sqrt{1 - \sqrt{x}}$

25. $(f + g)(x) = x^3 + x + 4$ $\qquad\qquad$ $(0 \leqq x < 2)$

$(f - g)(x) = x^3 - x - 4$ $\qquad\qquad$ $(0 \leqq x < 2)$

$(fg)(x) = x^3(x + 4) = x^4 + 4x^3$ $\qquad$ $(0 \leqq x < 2)$

$(f/g)(x) = \dfrac{x^3}{x + 4}$ $\qquad\qquad\qquad$ $(0 \leqq x < 2)$

$(g \circ f)(x) = g(x^3) = x^3 + 4$ $\qquad\;\,$ $(0 \leqq x < 2)$

29. $f(x) = 1/x$, $f(x)g(x) = x$, $g(x) = \dfrac{x}{1/x} = x^2$

1. $\lim\limits_{x \to 1} x^2 = 1$

5. $\lim\limits_{x \to 2} \dfrac{x^2 - 1}{x - 1} = \lim\limits_{x \to 2} (x + 1) = 3$

9. $\lim\limits_{x \to 2} |x| = 2$

13. $\lim\limits_{x \to 1} f(x) = \lim\limits_{x \to 1} (x + 1) = 2$

17. $\lim\limits_{h \to 0} \dfrac{(x + h + 1)^2 - (x + 1)^2}{h}$

    $= \lim\limits_{h \to 0} \dfrac{(x + 1)^2 + 2h(x + 1) + h^2 - (x + 1)^2}{h}$

    $= \lim\limits_{h \to 0} [2(x + 1) + h] = 2(x + 1)$

21. $\lim\limits_{h \to 0} \dfrac{\sqrt{x + h} - \sqrt{x}}{h} = \lim\limits_{h \to 0} \dfrac{x + h - x}{h(\sqrt{x + h} + \sqrt{x})} = \lim\limits_{h \to 0} \dfrac{h}{h(\sqrt{x + h} + \sqrt{x})}$

    $= \lim\limits_{h \to 0} \dfrac{1}{\sqrt{x + h} - \sqrt{x}} = \dfrac{1}{2\sqrt{x}}$

25. $\lim\limits_{h \to 0} \dfrac{|-1 + h| - 1}{h} = \lim\limits_{h \to 0} \dfrac{1 - h - 1}{h} \quad (h < 1)$

    $= \lim\limits_{h \to 0} (-1) = -1$

29. $\lim\limits_{x \to a} \dfrac{\sqrt{x} - \sqrt{a}}{x - a} = \lim\limits_{x \to a} \dfrac{x - a}{(x - a)(\sqrt{x} + \sqrt{a})} = \lim\limits_{x \to a} \dfrac{1}{\sqrt{x} + \sqrt{a}} = \dfrac{1}{2\sqrt{a}}$

Review 3, p. 54

1. (a) $D = \{x \mid x \text{ real}\}$, $R = \{y \mid y \text{ real}, y \geqq 2\}$

    (b) $D = \{x \mid x \text{ real}\}$, $R = \{y \mid y \text{ real}, y \geqq 0\}$

(c) $D = \{x \mid x \text{ real}, x \neq -5\}$, $R = \{y \mid y \text{ real}, y \neq 0\}$

(d) $f(x) = \dfrac{x^2}{x^2 + 1} = 1 - \dfrac{1}{x^2 + 1}$

$D = \{x \mid x \text{ real}\}$, $R = \{y \mid y \text{ real}, 0 \leqq y < 1\}$

(e) $D = \{x \mid x \text{ real}\}$, $R = \{y \mid y \text{ real}\}$

(f)

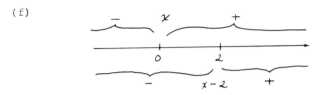

$D = \{x \mid x \text{ real}, x \leqq 0 \text{ or } x \geqq 2\}$, $R = \{y \mid y \text{ real}, y \geqq 0\}$

5. (a) $f(0) = 0$, (b) $f(4) = 2$

(c) $f(-1)$ does not exist, (d) $f(x^2) = \sqrt{x^2} = |x|$

(e) $\dfrac{f(1 + h) - f(1)}{h} = \dfrac{\sqrt{1 + h} - 1}{h} = \dfrac{1 + h - 1}{h(\sqrt{1 + h} + 1)} = \dfrac{1}{\sqrt{1 + h} + 1}$

9. $(f \circ g)(x) = f(g(x)) = f(-x) = -2x + 3$ $\qquad (x \leqq 0)$

$(g \circ f)(x) = g(f(x)) = g(2x + 3) = -2x - 3$

$2x + 3 \leqq 0$, $\qquad x \leqq -3/2$

But since the domain of f is $x \geqq 0$, the domain of $g \circ f$ is empty; $g \circ f$ does not exist.

Chapter Four
The Derivative

1. $A = (-1, 1)$, $P = (-1 + h, (-1 + h)^2)$

$$m_{AP} = \frac{(-1 + h)^2 - 1}{-1 + h + 1} = \frac{-2h + h^2}{h} = -2 + h \rightarrow -2$$

5. $A = (2, 4)$, $P = (2 + h, 3(2 + h)^2 - 5(2 + h) + 2)$

$$m_{AP} = \frac{12 + 12h + 3h^2 - 10 - 5h + 2 - 4}{2 + h - 2} = \frac{7h + 3h^2}{h} = 7 + 3h \rightarrow 7$$

9. $A = (0, -1)$, $P = (h, \frac{h + 1}{h - 1})$

$$m_{AP} = \frac{\frac{h + 1}{h - 1} + 1}{h} = \frac{h + 1 + h - 1}{h(h - 1)} = \frac{2}{h - 1} \rightarrow -2$$

13. $A = (2, 4)$, $P = (2 + h, (2 + h)^2)$

$$m_{AP} = \frac{(2 + h)^2 - 4}{2 + h - 2} = \frac{4h + h^2}{h} = 4 + h \rightarrow 4$$

Tangent:

$y - 4 = 4(x - 2)$
$y - 4 = 4x - 8$
$4x - y - 4 = 0$

Normal:

$y - 4 = -\frac{1}{4}(x - 2)$

$x + 4y - 18 = 0$

17. $A = (3, -7)$, $P = (3 + h, (h - 1)(2h + 7))$

$$m_{AP} = \frac{2h^2 + 5h - 7 + 7}{3 + h - 3} = \frac{2h^2 + 5h}{h} = 2h + 5 \rightarrow 5$$

Tangent:

$y + 7 = 5(x - 3)$

$y + 7 = 5x - 15$

$5x - y - 22 = 0$

Normal:

$y + 7 = -1/5(x - 3)$

$5y + 35 = -x + 3$

$x + 5y + 32 = 0$

21. $A = (x, x^2 - 4x)$, $P = (x + h, (x + h)^2 - 4(x + h))$

$$m_{AP} = \frac{x^2 + 2hx + h^2 - 4x - 4h - x^2 + 4x}{x + h - x}$$

$$= \frac{2hx + h^2 - 4h}{h} = 2x + h - 4 \rightarrow 2x - 4 = 0$$

$$x = 2, \quad y = -4$$

25. $f(x) = |x|$, $A = (0, 0)$

As P approaches A from the right $m_{AP} = 1 \rightarrow 1$
As P approaches A from the left $m_{AP} = -1 \rightarrow -1$
(see problem 19)

Section 4.2, pp. 65-66

1. $f'(x) = \lim_{h \to 0} \frac{f(x + h) - f(x)}{h}$

$$= \lim_{h \to 0} \frac{[2(x + h)^2 - 4(x + h) + 1] - [2x^2 - 4x + 1]}{h}$$

$$= \lim_{h \to 0} \frac{2x^2 + 4hx + 2h^2 - 4x - 4h + 1 - 2x^2 + 4x - 1}{h}$$

$$= \lim_{h \to 0} (4x + 2h - 4) = 4x - 4$$

5. $f(s) = s^4 + 2s^2 + 1$

$f'(s) = \lim_{h \to 0} \frac{f(s + h) - f(s)}{h}$

$$= \lim_{h \to 0} \frac{[(s + h)^4 + 2(s + h)^2 + 1] - [s^4 + 2s^2 + 1]}{h}$$

$$= \lim_{h \to 0} \frac{s^4 + 4hs^3 + 6h^2s^2 + 4h^3s + h^4 + 2s^2 + 4hs + 2h^2 + 1 - s^4 - 2s^2 - 1}{h}$$

$$= \lim_{h \to 0} (4s^3 + 6hs^2 + 4h^2s + h^3 + 4s + 2h) = 4s^3 + 4s$$

9. $f'(v) = \lim_{h \to 0} \frac{f(v + h) - f(v)}{h}$

$$= \lim_{h \to 0} \frac{\dfrac{v + h - 4}{v + h + 1} - \dfrac{v - 4}{v + 1}}{h}$$

$$= \lim_{h \to 0} \frac{(v + h - 4)(v + 1) - (v - 4)(v + h + 1)}{h(v + h + 1)(v + 1)}$$

$$= \lim_{h \to 0} \frac{v^2 + hv - 4v + v + h - 4 - v^2 - hv - v + 4v + 4h + 4}{h(v + h + 1)(v + 1)}$$

$$= \lim_{h \to 0} \frac{5}{(v + h + 1)(v + 1)} = \frac{5}{(v + 1)^2}$$

$$f'(0) = 5 \quad f'(2) = 5/9$$

13. $f'(x) = \lim_{h \to 0} \dfrac{f(x + h) - f(x)}{h}$

$$= \lim_{h \to 0} \frac{\dfrac{(x+h)^2 + 1}{x + h} - \dfrac{x^2 + 1}{x}}{h}$$

$$= \lim_{h \to 0} \frac{x^3 + 2hx^2 + h^2x + x - x^3 - x - hx^2 - h}{hx(x + h)}$$

$$= \lim_{h \to 0} \frac{x^2 + hx - 1}{x(x + h)} = \frac{x^2 - 1}{x^2}$$

$$\frac{x^2 - 1}{x^2} = 0, \quad x^2 = 1, \quad x = \pm 1, \quad (1, \ 2), \quad (-1, \ -2)$$

17. $f'(x) = \lim_{h \to 0} \dfrac{f(x + h) - f(x)}{h} = \lim_{h \to 0} \dfrac{(x+h)^3 - x^3}{h}$

$$= \lim_{h \to 0} \frac{x^3 + 3hx^2 + 3h^2x + h^3 - x^3}{h}$$

$$= \lim_{h \to 0} (3x^2 + 3hx + h^2) = 3x^2$$

$$y' = 3x^2$$

$$3x^2 = 1, \quad x = \pm \frac{1}{\sqrt{3}} \ \left(\pm \frac{1}{\sqrt{3}}, \ \frac{1}{3\sqrt{3}} \right)$$

21. $s = t^2 + t - 1$

$$v = \frac{ds}{dt} = \lim_{h \to 0} \frac{f(t+h) - f(t)}{h} = \lim_{h \to 0} \frac{[(t+h)^2 + (t+h) - 1] - [t^2 + t - 1]}{h}$$

$$= \lim_{h \to 0} \frac{t^2 + 2ht + h^2 + t + h - 1 - t^2 - t + 1}{h} = \lim_{h \to 0} \frac{2ht + h^2 + h}{h}$$

$$= \lim_{h \to 0} (2t + h + 1) = 2t + 1$$

At $t = 4$, $s = 19$ and $v = 9$

30   *Section 4.2*

25. $s = t^4 - 1$

$v = \dfrac{ds}{dt} = \lim_{h \to 0} \dfrac{f(t+h) - f(t)}{h} = \lim_{h \to 0} \dfrac{[(t+h)^4 - 1] - [t^4 - 1]}{h}$

$= \lim_{h \to 0} \dfrac{t^4 + 4ht^3 + 6h^2t^2 + 4h^3t + h^4 - 1 - t^4 + 1}{h}$

$= \lim_{h \to 0} \dfrac{4ht^3 + 6h^2t^2 + 4h^3t + h^4}{h} = \lim_{h \to 0} (4t^3 + 6ht^2 + 4h^2t + h^3)$

$= 4t^3$

At $t = 2$, $s = 15$ and $v = 32$

29. $s = t^2(t+1)^2 = t^4 + 2t^3 + t^2$

$v = \dfrac{ds}{dt} = \lim_{h \to 0} \dfrac{f(t+h) - f(t)}{h}$

$= \lim_{h \to 0} \dfrac{[(t+h)^4 + 2(t+h)^3 + (t+h)^2] - [t^4 + 2t^3 + t^2]}{h}$

$= \lim_{h \to 0} \dfrac{\begin{array}{c} t^4 + 4ht^3 + 6h^2t^2 + 4h^3t + h^4 + 2t^3 + 6ht^2 + 6h^2t + 2h^3 + t^2 \\ + 2ht + h^2 - t^4 - 2t^3 - t^2 \end{array}}{h}$

$= \lim_{h \to 0} (4t^3 + 6ht^2 + 4h^2t + h^3 + 6t^2 + 6ht + 2h^2 + 2t)$

$= 4t^3 + 6t^2 + 2t$

At $t = 3$, $s = 144$ and $v = 168$

33. $s = -At^2 + Bt + C$ $\qquad\qquad 0 \leqq t \leqq ?$

$v = -2At + B = 0$, $\quad t = \dfrac{B}{2A}$

$s = -\dfrac{AB^2}{4A^2} + \dfrac{B^2}{2A} + C = \dfrac{B^2 + 4AC}{4A}$ ft

$-AT^2 + Bt + C = 0$

$t = \dfrac{-B \pm \sqrt{B^2 + 4AC}}{-2A} = \dfrac{B + \sqrt{B^2 + 4AC}}{2A}$ sec

(only the - is used since $t \geqq 0$)

$v = -B - \sqrt{B^2 + 4AC} + B = -\sqrt{B^2 + 4AC}$ ft/sec

37. 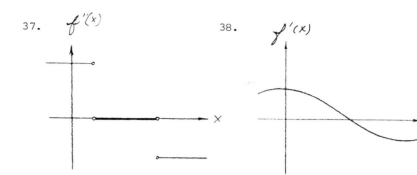   38.

41. $P = R - C = -7000 - 30x + 0.6x^2 - 0.001x^3$

$\dfrac{dP}{dx} = -30 + 1.2x - 0.003x^2$

At $x = 300$:

$P = -7000 - 9000 + 54,000 - 27,000 = \$11,000$

$\dfrac{dP}{dx} = -30 + 360 - 270 = \$60$

At $x = 400$:

$P = -7000 - 12,000 + 96,000 - 64,000 = \$13,000$

$\dfrac{dP}{dx} = -30 + 480 - 480 = -\$30$

Increase production in the first case, decrease in the second.

Section 4.3, pp. 70-71

1. $y' = 6x + 5$ 5. $y' = 35x^4 + 15x^2 + 1$

9. $y' = (x^2 + 2x - 1)(2x - 2) + (x^2 - 2x + 1)(2x + 2)$

$= 2(x^3 + 2x^2 - x - x^2 - 2x + 1 + x^3 - 2x^2 + x + x^2 - 2x + 1)$

$= 4(x^3 - 2x + 1)$

32   *Section 4.3*

13. $y' = 6x - 5 = 1$ 17. $y' = (x^4 + 1)2x + (x^2 - 1)4x^3 = 4$

21. $y' = 3x^2 - 24x + 45 = 0$, $x^2 - 8x + 15 = 0$

     $(x - 3)(x - 5) = 0$, $x = 3$ $x = 5$

     $(3, -1)(5, -5)$

25. $s = 5 - 3t + 4t^2 - t^3$, $v = -3 + 8t - 3t^2$

     At $t = 1$: $s = 5 - 3 + 4 - 1 = 5$ ft,

        $v = -3 + 8 - 3 = 2$ ft/sec, $|v| = 2$ ft/sec

     At $t = 4$: $s = 5 - 12 + 64 - 64 = -7$ ft,

        $v = -3 + 32 - 48 = -19$ ft/sec, $|v| = 19$ ft/sec

29. $y' = (x + 1)(x + 2) \cdot 1 + (x + 1) \cdot 1(x + 3) + 1 \cdot (x + 2)(x + 3)$

     $= x^2 + 3x + 2 + x^2 + 4x + 3 + x^2 + 5x + 6 = 3x^2 + 12x + 11$

33. $f(x) = u(x) - v(x)$

     $f'(x) = \lim_{h \to 0} \dfrac{f(x + h) - f(x)}{h}$

          $= \lim_{h \to 0} \dfrac{[u(x + h) - v(x + h)] - [u(x) - v(x)]}{h}$

          $= \lim_{h \to 0} \left(\dfrac{u(x + h) - u(x)}{h} - \dfrac{v(x + h) - v(x)}{h} \right)$

          $= u'(x) - v'(x)$

Section 4.4, pp. 79-80

1. $y = x^{-2}$ $y' = -2x^{-3} = -2/x^3$

5. $y = x^{2/3} - a^{2/3}$ $y' = \dfrac{2}{3} x^{-1/3} = \dfrac{2}{3x} 1/3$

9. $s = t^{2/3} - t^{-1/3}$ $s' = \dfrac{2}{3}t^{-1/3} + \dfrac{1}{3}t^{-4/3} = \dfrac{2t + 1}{3t^{4/3}}$

13. $p = \dfrac{q^{2/3} + a^{2/3}}{q^{2/3} - a^{2/3}}$

$p' = \dfrac{(q^{2/3} - a^{2/3})\, \frac{2}{3}\, q^{-1/3} - (q^{2/3} + a^{2/3})\, \frac{2}{3}\, q^{-1/3}}{(q^{2/3} - a^{2/3})^2}$

$ = \dfrac{-4a^{2/3}}{3q^{1/3}(q^{2/3} - a^{2/3})^2}$

17. $\dfrac{dy}{dx} = \dfrac{dy}{du} \cdot \dfrac{du}{dx} = \dfrac{1}{2\sqrt{u}} \cdot 4x = \dfrac{2x}{\sqrt{u}}$

21. $y = \sqrt{u}, \quad u = x^2 + 2x - 5$

$\dfrac{dy}{dx} = \dfrac{dy}{du} \cdot \dfrac{du}{dx} = \dfrac{1}{2\sqrt{u}}\,(2x + 2) = \dfrac{x + 1}{\sqrt{x^2 + 2x - 5}}$

25. $y = \dfrac{x^2 - 3}{3x^2 - 1}$ at $x = 2$

$y' = \dfrac{(3x^2 - 1)\,2x - (x^2 - 3)\,6x}{(3x^2 - 1)^2} = \dfrac{11 \cdot 4 - 1 \cdot 12}{11^2} = \dfrac{32}{121}$

29. $y = 2x^3 - 3x^2 + 1, \quad y' = 6x^2 - 6x = 12$

$x^2 - x - 2 = 0, \qquad (x - 2)(x + 1) = 0$

$x = 2, \; x = -1, \qquad (2, 5), \; (-1, -4)$

33. $y' = \dfrac{1}{2\sqrt{x}} = \dfrac{1}{4}, \qquad y - 2 = \dfrac{1}{4}\,(x - 4),$

$x - 4y + 4 = 0$

37. $PV = k, \quad 1 \cdot 10 = k, \quad PV = 10, \quad (1 + t)V = 10$

$V = \dfrac{10}{t + 1}, \quad \dfrac{dV}{dt} = \dfrac{-10}{(t + 1)^2} = \dfrac{-10}{10^2} = -\dfrac{1}{10}$ atmos/min

Section 4.5, pp. 82-83

1. $y' = 4(x + 1)^3$

5. $y' = \dfrac{4}{2\sqrt{4x + 2}} = \dfrac{2}{\sqrt{4x + 2}}$

9. $y' = \frac{1}{2}\left(\frac{x+1}{x-1}\right)^{-1/2} \frac{(x-1)-(x+1)}{(x-1)^2} = \frac{-1}{(x+1)^{1/2}(x-1)^{3/2}}$

13. $y' = \frac{2}{3}\left(\frac{2x-1}{2x+1}\right)^{-1/3} \frac{(2x+1)2-(2x-1)2}{(2x+1)^2}$

$= \frac{8}{3(2x-1)^{1/3}(2x+1)^{5/3}}$

17. $y' = -3(x^{-2}+x)^{-4}(-2x^{-3}+1) = \frac{-3(1-2/x^3)}{(1/x^2+x)^4} = \frac{3x^5(2-x^3)}{(1+x^3)^4}$

21. $y' = \frac{(x^2-4)\left(\frac{x}{2\sqrt{x+1}}+\sqrt{x+1}\right) - x\sqrt{x+1}\cdot 2x}{(x^2-4)^2}$

$= \frac{(x^2-4)(x+2x+2) - 4x^2(x+1)}{2\sqrt{x+1}(x^2-4)^2}$

$= \frac{3x^3+2x^2-12x-8-4x^3-4x^2}{2\sqrt{x+1}(x^2-4)^2}$

$= \frac{-x^3-2x^2-12x-8}{2\sqrt{x+1}(x^2-4)^2} = -\frac{x^3+2x^2+12x+8}{2\sqrt{x+1}(x^2-4)^2}$

25. $y' = \frac{(x+2)^2 3(x-1)^2 - (x-1)^3 2(x+2)}{(x+2)^4}$

$= \frac{1\cdot 3(-2)^2 - (-2)^3 2}{1^3} = 12 + 16 = 28$

29. $y' = \frac{(x-3)-(x+2)}{(x-3)^2} = \frac{-5}{(x-3)^2} = -5$

$(x-3)^2 = 1, \quad x-3 = \pm 1, \quad x = 3 \pm 1 = 2, 4$

33. $y' = 2(x-4) = 2(-3) = -6$

$y - 9 = -6(x-1), \qquad 6x + y - 15 = 0$

37. $f(x) = (x-a)^n \cdot P(x)$

$f'(x) = (x-a)^n \cdot P'(x) + n(x-a)^{n-1} \cdot P(x)$

$= (x-a)^{n-1}[(x-a)P'(x) + nP(x)]$

$= (x-a)^{n-1}Q(x)$

1. $y = \dfrac{1 - x^2}{2}$ $\qquad\qquad$ $2x + 2y' = 0$

$y' = -x = -1$ $\qquad\qquad$ $y' = -x = -1$

5. $y = \dfrac{-2 \pm \sqrt{4 - 4x^2}}{2}$ $\qquad$ $x^2 + y^2 + 2y = 0$

$y = -1 \pm \sqrt{1 - x^2}$ $\qquad\qquad$ $2x + 2yy' + 2y' = 0$

$y = -1 - \sqrt{1 - x^2}$ $\qquad\qquad$ $y'(y + 1) = -x$

$y' = -\dfrac{-2x}{2\sqrt{1 - x^2}} = \dfrac{x}{\sqrt{1 - x^2}}$ $\qquad$ $y' = \dfrac{-x}{y + 1}$

$\qquad = 0$ $\qquad\qquad\qquad\qquad$ $= 0$

9. $3x^2 + 3y^2y' = 0,$ $\qquad\qquad$ $y' = -\dfrac{x^2}{y^2}$

13. $\dfrac{1}{3} x^{-2/3} - \dfrac{1}{3} y^{-2/3} y' = 0$ $\quad$ $y' = \dfrac{y^{2/3}}{x^{2/3}} = (\dfrac{y}{x})^{2/3}$

17. $2xy' + 2y - 2yy' = 0,$ $\qquad$ $y'(x - y) = -y$

$y' = \dfrac{y}{y - x}$

21. $3x^2 + 3x^2y' + 6xy + 3y^2y' = 0$

$y'(x^2 + y^2) = -(x^2 + 2xy),$ $\quad$ $y' = -\dfrac{x^2 + 2xy}{x^2 + y^2}$

25. $\dfrac{(x - y)(1 + y') - (x + y)(1 - y')}{(x - y)^2} = 2x + 2yy'$

$[(x - y) - (x + y)] + y'(x - y + x + y) = 2x(x - y)^2 + 2yy'(x - y)^2$

$y'[2x - 2y(x - y)^2] = 2x(x - y)^2 + 2y$

$y' = \dfrac{y + x \ (x - y)^2}{x - y \ (x - y)^2}$

29. $2x + 2yy' - 6 - 8y' = 0,$ $y'(y - 4) = 3 - x,$

$$y' = \frac{3 - x}{y - 4} = \frac{3 - 6}{0 - 4} = \frac{-3}{-4} = \frac{3}{4}$$

$$y - 0 = \frac{3}{4}(x - 6), \qquad\qquad 3x - 4y - 18 = 0$$

33. $2x + 2yy' - 6 - 4y' = 0$ $y'(y - 2) = 3 - x$

$$y' = \frac{3 - x}{y - 2}$$

Let (x_o, y_o) be the point
of tangency.

$$y' = \frac{3 - x_o}{y_o - 2}$$

$y - 3 = \dfrac{3 - x_o}{y_o - 2} (x + 4)$ is the equation of tangent line.

(x_o, y_o) is on the tangent line.

$$y_o - 3 = \frac{3 - x_o}{y_o - 2} (x_o + 4)$$

$$y_o^2 - 5y_o + 6 = - x_o^2 - x_o + 12$$

$x_o^2 + y_o^2 + x_o - 5y_o - 6 = 0.$ (x_o, y_o) is on the
original curve.

$$x_o^2 + y_o^2 - 6x_o - 4y_o - 12 = 0$$

$7x_o - y_o + 6 = 0,$ $y_o = 7x_o + 6$

$$x_o^2 + 49x_o^2 + 84x_o + 36 + x_o - 35x_o - 30 - 6 = 0$$

$50x_o^2 + 50x_o = 0,$ $50x_o(x_o + 1) = 0$

$x_o = 0, \; y_o = 6$ $x_o = -1, \; y_o = -1$

$y - 3 = \dfrac{3}{4}(x + 4)$ $y - 3 = - \dfrac{4}{3}(x + 4)$

$3x - 4y + 24 = 0$ $4x + 3y + 7 = 0$

(The fact that the two tangent lines are perpendicular
is accidental.)

37. $4x + xy' + y - 2yy' - 6 + 3y' = 0$

$y'(x - 2y + 3) = -4x - y + 6$

$y' = \dfrac{6 - 4x - y}{3 + x - 2y}$

At $(2, 4)$, $y' = \dfrac{6 - 8 - 4}{3 + 2 - 8} = \dfrac{-6}{-3} = 2$

At $(2, 1)$, $y' = \dfrac{6 - 8 - 1}{3 + 2 - 2} = \dfrac{-3}{3} = -1$

At $(3, 6)$, $y' = \dfrac{6 - 12 - 6}{3 + 3 - 12} = \dfrac{-12}{-6} = 2$

At $(3, 0)$, $y' = \dfrac{6 - 12 - 0}{3 + 3 - 0} = \dfrac{-6}{6} = -1$

At $(1, 2)$, $y' = \dfrac{6 - 4 - 2}{3 + 1 - 4} = \dfrac{0}{0}$ no derivative

$2x^2 + xy - y^2 - 6x + 3y = (2x - y)(x + y) - 3(2x - y)$

$\qquad = (2x - y)(x + y - 3) \qquad 2x - y = 0, \ x + y = 3$

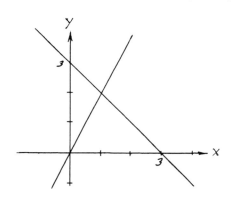

Section 4.7, pp. 89-90

1. $y' = 14x + 2$, $\qquad\qquad\qquad y'' = 14$, $\quad y''' = 0$

5. $f(x) = \dfrac{1}{x^2 - 1}$,$\qquad\qquad f'(x) = \dfrac{-2x}{(x^2 - 1)^2}$

$f''(x) = \dfrac{(x^2 - 1)^2(-2) + 2x(x^2 - 1)4x}{(x^2 - 1)^4}$

$\qquad = \dfrac{-2x^2 + 2 + 8x^2}{(x^2 - 1)^3} = \dfrac{6x^2 + 2}{(x^2 - 1)^3}$

9. $\frac{2}{3}x^{-1/3} + \frac{2}{3}y^{-1/3}\, y' = 0$

$y' = -\dfrac{y^{1/3}}{x^{1/3}}$,

$y'' = -\dfrac{x^{1/3}\cdot \frac{1}{3}y^{-2/3}\, y' - y^{1/3}\, \frac{1}{3}y^{-2/3}}{x^{2/3}}$

$= -\dfrac{-x^{1/3}\, y^{-2/3}\, y^{1/3}\, x^{-1/3} - y^{1/3}\, x^{-2/3}}{3x^{2/3}}$

$= \dfrac{y^{-1/3} + y^{1/3}\, x^{-2/3}}{3x^{2/3}} = \dfrac{x^{2/3} + y^{2/3}}{3x^{4/3}\, y^{1/3}} = \dfrac{1}{3x^{4/3}\, y^{1/3}}$

13. $f(x) = 4x^2 - 2x + 1,$ $\qquad f(1) = 3$

$f'(x) = 8x - 2,$ $\qquad f'(1) = 6$

$f''(x) = 8$ $\qquad f''(1) = 8$

17. $2x + 2yy' = 0$ $\qquad y' = -\dfrac{x}{y} = -\dfrac{3}{-4} = \dfrac{3}{4}$

$y'' = -\dfrac{y\cdot 1 - xy'}{y^2} = -\dfrac{-4 - 3\cdot \frac{3}{4}}{16} = \dfrac{4 + 9/4}{16} = \dfrac{25}{64}$

21. $s = t^3 - 2t^2 - 4t - 8$

$v = 3t^2 - 4t - 4 = (3t+2)(t-2), \quad a = 6t - 4$

t	s	v	a
0	-8	-4	-4
2/3	-304/27	-16/3	0
2	-16	0	8
>2		+	+

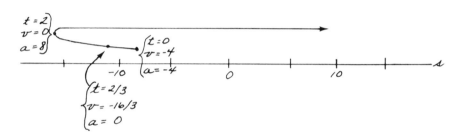

25. $s = 2t^3 - 15t^2 + 24t$ $v = 6t^2 - 30t + 24 = 6(t-1)(t-4)$

$a = 12t - 30 = 6(2t - 5)$

t	s	v	a
0	0	24	-30
1	11	0	-18
5/2	-5/2	-27/2	0
4	-16	0	18
>4		+	+

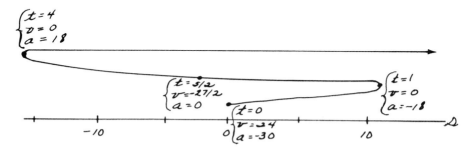

29.

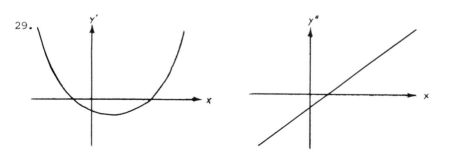

33. $f(x) = u(x)v(x)$ $f'(x) = u(x)v'(x) + u'(x)v(x)$

$f''(x) = [u(x)v''(x) + u'(x)v'(x)] + [u'(x)v'(x) + u''(x)v(x)]$

$= u(x)v''(x) + 2u'(x)v'(x) + u''(x)v(x)$

$f'''(x) = [u(x)v'''(x) + u'(x)v''(x)] + [2u'(x)v''(x)$

$+ 2u''(x)v'x)] + [u''(x)v'(x) + u'''(x)v(x)]$

$= u(x)v'''(x) + 3u'(x)v''(x) + 3u''(x)v'(x)$

$+ u'''(x)v(x)$

$$f^{(4)}(x) = u(x)v^{(4)}(x) + 4u'(x)v'''(x) + 6u''(x)v''(x)$$
$$+ 4u'''(x)v'(x) + u^{(4)}(x)v(x)$$

Review 4, p. 91

1. $y = 4x^{-3}$ $\qquad y' = -12x^{-4} = -\dfrac{12}{x^4}$

5. $y' = \dfrac{x\dfrac{2x}{2\sqrt{x^2+1}} - \sqrt{x^2+1}}{x^2} = \dfrac{x^2 - (x^2+1)}{x^2\sqrt{x^2+1}} = \dfrac{-1}{x^2\sqrt{x^2+1}}$

9. $y = \dfrac{2x^2 - 5x - 3}{x+4}$ $\qquad y' = \dfrac{(x+4)(4x-5) - (2x^2-5x-3)}{(x+4)^2}$

$= \dfrac{4x^2 + 11x - 20 - 2x^2 + 5x + 3}{(x+4)^2} = \dfrac{2x^2 + 16x - 17}{(x+4)^2}$

13. $y' = \dfrac{(2x-1)2(x+1) - (x+1)^2 2}{(2x-1)^2}$

$= \dfrac{2(x+1)(2x-1-x-1)}{(2x-1)^2} = \dfrac{2(x+1)(x-2)}{(2x-1)^2}$

17. $\dfrac{dy}{du} = 2u$

$\dfrac{dy}{dx} = \dfrac{dy}{du} \cdot \dfrac{du}{dx} = 2u(1 + \dfrac{1}{2\sqrt{x+1}}) = 2u\dfrac{1 + 2\sqrt{x+1}}{2\sqrt{x+1}}$

$= \dfrac{u(1 + 2\sqrt{x+1})}{\sqrt{x+1}}$

21. $y' = \lim_{h \to 0} \dfrac{f(x+h) - f(x)}{h} = \lim_{h \to 0} \dfrac{\dfrac{x+h+1}{2x+2h-3} - \dfrac{x+1}{2x-3}}{h}$

$= \lim_{h \to 0} \dfrac{(2x-3)(x+h+1) - (x+1)(2x+2h-3)}{h(2x+2h-3)(2x-3)}$

$= \lim_{h \to 0} \dfrac{2x^2+2hx+2x-3x-3h-3-2x^2-2hx+3x-2x-2h+3}{h(2x+2h-3)(2x-3)}$

$= \lim_{h \to 0} \dfrac{-5h}{h(2x+2h-3)(2x-3)} = \lim_{h \to 0} \dfrac{-5}{(2x+2h-3)(2x-3)}$

$$= \frac{-5}{(2x - 3)^2}$$

25. $y' = 3x^2 - 1 = 2,$ $x^2 = 1,$ $x = \pm 1,$ $(1, 0)$ and $(-1, 0)$

Chapter Five
Curve Sketching

1. $y = x(x + 3)$ $(0, 0)$, $(-3, 0)$

5. $(-1/4, 0)$, $(2, 0)$, $(-3/2, 0)$, $(0, -18)$

9. $(1/3, 0)$, $(-1/2, 0)$, $(0, 8)$ 13. $(3, 0)$, $(0, 9)$

17. No asymptotes   21. $x = -3$, $y = 1$

25. $x = -1$, $x = 3$, $y = 0$ 29. $x = -3/2$, $x = -1$, $y = 0$

33. $y = \dfrac{4x^2 - (4x^2 + 3)}{2x + \sqrt{4x^2 + 3}} = \dfrac{-3}{2x + \sqrt{4x^2 + 3}}$ $y = 0$

    (The curve approaches it at the positive end only.)

37. $y = \dfrac{x^2 + x - 2}{x} = x + 1 - \dfrac{2}{x}$, $x = 0$, $y = x + 1$

41. (a) $n < m$; $y = 0$ (b) $n = m$; $y = a_n/b_m$

    (c) $n > m$; no horizontal asymptote

1. Replacing x by $-x$,   we have

    $y = (-x)^4 - (-x)^2 = x^4 - x^2$

    Symmetric about the y axis

5. Replacing y by $-y$, we have

    $(-y)^2 = \dfrac{x + 1}{x}$, $y^2 = \dfrac{x + 1}{x}$

    Symmetric about the x axis

9. Replacing x by -x and y by -y, we have

$$-y = \frac{-x}{(-x)^2 + 1} = \frac{-x}{x^2 + 1}, \quad y = \frac{x}{x^2 + 1}$$

Symmetric about the origin

13. Intercepts: $(-1, 0)$ odd, $(6, 0)$ odd, $(0, -6)$

No symmetry $\qquad y = (x + 1)(x - 6)$

No asymptotes

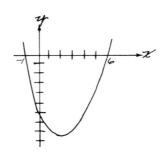

17. Intercepts: $(-1, 0)$ odd

Asymptotes: $x = 0$ odd, $y = 1$

No symmetry

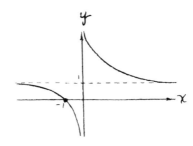

21. Intercepts: $(-2, 0)$ odd, $(4, 0)$ even, $(0, 8)$

Asymptotes: $x = 1$ odd, $y = x - 1$

No symmetry

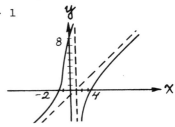

25. No intercepts

Asymptotes: $x = 0$ odd, $y = x$

Symmetry about the origin

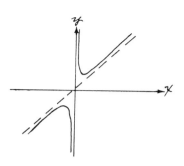

29. Intercepts: $(0, 0)$ odd

Asymptotes: $x = \pm 1$ odd, $y = x$

Symmetry about the origin

$$y = \frac{x^3}{x^2 - 1}$$

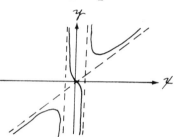

33. Yes. Every point of a line is a point of symmetry. Every x-intercept of $y = \sin x$ is a point of symmetry. A curve cannot have <u>exactly</u> two points of symmetry--if it has two it must have infinitely many.

37. No. Suppose $f(a) = b \neq 0$. Then (a, b) and $(a, -b)$ are both on the graph. But no function can have more than one point on its graph with a given x coordinate.

Section 5.3, p. 112

1. Intercepts: $(0, 0)$, $(\pm 1, 0)$

No asymptotes

Symmetry about the origin

$D = \{x \mid x = 0, \ x \geq 1, \text{ or } x \leq -1\}$

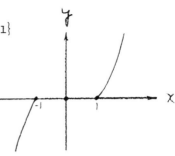

5. No intercepts

Asymptotes: $y = 0, \ y = 2x$

No symmetry

$D = \{x \mid x \geq 1 \text{ or } x \leq -1\}$

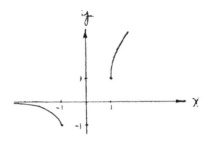

9. $z = \dfrac{2x}{(x-1)^2}$

Intercepts: $(0, 0)$ odd

Asymptotes: $x = 1$ even, $z = 0$

No symmetry

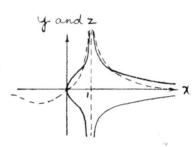

13. $z = \dfrac{x(x-1)}{(x+1)^2}$

Intercepts: $(0, 0)$ odd, $(1, 0)$ even

Asymptotes: $x = -1$ even, $z = 1$

No symmetry

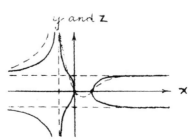

17. $z = (x - 1)(x - 3)^2$

   Intercepts: $(1, 0)$ odd, $(3, 0)$ even, $(0, -9)$

   No asymptotes

   No symmetry

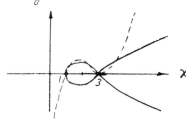

21. Intercepts: $(-1, 0)$ odd

   Asymptotes: $x = 0$ odd, $y = 1$

   No symmetry

$$y = \frac{x + 1}{x}$$

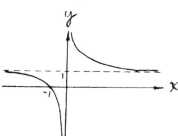

25. $y = -2 - h \ (h \neq 0)$

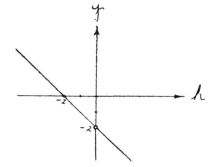

1. $y' = 2x + 2$

    $(-1, -4)$ absol. min.

5. $y' = 3x^2 - 6x = 3x(x - 2)$

    $(0, 1)$ rel. max., $(2, -3)$ rel. min.

9. $y' = 12x^3 + 12x^2 = 12x^2(x + 1)$

    $(-1, -1)$ absol. min., $(0, 0)$ neither

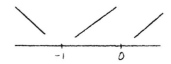

13. $y' = \dfrac{x + 1 - x}{(x + 1)^2} = \dfrac{1}{(x + 1)^2}$ No critical point

17. $y' = \dfrac{-2x}{(x^2 - 1)^2}$

    $(0, -1)$ rel. max.

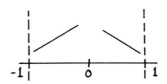

21. $y' = (x+1)^2 + (x-2)2(x+1) = 3(x+1)(x-1)$

   $(-1, 0)$ rel. max., $(1, -4)$ rel. min.

25. $y' = \dfrac{1}{3x^{2/3}}$

   $(0, 0)$ neither

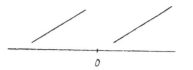

29. $y' = (x-1)^{1/3}\dfrac{2}{3}(x+2)^{-1/3} + (x+2)^{2/3}\dfrac{1}{3}(x-1)^{-2/3}$

   $= \dfrac{2(x-1) + (x+2)}{3(x+2)^{1/3}(x-1)^{2/3}} = \dfrac{x}{(x+2)^{1/3}(x-1)^{2/3}}$

   $(0, -\sqrt[3]{4})$ rel. min., $(-2, 0)$ rel. max., $(1,0)$ neither

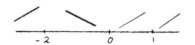

33. $y' = \dfrac{4}{3}x^{1/3} - \dfrac{2}{3}x^{-1/3} = \dfrac{2(2x^{2/3}-1)}{3x^{1/3}}$

   $(0, 0)$ rel. max., $(1/2\sqrt{2}, -1/4)$ absol. min.
   $(-1/2\sqrt{2}, -1/4)$ absol. min.

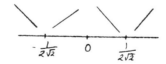

Section 5.5, pp. 121-122

1. $y' = 2x - 1$, $\quad y'' = 2$, $\quad (1/2, -25/4)$ absol. min.

5. $y' = 12x^3 - 12x^2 - 24x = 12x(x-2)(x+1)$

   $y'' = 36x^2 - 24x - 24$

$(0, 24)$	$(2, -8)$	$(-1, 19)$
$y'' = -24$	$y'' = 72$	$y'' = 36$
rel. max.	absol. min.	rel. min.

9. $y' = \dfrac{(x+1)\ 2x - x^2}{(x+1)^2} = \dfrac{x(x+2)}{(x+1)^2}$

$y'' = \dfrac{(x+1)^2(2x+2) - (x^2+2x)\ 2(x+1)}{(x+1)^4} = \dfrac{2}{(x+1)^3}$

(0, 0)   (-2, -4)

$y'' = 2$ $y'' = -2$

rel. min.   rel. max.

13. $y' = 2x - 2$

$y'' = 2$

17. $y' = 4x^3 - 18x^2$

$y'' = 12x^2 - 36x = 12x(x-3)$

(0, 0), (3, -81)

21. $y' = \dfrac{x^2 + 1 - 2x^2}{(x^2+1)^2} = \dfrac{1 - x^2}{(x^2+1)^2}$

$y'' = \dfrac{(1+x^2)^2(-2x) - (1-x^2)2(1+x^2)2x}{(1+x^2)^4} = \dfrac{2x(x^2-3)}{(1+x^2)^3}$

(0, 0), $(\sqrt{3}, \sqrt{3}/4)$,

$(-\sqrt{3}, -\sqrt{3}/4)$

25. $y' = 12x^2 - 30x - 18 = 6(2x+1)(x-3)$

$y'' = 24x - 30 = 6(4x-5)$

(-1/2, 59/4)   (3, -71)   (5/4, -225/8)

$y'' = -42$ $y'' = 42$ point of
inflection

rel. max.   rel. min.

29. $y' = 5x^4 - 20x^3 = 5x^3(x-4)$, $\quad y'' = 20x^3 - 60x^2 = 20x^2(x-3)$

(0, 0)   (4, -256)   (3, -162)

$y'' = 0$ $y'' = 320$ point of inflection

rel. max.   rel. min.

33. $y' = \dfrac{(cx+d)\,a - (ax+b)c}{(cx+d)^2} = \dfrac{ad - bc}{(cx+d)^2}$ (no critical point)

$y'' = \dfrac{(ad-bc)(-2)(c)}{(cx+d)^3} = \dfrac{-2c(ad-bc)}{(cx+d)^3}$ (no point of inflection)

Section 5.6, pp. 125–126

1. $y = x^2 - 2$, $-1 \leqq x \leqq 2$

$y' = 2x$ $y'' = 2$

(0, -2)   (-1, -1)   (2, 2)

absol. min.   rel. max.   absol. max.

5. $y = 1/x^2$, $-1 \leqq x \leqq 1$

$y' = -2/x^3$

No critical point   (-1, 1)   (1, 1)

                         absol. min.   absol. min.

9. $y = x^3 - 3x$, $-\sqrt{3} \leqq x \leqq 2$

$y' = 3x^2 - 3$ $y'' = 6x$

(1, -2)   (-1, 2) $(-\sqrt{3}, 0)$ (2, 2)

$y'' > 0$ $y'' < 0$ rel. min.   absol. max.

absol. min.   absol. max.

13. $y = \begin{cases} x^2, & \text{if } x < 0 \\ y^3, & \text{if } x \geqq 0 \end{cases}$

$y' = \begin{cases} 2x, & \text{if } x < 0 \\ 3x^2, & \text{if } x \geqq 0 \end{cases}$

(0, 0) absol. min.

17. $y' = \begin{cases} -1, & \text{if } x < -1 \\ -2x, & \text{if } -1 < x < 1 \\ 1, & \text{if } x > 1 \end{cases}$

(0, 1)   (−1, 0)   (1, 0)

$y'' = -2$ absol. min.   absol. min.

rel. max.

21. $y' = \begin{cases} 2, & \text{if } x < -2 \\ -1, & \text{if } -2 < x < 0 \\ 2 - 2x, & \text{if } x > 0 \end{cases}$

(1, 1)   (−2, 2)   (0, 0)

$y'' = -2$ absol. max.   rel. min.

rel. max.

Section 5.7, pp. 129-130

1. $y = x^2(x + 6)$, $y' = 3x^2 + 12x = 3x(x + 4)$

$y'' = 6x + 12 = 6(x + 2)$

(0, 0)   (−4, 32)   (−2, 16)

$y'' = 12$ $y'' = -12$ point of
                                              inflection

min.   max.

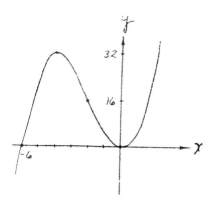

5. $y' = (x + 1)^2 \ 3(x - 3)^2 + (x - 3)^3 \ 2(x + 1)$

$\quad = (x + 1)(x - 3)^2 \ (5x - 3)$

$y'' = (x + 1)(x - 3)^2 \ 5 + (x + 1) \ 2(x - 3)(5x - 3)$

$\quad + (x - 3)^2 \ (5x - 3)$

$\quad = (x - 3)(20x^2 - 24x - 12) = 4(x - 3)(5x^2 - 6x - 3)$

(-1, 0)　　(3, 0)　　(3/5, -35.4)　　(3,0), (1,6, -18.5)
　　　　　　　　　　　　　　　　　　　(-0.4, -14.1)

y'' is -　　$y'' = 0$　　y'' is +

max.　　　neither　min.　　　points of inflection

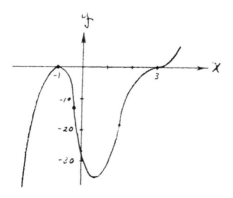

9. $y' = (x-1)^{1/3} \frac{2}{3}(x+3)^{-1/3} + (x+3)^{2/3} \frac{1}{3}(x-1)^{-2/3}$

$\quad = \dfrac{2(x-1)+(x+3)}{3(x+3)^{1/3}(x-1)^{2/3}} = \dfrac{3x-1}{3(x+3)^{1/3}(x-1)^{2/3}}$

$y'' = \dfrac{3(x+3)^{1/3}(x-1)^{2/3} \cdot 3 - (3x+1)3[(x+3)^{1/3}}{9(x+3)^{2/3}(x-1)^{4/3}}$

$\qquad \dfrac{\frac{2}{3}(x-1)^{-1/3} + (x-1)^{2/3}\frac{1}{3}(x+3)^{-2/3}]}{\phantom{9(x+3)^{2/3}(x-1)^{4/3}}}$

$\quad = \dfrac{9(x+3)(x-1) - (3x+1)[2(x+3)+(x-1)]}{9(x+3)^{4/3}(x-1)^{5/3}}$

$\quad = \dfrac{-32}{9(x+3)^{4/3}(x-1)^{5/3}}$

$(1, 0) \qquad (-3, 0) \qquad (-1/3, -4\sqrt[3]{4}/3) \qquad (1, 0)$

neither   max y'' is +   point of
                                              inflection

                    min

13. $y' = \dfrac{(x+1)2x - x^2}{(x+1)^2} = \dfrac{x(x+2)}{(x+1)^2}$

$\quad y'' = \dfrac{(x+1)^2(2x+2) - (x^2+2x)2(x+1)}{(x+1)^4} = \dfrac{2}{(x+1)^3}$

$(0, 0)$ $(-2, -4)$

y'' is $+$ y'' is $-$

min   max

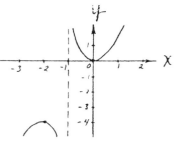

17. $y^2 = x^4 - 4x^3 = x^3(x - 4)$

      $z = x^4 - 4x^3$

      $z' = 4x^3 - 12x^2 = 4x^2(x - 3)$

      $z'' = 12x^2 - 24x = 12x(x - 2)$

      $x = 0$ $x = 3$

      $z = 0$ $z = -27$

      $y = 0$ $y = \underline{\quad\quad}$

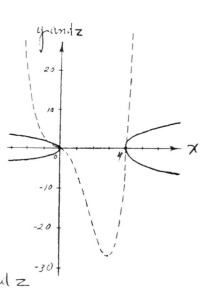

21. $z = \dfrac{x - 1}{x^3}$

      $z' = \dfrac{x^3 - (x - 1)3x^2}{x^6}$

        $= \dfrac{3 - 2x}{x^4}$

      $x = 3/2$

      $z = 4/27$

      $y = \pm 2/3\sqrt{3}$

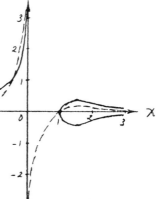

1. Intercepts: ($\pm$1, 0) odd, (0, -1)

    Asymptote: y = 1

    Symmetry about the y axis

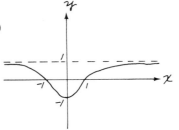

5. Intercepts: (1, 0) even, (-3, 0) odd, (3/2, 0) odd,

    (0, -81)

    No asymptote

    No symmetry

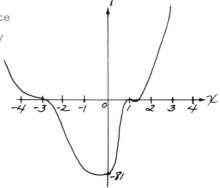

9. $z = \dfrac{x(x - 2)}{(x + 1)^2}$

    Intercepts: (0, 0) odd, (2, 0) odd

    Asymptotes: x = -1 even, y = 1

    No symmetry

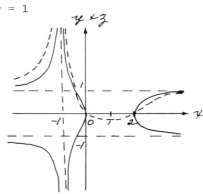

13. $y' = 6x^2 - 6x - 12 = 6(x - 2)(x + 1)$

$y'' = 12x - 6 = 6(2x - 1)$

$(-1,\ 15)$	$(2,\ -12)$
$y'' < 0$	$y'' > 0$
rel. max.	rel. min.

17. $y' = x^3 \cdot 2(3x - 1)3 + 3x^2(3x - 1)^2$

$\quad = 3x^2(3x - 1)(5x - 1) = 3(15x^4 - 8x^3 + x^2)$

$y'' = 3(60x^3 - 24x^2 + 2x) = 6x(30x^2 - 12x + 1)$

$(1/3,\ 0)$	$(1/5,\ 4/3125)$	$(0,\ 0)$
$y'' > 0$	$y'' < 0$	$y'' = 0$
rel. min.	rel. max.	Neither, since $(0,\ 0)$ is an odd intercept, the graph crosses the x axis.

21. $y' = 2(x^2 - 9)2x = 4x(x^2 - 9)$

$y'' = 4x \cdot 2x + 4(x^2 - 9) = 12(x^2 - 3)$

Critical points:	$(0,\ 81)$	$(\pm 3,\ 0)$
	$y'' < 0$	$y'' > 0$
	rel. max.	absol. min.

Points of inflection: $(\pm \sqrt{3},\ 36)$

25. $y' = \dfrac{(3x^2 - 1)6x - 3x^2 \cdot 6x}{(3x^2 - 1)^2} = \dfrac{-6x}{(3x^2 - 1)^2}$

$y'' = \dfrac{(3x^2 - 1)^2(-6) + 6x \cdot 2(3x^2 - 1)6x}{(3x^2 - 1)^4}$

$\quad = \dfrac{-18x^2 + 6 - 72x^2}{(3x^2 - 1)^3} = \dfrac{6(9x^2 + 1)}{(3x^2 - 1)^3}$

Critical point:

(0, 0)

$y'' < 0$

rel. max.

No point of
inflection

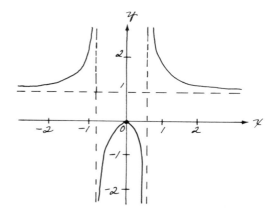

29. $f'(x) = 3ax^2 + 2bx + c,$ $f''(x) = 6ax + 2b$

If x_1 and x_2 are critical values, then

$3ax_1^2 + 2bx_1 + c = 0,$ $3ax_2^2 + 2bx_2 + c = 0$

Subtracting and dividing by $x_1 - x_2$, we have

$3a(x_1^2 - x_2^2) + 2b(x_1 - x_2) = 0,$ $3a(x_1 + x_2) + 2b = 0$

$6a(x_1 + x_2)/2 + 2b = 0$

Therefore $x = (x_1 + x_2)/2$ is a point of inflection.

$f(x) = ax^4 + bx^3 + cx^2 + dx + e$

$f'(x) = 4ax^3 + 3bx^2 + 2cx + d,$ $f''(x) = 12ax^2 + 6bx + 2c$

If x_1 and x_2 are critical values, then

$4ax_1^3 + 3bx_1^2 + 2cx_1 + d = 0$

$4ax_2^3 + 3bx_2^2 + 2cx_2 + d = 0$

Subtracting and dividing by $x_1 - x_2$, we have

$4a(x_1^3 - x_2^3) + 3b(x_1^2 - x_2^2) + 2c(x_1 - x_2) = 0$

$4a(x_1^2 + x_1 x_2 + x_2^2) + 3b(x_1 + x_2) + 2c = 0$

It is now clear that $x = (x_1 + x_2)/2$ is not a point
of inflection.

Chapter Six
Further Applications of the Derivative

1. $P = xy$, $x + y = 48$, $P = x(48 - x) = 48x - x^2$

$\dfrac{dP}{dx} = 48 - 2x = 0$, $x = 24$, $y = 24$

$360,000 \, x^{-1}$

$-360,000$
$\overline{ x^2}$

5. $xy = 180,000$, $y = \dfrac{180,000}{x}$

$L = 4x + 2y = 4x + \dfrac{360,000}{x}$

$L' = 4 - \dfrac{360,000}{x^2} = 0$

$x^2 = 90,000$

$x = 300$ ft, $y = 600$ ft,

$L = 2400$ ft

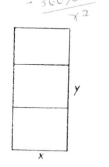

9. $\dfrac{H}{R} = \dfrac{h}{R - r}$, $h = \dfrac{H}{R}(R - r)$, $V = \pi r^2 h = \dfrac{\pi H}{R}(Rr^2 - r^3)$

$V' = \dfrac{\pi H}{R}(2Rr - 3r^2) = \dfrac{\pi H r}{R}(2R - 3r) = 0$

$r = \dfrac{2R}{3}$, $V = \dfrac{4\pi R^2 H}{27}$

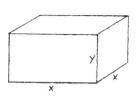

13. $4000 = x^2 y$, $y = \dfrac{4000}{x^2}$

$A = 4xy + x^2 = \dfrac{16,000}{x} + x^2$

$A' = -\dfrac{16,000}{x^2} + 2x = 0$

$x^3 = 8000$, $x = 20$ in, $y = 10$ in

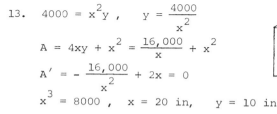

17. $x^2 + y^2 = R^2$

$A = \dfrac{1}{2} \cdot 2x(R - y) = x(R - y)$

$ = \sqrt{R^2 - y^2}\,(R - y)$

$A' = \sqrt{R^2 - y^2}(-1)$

$ + (R - y)\,\dfrac{-2y}{2\sqrt{R^2 - y^2}}$

$ = \dfrac{2y^2 - Ry - R^2}{\sqrt{R^2 - y^2}} = \dfrac{(2y + R)(y - R)}{\sqrt{R^2 - y^2}}$

$y = -\dfrac{R}{2}$, Base $= 2x = 2\sqrt{R^2 - y^2} = \sqrt{3}\,R$

Height $= R - y = \dfrac{3R}{2}$

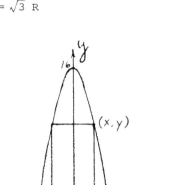

21. $A = 2xy = 2x(16 - x^2) = 32x - 2x^3$

$A' = 32 - 6x^2 = 0$

$x^2 = \dfrac{16}{3}$, $x = \dfrac{4}{\sqrt{3}}$

$y = 16 - \dfrac{16}{3} = \dfrac{32}{3}$

Dimensions: $\dfrac{8}{\sqrt{3}}$ x $\dfrac{32}{3}$

25. $A = \dfrac{4b}{a}\,x\sqrt{a^2 - x^2}$, $0 \leqq x \leqq a$

$A^2 = \dfrac{16b^2}{a^2}\,x^2(a^2 - x^2) = \dfrac{16b^2}{a^2}(a^2 x^2 - x^4)$

$\dfrac{dA^2}{dx} = \dfrac{16b^2}{a^2}(2a^2 x - 4x^3) = \dfrac{32b^2}{a^2}\,x(a^2 - 2x^2)$

This is 0 if $x = 0$ or $x = \pm\, a/\sqrt{2}$. Since $x = -a/\sqrt{2}$
is meaningless and $x = 0$ gives $A = 0$, the maximum
value of A occurs when $x = a/\sqrt{2}$.

29. $x = \dfrac{k(t+a)}{t^2+b}$, $\quad \dfrac{dx}{dt} = k\,\dfrac{t^2+b-2t(t+a)}{(t^2+b)^2} = k\,\dfrac{b-2at-t^2}{(t^2+b)^2}$

$\dfrac{k(b-2a-1)}{(1+b)^2} = 0$, $\qquad b-2a-1=0$, $\qquad b = 2a+1$

When $t = 0$, $x = 20,000$; when $t = 1$, $x = 25,000$.
This gives:

$\dfrac{ak}{b} = 20,000$ $\qquad\qquad\qquad \dfrac{k(1+a)}{1+b} = 25,000$

$ak = 20,000b = 20,000(2a+1)$ $\quad k(a+1) = 25,000(b+1)$

$\qquad\qquad\qquad\qquad\qquad\qquad\qquad\qquad = 25,000(2a+2)$

$k = \dfrac{20,000(2a+1)}{a} = \dfrac{25,000(2a+2)}{a+1}$

$\dfrac{8a+4}{a} = \dfrac{10a+10}{a+1}$, $\qquad 10a^2+10a = 8a^2+12a+4$

$2a^2 - 2a - 4 = 0$, $\qquad a^2 - a - 2 = 0$, $\qquad (a-2)(a+1) = 0$

$a = 2$ $\quad$ or $\quad a = -1$

$a \neq -1$, since then $\dfrac{k(1+a)}{1+b} \neq 25,000$. Therefore $a = 2$,
$b = 5$, $k = 50,000$; and the original formula is
$x = 50,000\,\dfrac{t+2}{t^2+5}$. When $t = 30$, $x = 50,000\,\dfrac{32}{905} = 1767$.

33. $t_1 = \dfrac{d_1}{r_1} = \dfrac{\sqrt{x^2+16}}{4}$

$t_2 = \dfrac{d_2}{r_2} = \dfrac{10-x}{5}$

$t = t_1 + t_2 = \dfrac{\sqrt{x^2+16}}{4} + \dfrac{10-x}{5}$

$t' = \dfrac{x}{4\sqrt{x^2+16}} - \dfrac{1}{5}$, $\quad 5x = 4\sqrt{x^2+16}$

$25x^2 = 16x^2 + 256$, $\quad x^2 = \dfrac{256}{9}$, $\quad x = \dfrac{16}{3}$

37. $I = \dfrac{E}{R+r}$, $\qquad P = RI^2 = \dfrac{RE^2}{(R+r)^2}$

$\dfrac{dP}{dR} = \dfrac{(R+r)^2E^2 - RE^2 \cdot 2(R+r)}{(R+r)^4} = \dfrac{E^2(r-R)}{(R+r)^3}$, $\qquad R = r$

41. Number sold $= x = \dfrac{k}{p^2}$, Profit per widget $= p - 50$

Total profit $= P = \dfrac{k}{p^2}\ (p - 50)$

$\dfrac{dP}{dp} = k\ \dfrac{p^2 - (p - 50)2p}{p^4}$ $k\ \dfrac{100 - p}{p^3} = 0$, $p = 100$

45. $x^2 = z^2 + (k - x)^2$, $z = \sqrt{2kx - k^2}$, $w = v - \sqrt{2kx - k^2}$

$v^2 = w^2 + k^2 = v^2 - 2v\sqrt{2kx - k^2} + 2kx$, $kx = v\sqrt{2kx - k^2}$

$v = \dfrac{kx}{\sqrt{2kx - k^2}}$, $A = \dfrac{1}{2}xy = \dfrac{k}{2}\ \dfrac{x^2}{\sqrt{2kx - k^2}} (\dfrac{k}{2} < x < k)$

$A^2 = \dfrac{k^2}{4}\ \dfrac{x^4}{2kx - k^2} = \dfrac{k}{4}\ \dfrac{x^4}{2x - k}$

$\dfrac{dA^2}{dx} = \dfrac{k}{4}\ \dfrac{2x^3(3x - 2k)}{(2x - k)^2}$, $3x - 2k = 0$, $x = \dfrac{2k}{3}$

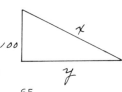

Section 6.2, pp. 143-144

1. $V = xyz$

$\dfrac{dV}{dt} = xy\ \dfrac{dz}{dt} + xz\ \dfrac{dy}{dt} + yz\ \dfrac{dx}{dt} = 10 \cdot 10 \cdot 2 + 10 \cdot 10 \cdot 7 + 10 \cdot 10 \cdot 5$

$= 1400$ in^3/min

$s = 2xy + 2xz + 2yz$

$\dfrac{ds}{dt} = 2(x\ \dfrac{dy}{dt} + y\ \dfrac{dx}{dt} + x\ \dfrac{dz}{dt} + z\ \dfrac{dx}{dt} + y\ \dfrac{dz}{dt} + z\ \dfrac{dy}{dt})$

$= 2(10 \cdot 7 + 10 \cdot 5 + 10 \cdot 2 + 10 \cdot 5 + 10 \cdot 2 + 10 \cdot 7) =$

560 in^2/min

5. $x^2 = 10,000 + y^2$

When $x = 260$,

$y = \sqrt{67,600 - 10,000} = \sqrt{57,600} = 240$

$2x\ \dfrac{dx}{dt} = 2y\ \dfrac{dy}{dt}$, $\dfrac{dy}{dt} = \dfrac{x}{y}\ \dfrac{dx}{dt} = \dfrac{260}{240} \cdot 5 = \dfrac{65}{12}$ ft/sec

9. $\dfrac{dx}{dt} = -1$ ft/sec

$\dfrac{x}{y} = \dfrac{10 - x}{6}$

$6x = y(10 - x)$

$6\dfrac{dx}{dt} = -y\dfrac{dx}{dt} + (10 - x)\dfrac{dy}{dt}$

$6(-1) = -\dfrac{3}{2}(-1) + 8\dfrac{dy}{dt}$

$8\dfrac{dy}{dt} = -6 - \dfrac{3}{2} = -\dfrac{15}{2}$, $\dfrac{dy}{dt} = -\dfrac{15}{16}$ ft/sec

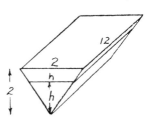

13. $pv^{1.4} = k$, $1.4pv^{0.4}\dfrac{dv}{dt} + v^{1.4}\dfrac{dp}{dt} = 0$

$1.4(40)28^{0.4}(-2) + 28^{1.4}\dfrac{dp}{dt} = 0$

$-28.4 + 28\dfrac{dp}{dt} = 0$, $\dfrac{dp}{dt} = 4$ lb/in^2/min

17. $y = \sqrt{x}$, $\dfrac{dy}{dt} = \dfrac{1}{2\sqrt{x}}\dfrac{dx}{dt}$

(a) $\dfrac{dy}{dt} = \dfrac{3}{2\sqrt{1}} = \dfrac{3}{2}$ units/min , (b) $\dfrac{dy}{dt} = \dfrac{3}{2\sqrt{4}} = \dfrac{3}{2}$ units/min

21. $V = 12 \cdot \dfrac{1}{2}h^2 = 6h^2$

$\dfrac{dV}{dt} = 12h\dfrac{dh}{dt}$

$1 = 12 \cdot \dfrac{3}{2}\dfrac{dh}{dt}$

$\dfrac{dh}{dt} = \dfrac{1}{18}$ ft $= \dfrac{2}{3}$ in

Section 6.3, pp. 149-150

1. $dy = (3x^2 - 2x)\ dx = x(3x - 2)\ dx$

5. $dy = 4(x^2 + 1)^3 \cdot 2x\ dx = 8x(x^2 + 1)^3\ dx$

9. $3x^2\ dx - 3x^2\ dy - 6xy\ dx + 6\ xy\ dy + 3y^2\ dx = 0$

$(x^2 - 2xy)\ dy = (x^2 - 2xy + y^2)\ dx$, $dy = \dfrac{(x - y)^2}{x(x - 2y)}\ dx$

13. $\quad C = \pi x \qquad\qquad\qquad A = \pi r^2$

$\quad dC = \pi\,dx \qquad\qquad\quad dA = \dfrac{\pi x\ dx}{2}$

$\quad \left|dC\right| \leq \left|\pi(.03)\right| = 0.03\pi\quad \left|dA\right| \leq \left|\dfrac{\pi}{2}(7.38)(0.03)\right| = 0.11\pi$

17. $\quad R = r\,\dfrac{x}{1 - x}\ ,\quad dR = r\,\dfrac{1 - x + x}{(1 - x)^2}\,dx = \dfrac{rl}{(1 - x)^2}\,dx$

$\quad R = \dfrac{40 \cdot 50}{50} = 40\ \text{ohm}\ ,\quad \left|dR\right| \leq \dfrac{40 \cdot 100}{(100 - 50)^2}\,0.1 = 0.16\ \text{ohm}$

21. $\quad y = \sqrt[3]{x}\ ,\qquad\quad dy = \dfrac{dx}{3x^{2/3}}$

$\quad x = 27 \qquad\qquad\qquad\qquad dx = -2$

$\quad y = 3 \qquad\qquad\qquad\qquad dy = \dfrac{-2}{3 \cdot 27^{2/3}} = -0.074$

$\quad \sqrt[3]{25} \doteq y + dy = 2.926$

25. $\quad y = \dfrac{1}{\sqrt{x}}\ ,\qquad\quad dy = -\dfrac{dx}{2x^{3/2}}$

$\quad x = 25 \qquad\qquad\qquad\qquad dx = 2$

$\quad y = 0.2 \qquad\qquad\qquad\qquad dy = -\dfrac{2}{2 \cdot 25^{3/2}} = -0.008$

$\quad \dfrac{1}{\sqrt{27}} \doteq y + dy = 0.192$

29. $\quad V = x^3\ ,\qquad\qquad\qquad dV = 3x^2\,dx$

$\quad \dfrac{dV}{V} = \dfrac{3x^2\,dx}{x^3} = 3\,\dfrac{dx}{x}\ ,\quad 0.03 = 3\,\dfrac{dx}{x}\ ,\quad \dfrac{dx}{x} = 0.01 = 1\%$

Section 6.4, p. 153

1. $\quad Q' = 0.000\ 008 - \dfrac{3200}{x^2} = 0\ ,\quad \dfrac{3200}{x^2} = 0.000\ 008$

$\quad x^2 = \dfrac{3200}{0.000\ 008} = 400,000,000$

$\quad x = 20,000$

5. $P = px - C = 60x - 0.01x^2 - (0.004x^2 + 25x + 4000)$

$\quad = -0.014x^2 + 35x - 4000$

$\dfrac{dP}{dx} = -0.028x + 35 = 0, \quad P'' = -0.028 < 0$

$0.028x = 35, \quad\quad\quad\quad x = \dfrac{35}{0.028} = 1250$

$p = 60 - 0.01(1250) = \$47.50$

9. $P = px - C = 70x - 0.1x^2 - 30x - 1000$

$\quad = -0.1x^2 + 40x - 1000$

$P' = -0.2x + 40 = 0 \quad\quad P'' = -0.2 < 0$

$0.2x = 40 \quad\quad\quad\quad x = 200$

$p = 70 - 0.1(200) = \$50$

$P = -0.1(40,000) + 40(200) - 1000 = \$3000$

13. $R' = 80 - 0.2x$

At $x = 300$, $R' = 80 - 0.2(300) = 20$

At $x = 500$, $R' = 80 - 0.2(500) = -20$

17. $P = R - C$, $\quad P' = R' - C' = 0$, $\quad R' = C'$

Section 6.5, pp. 158-159

1. $f(x) = 3x^4 - 2x^3 + 3x + c$

5. $f'(x) = \dfrac{x^{-1}}{-1} + C = -\dfrac{1}{x} + C$

9. $f'(x) = (x-1)^2 = x^2 - 2x + 1$

$f(x) = \dfrac{x^3}{3} - x^2 + x + C$

13. $f(x) = \dfrac{3}{2}x^{2/3} + C$, $\quad 2 = \dfrac{3}{2} + C$, $\quad C = \dfrac{1}{2}$

$f(x) = \dfrac{3x^{2/3} + 1}{2}$

17. $f'(x) = 12x + C_1$ $3 = 12 + C_1$

   $C_1 = -9$ $f'(x) = 12x - 9$

   $f(x) = 6x^2 - 9x + C_2$ $4 = 6 - 9 + C_2$

   $C_2 = 7$ $f(x) = 6x^2 - 9x + 7$

21. $f''(x) = 6x + C_1$ $6 = 6 + C_1$

   $C_1 = 0$ $f''(x) = 6x$

   $f'(x) = 3x^2 + C_2$ $4 = 3 + C_2$

   $C_2 = 1$ $f'(x) = 3x^2 + 1$

   $f(x) = x^3 + x + C_3$ $0 = 1 + 1 + C_3$

   $C_3 = -2$ $f(x) = x^3 + x - 2$

25. $a = 4t$ $v = 2t^2 + c_1$

   Since $v = 16$ when $t = 0$, $c_1 = 16$

   $v = 2t^2 + 16$ $s = \dfrac{2t^3}{3} + 16t + c_2$

   Since $s = 4$ when $t = 0$, $c_2 = 4$

   $s = \dfrac{2t^3}{3} + 16t + 4$

29. $a = -32$ $v = -32t + v_o$

   $s = -16t^2 + v_o t + s_o$

33. $a = kt$ $v = \dfrac{kt^2}{2} + c$

   Since $v = 0$ when $t = 0$, $c = 0$

   $v = \dfrac{kt^2}{2}$

   Since $v = 88$ when $t = 30$, $k = \dfrac{44}{225}$

   $v = \dfrac{22t^2}{225} = \dfrac{22 \cdot 60^2}{225} = 352$ ft/sec $= 240$ mph

37. $C' = 3x^2 - 4x + 5$ $\qquad$ $C = x^3 - 2x^2 + 5x + k$

$1000 = 1000 - 200 + 50 + k$

$k = 150$ $\qquad\qquad$ $C = x^3 - 2x^2 + 5x + 150$

Review 6, pp. 159-161

1. $P = x + 2y + \dfrac{\pi x}{2}$, $\quad y = \dfrac{2P - 2x - \pi x}{4}$

$A = xy + \dfrac{\pi x^2}{8} = \dfrac{2Px - 2x^2 - \pi x^2}{4} + \dfrac{\pi x^2}{8} = \dfrac{4Px - 4x^2 - \pi x^2}{8}$

$A' = \dfrac{4P - 8x - 2\pi x}{8}$

$2P - 4x - \pi x = 0,\quad x = \dfrac{2P}{4 + \pi}$, $y = \dfrac{2P - \dfrac{(2+\pi)\,2P}{4 + \pi}}{4} = \dfrac{P}{4 + \pi}$

$y + \dfrac{x}{2} = \dfrac{2P}{4 + \pi}$

5. C is total cost; Q is cost per widget

$C = 0.1x^2 + 20x + 9000$ $\qquad$ $Q = 0.1x + 20 + \dfrac{9000}{x}$

$Q' = 0.1 - \dfrac{9000}{x^2} = 0$ $\qquad$ $Q'' = \dfrac{18,000}{x^3} > 0$

$\dfrac{9000}{x^2} = 0.1$ $\qquad\qquad$ $x^2 = \dfrac{9000}{0.1} = 90,000$

$x = 300$

9. $V = \dfrac{4}{3}\pi r^3$ $\qquad\qquad\qquad$ $\dfrac{dV}{dt} = 4\pi r^2\,\dfrac{dr}{dt}$

$5 = 4\pi \cdot 9\,\dfrac{dr}{dt}$ $\qquad\qquad$ $\dfrac{dr}{dt} = \dfrac{5}{36\pi}$ ft/min

13. $f(x) = \dfrac{x^3}{3} - \dfrac{1}{x} + C$ $\qquad\qquad$ $1 = \dfrac{1}{3} - 1 + C$

$C = \dfrac{5}{3}$ $\qquad\qquad$ $f(x) = \dfrac{x^3}{3} - \dfrac{1}{x} + \dfrac{5}{3} = \dfrac{x^4 + 5x - 3}{3x}$

17. If y is the distance in miles, n is the number of
    steps and ℓ is the length of a stride, then

$$y = \frac{n\ell}{12 \cdot 5280} \qquad \text{When } \ell = 28, \quad y = 10.3.$$

$$10.3 = \frac{28n}{12 \cdot 5280} \qquad n = 23,300$$

$$y = \frac{23,300 \ \ell}{12 \cdot 5280} = 0.368 \ \ell$$

$$dy = 0.368 \ d\ell = 0.368(2) = 0.7 \text{ mi}$$

21. $y = x^3 + C$ $\qquad\qquad\qquad y' = 3x^2$

$$y' = -\frac{1}{3x^2} = -\frac{1}{3}x^{-2} \qquad y = \frac{1}{3x} + k$$

Chapter Seven
The Integral

1. $\displaystyle\sum_{i=1}^{7} i^3 = 1^3 + 2^3 + 3^3 + 4^3 + 5^3 + 6^3 + 7^3$

   $\qquad = 1 + 8 + 27 + 64 + 125 + 216 + 343$

5. $\displaystyle\sum_{i=0}^{7} (2i + 1) = (2\cdot 0 + 1) + (2\cdot 1 + 1) + (2\cdot 2 + 1) + (2\cdot 3 + 1)$

   $\qquad\qquad + (2\cdot 4 + 1) + (2\cdot 5 + 1) + (2\cdot 6 + 1) + (2\cdot 7 + 1)$

   $\qquad\quad = 1 + 3 + 5 + 7 + 9 + 11 + 13 + 15$

9. $\displaystyle\sum_{i=1}^{n} (2i + 1) = (2\cdot 1 + 1) + (2\cdot 2 + 1) + (2\cdot 3 + 1) + \cdots$

   $\qquad\qquad + (2\cdot n + 1) = 3 + 5 + 7 + \cdots + (2n + 1)$

13. $2 + 4 + 6 + \cdots + 22 = \displaystyle\sum_{i=1}^{11} 2i$

17. $1 + 3 + 5 + \cdots + (2n - 1) = \displaystyle\sum_{i=1}^{n} (2i - 1)$

21. $\displaystyle\sum_{i=1}^{n} (2i - 1) = 1 + 3 + 5 + \cdots + (2n - 1)$

    $\displaystyle\sum_{i=0}^{n-1} (2i + 1) = 1 + 3 + 5 + \cdots + (2n - 1)$

    Equal

25. $\displaystyle\sum_{i=0}^{n} (i^2 + 1) = 1 + 2 + 5 + \cdots + (n^2 + 1)$

    $\displaystyle\sum_{i=1}^{n+1} [(i - 1)^2 + 1] = 1 + 2 + 5 + \cdots + (n^2 + 1)$

    Equal

29. $\displaystyle\sum_{i=1}^{n} (\frac{i}{n})^2 \frac{1}{n} = \frac{1}{n^3} \sum_{i=1}^{n} i^2 = \frac{1}{n^3} \frac{n(n+1)(2n+1)}{6}$

    $\qquad\qquad = \frac{(n+1)(2n+1)}{6n^2}$

33. $\displaystyle\sum_{i=1}^{n} (\frac{i-1}{n})^2 \frac{1}{n} = \frac{1}{n^3} \sum_{i=1}^{n} (i-1)^2 = \frac{1}{n^3} \sum_{i=0}^{n-1} i^2$

$\qquad = \dfrac{1}{n^3} \displaystyle\sum_{i=1}^{n-1} i^2$

$\qquad = \dfrac{1}{n^3} \dfrac{(n-1)n(2n-1)}{6} = \dfrac{(n-1)(2n-1)}{6n^2}$

37. $\displaystyle\sum_{i=1}^{n} 1 = \underbrace{1 + 1 + 1 + \cdots + 1}_{n \text{ terms}} = n$

41.

$$\begin{array}{ccccccccccccc}
1 & + & 2 & + & 3 & + \cdots + & (n-2) & + & (n-1) & + & n \\
n & + & (n-1) & + & (n-2) & + \cdots + & 3 & + & 2 & + & 1 \\
\hline
(n+1) & + & (n+1) & + & (n+1) & + \cdots + & (n+1) & + & (n+1) & + & (n+1)
\end{array}$$

Hence $\quad 2\displaystyle\sum_{i=1}^{n} i = n(n+1), \quad \sum_{i=1}^{n} i = \dfrac{n(n+1)}{2}$

Section 7.2, pp. 173–174

1. $A(R_1) = A(R_2)$

$ab = A(R_1 \cup R_2) = A(R_1) + A(R_2) = 2A(R_1)$

$A(R_1) = \dfrac{1}{2} ab$

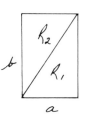

5. $\begin{aligned} x_0 &= 0 \\ x_1 &= \frac{1}{n} \\ x_2 &= \frac{2}{n} \\ &\ \ \vdots \\ x_n &= \frac{n}{n} \end{aligned} \qquad\qquad \begin{aligned} x_1^* &= \frac{1}{n} \\[4pt] x_2^* &= \frac{2}{n} \\ &\ \ \vdots \\ x_n^* &= \frac{n}{n} \end{aligned}$

$A = \displaystyle\lim_{n\to+\infty} \sum_{i=1}^{n} \frac{1}{n} f(\frac{i}{n}) = \lim_{n\to+\infty} \sum_{i=1}^{n} \frac{1}{n} \frac{i}{n} = \lim_{n\to+\infty} \frac{1}{n^2} \sum_{i=1}^{n} i$

$\qquad = \displaystyle\lim_{n\to+\infty} \frac{1}{n^2} \frac{n(n+1)}{2} = \lim_{n\to+\infty} \frac{1/n + 1}{2} = \frac{1}{2}$

9. $x_0 = 1$

    $x_1 = 1 + \dfrac{1}{n}$ $x_1^* = 1 + \dfrac{1}{n}$

    $x_2 = 1 + \dfrac{2}{n}$ $x_2^* = 1 + \dfrac{2}{n}$

    $\vdots$ $\vdots$

    $x_n = 1 + \dfrac{n}{n}$ $x_n^* = 1 + \dfrac{n}{n}$

$$A = \lim_{n \to +\infty} \sum_{i=1}^{n} \frac{1}{n} f(1 + \frac{i}{n}) = \lim_{n \to +\infty} \sum_{i=1}^{n} \frac{1}{n}(1 + \frac{i}{n})^2$$

$$= \lim_{n \to +\infty} \sum_{i=1}^{n} (\frac{1}{n} + \frac{2i}{n^2} + \frac{i^2}{n^3})$$

$$= \lim_{n \to +\infty} (\frac{1}{n} \cdot n + \frac{2}{n^2} \frac{n(n+1)}{2} + \frac{1}{n^3} \frac{n(n+1)(2n+1)}{6})$$

$$= \lim_{n \to +\infty} [1 + (1 + \frac{1}{n}) + \frac{(1 + 1/n)(2 + 1/n)}{6}] = 1 + 1 + \frac{1}{3} = \frac{7}{3}$$

13. $x_0 = 0$

    $x_1 = \dfrac{1}{n}$ $x_1^* = \dfrac{1}{n}$

    $x_2 = \dfrac{2}{n}$ $x_2^* = \dfrac{2}{n}$

    $\vdots$ $\vdots$

    $x_n = \dfrac{n}{n}$ $x_n^* = \dfrac{n}{n}$

$$A = \lim_{n \to +\infty} \sum_{i=1}^{n} \frac{1}{n} f(\frac{i}{n}) = \lim_{n \to +\infty} \sum_{i=1}^{n} \frac{1}{n} \frac{3i^2}{n^2} = \lim_{n \to +\infty} \frac{3}{n^3} \frac{n(n+1)(2n+1)}{6}$$

$$= \lim_{n \to +\infty} \frac{(1 + 1/n)(2 + 1/n)}{2} = 1$$

17. $x_0 = 0$

    $x_1 = \dfrac{1}{n}$ $x_1^* = \dfrac{1}{n}$

    $x_2 = \dfrac{2}{n}$ $x_2^* = \dfrac{2}{n}$

    $\vdots$ $\vdots$

    $x_n = \dfrac{n}{n}$ $x_n^* = \dfrac{n}{n}$

$$A = \lim_{n \to +\infty} \sum_{i=1}^{n} \frac{1}{n} f(\frac{i}{n}) = \lim_{n \to +\infty} \sum_{i=1}^{n} \frac{1}{n}(\frac{i}{n} - \frac{i^2}{n^2})$$

$$= \lim_{n \to +\infty} (\frac{1}{n^2} \frac{n(n+1)}{2} - \frac{1}{n^3} \frac{n(n+1)(2n+1)}{6})$$

$$= \lim_{n \to +\infty} (\frac{1+1/n}{2} - \frac{(1+1/n)(2+1/n)}{6}) = \frac{1}{6}$$

Section 7.3, pp. 178-179

1. $x_0 = 0$
    $x_1 = \frac{3}{n}$
    $x_2 = \frac{6}{n}$
    $\vdots$
    $x_n = \frac{3n}{n}$

    $x_1^* = \frac{3}{n}$
    $x_2^* = \frac{6}{n}$
    $\vdots$
    $x_n^* = \frac{3n}{n}$

$$\int_0^3 (x^2 - 1)\, dx = \lim_{n \to +\infty} \sum_{i=1}^{n} \frac{3}{n} f(\frac{3i}{n}) = \lim_{n \to +\infty} \sum_{i=1}^{n} \frac{3}{n}(\frac{9i^2}{n^2} - 1)$$

$$= \lim_{n \to +\infty} (\frac{27}{n^3} \frac{n(n+1)(2n+1)}{6} - \frac{3}{n} \cdot n) = \lim_{n \to +\infty} (\frac{9(1+1/n)(2+1/n)}{2} - 3)$$

$$= 6$$

5. $x_0 = 0$
    $x_1 = \frac{1}{n}$
    $x_2 = \frac{2}{n}$
    $\vdots$
    $x_n = \frac{n}{n}$

    $x_1^* = \frac{1}{n}$
    $x_2^* = \frac{2}{n}$
    $\vdots$
    $x_n^* = \frac{n}{n}$

$$\int_0^1 x^3\, dx = \lim_{n \to +\infty} \sum_{i=1}^{n} \frac{1}{n} \frac{i^3}{n^3} = \lim_{n \to +\infty} \frac{1}{n^4} \frac{n^2(n+1)^2}{4} = \lim_{n \to +\infty} \frac{(1+1/n)^2}{4} = \frac{1}{4}$$

9. $x_0 = 0$ $x_1^* = \frac{1}{n}$

$x_1 = \frac{1}{n}$ $x_2^* = \frac{2}{n}$

$x_2 = \frac{2}{n}$

$\vdots$ $\vdots$

$x_n = \frac{n}{n}$ $x_n^* = \frac{n}{n}$

$$\int_0^1 (x^2 - x)\,dx = \lim_{n \to +\infty} \sum_{i=1}^{n} \frac{1}{n}\left(\frac{i^2}{n^2} - \frac{i}{n}\right)$$

$$= \lim_{n \to +\infty} \left(\frac{1}{n^3} \frac{n(n+1)(2n+1)}{6} - \frac{1}{n^2} \frac{n(n+1)}{2}\right)$$

$$= \lim_{n \to +\infty} \left(\frac{(1 + 1/n)(2 + 1/n)}{6} - \frac{1 + 1/n}{2}\right) = -\frac{1}{6}$$

13. $x_0 = 0$ $x_1^* = \frac{1}{n}$ $x_0 = 1$ $x_1^* = 1 + \frac{1}{n}$

$x_1 = \frac{1}{n}$ $x_2^* = \frac{2}{n}$ $x_1 = 1 + \frac{1}{n}$ $x_2^* = 1 + \frac{2}{n}$

$x_2 = \frac{2}{n}$ $x_2 = 1 + \frac{2}{n}$

$\vdots$ $\vdots$ $\vdots$ $\vdots$

$x_n = \frac{n}{n}$ $x_n^* = \frac{n}{n}$ $x_n = 1 + \frac{n}{n}$ $x_n^* = 1 + \frac{n}{n}$

$$A = \int_0^2 \left|x^2 - x\right|\,dx = \int_0^1 (x - x^2)\,dx + \int_1^2 (x^2 - x)\,dx$$

$$\int_0^1 (x - x^2)\,dx = \lim_{n \to +\infty} \sum_{i=1}^{n} \frac{1}{n}\left(\frac{i}{n} - \frac{i^2}{n^2}\right)$$

$$= \lim_{n \to +\infty} \left(\frac{1}{n^2} \frac{n(n+1)}{2} - \frac{1}{n^3} \frac{n(n+1)(2n+1)}{6}\right)$$

$$= \lim_{n \to +\infty} \left(\frac{1 + 1/n}{2} - \frac{(1 + 1/n)(2 + 1/n)}{6}\right) = \frac{1}{6}$$

$$\int_1^2 (x^2 - x)\,dx = \lim_{n \to +\infty} \sum_{i=1}^{n} \frac{1}{n}\left[\left(1 + \frac{i}{n}\right)^2 - \left(1 + \frac{i}{n}\right)\right]$$

$$= \lim_{n \to +\infty} \sum_{i=1}^{n} \frac{1}{n}\left(\frac{i}{n} + \frac{i^2}{n^2}\right)$$

$$= \lim_{n \to +\infty} \left(\frac{1}{n^2} \frac{n(n+1)}{2} + \frac{1}{n^3} \frac{n(n+1)(2n+1)}{6} \right)$$

$$= \lim_{n \to +\infty} \left(\frac{1+1/n}{2} + \frac{(1+1/n)(2+1/n)}{6} \right) = \frac{5}{6}$$

$$A = \frac{1}{6} + \frac{5}{6} = 1$$

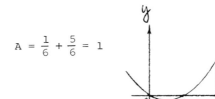

17. $x_0 = 0$ $\qquad x_n^* = \frac{1}{n}$

$x_1 = \frac{1}{n}$ $\qquad x_2^* = \frac{2}{n}$

$x_2 = \frac{2}{n}$

$\vdots$ $\qquad\qquad \vdots$

$x_n = \frac{n}{n}$ $\qquad x_n^* = \frac{n}{n}$

$$\int_0^1 (x^2 - x^3)\, dx = \lim_{n \to +\infty} \sum_{i=1}^{n} \frac{1}{n} \left(\frac{i^2}{n^2} - \frac{i^3}{n^3} \right)$$

$$= \lim_{n \to +\infty} \left(\frac{1}{n^3} \frac{n(n+1)(2n+1)}{6} - \frac{1}{n^4} \frac{n^2(n+1)^2}{4} \right)$$

$$= \lim_{n \to +\infty} \left(\frac{(1+1/n)(2+1/n)}{6} - \frac{(1+1/n)^2}{4} \right) = \frac{1}{12}$$

21. The graph of an odd function is symmetric about the origin. Thus

$$\int_{-a}^{0} f(x)\, dx = - \int_0^a f(x)\, dx$$

or

$$\int_{-a}^{a} f(x)\, dx = \int_{-a}^{0} f(x)\, dx + \int_0^a f(x)\, dx = 0$$

The graph of an even function is symmetric about the y axis. Thus

$$\int_{-a}^{0} f(x)\,dx = \int_{0}^{a} f(x)\,dx$$

or

$$\int_{-a}^{a} f(x)\,dx = \int_{-a}^{0} f(x)\,dx + \int_{0}^{a} f(x)\,dx = 2\int_{0}^{a} f(x)\,dx$$

Section 7.4, pp. 182–183

1. $\displaystyle\int_{0}^{2} x^2\,dx = \frac{x^3}{3}\Big|_{0}^{2} = \frac{8}{3}$

5. $\displaystyle\int_{1}^{2} (3x^2 + 2x)\,dx = (x^3 + x^2)\Big|_{1}^{2} = 8 + 4 - 2 = 10$

9. $\displaystyle\int_{1}^{2} x^2(x-1)\,dx = \int_{1}^{2} (x^3 - x^2)\,dx = \left(\frac{x^4}{4} - \frac{x^3}{3}\right)\Big|_{1}^{2}$

$\displaystyle\quad = \left(4 - \frac{8}{3}\right) - \left(\frac{1}{4} - \frac{1}{3}\right) = \frac{17}{12}$

13. $\displaystyle\int_{1}^{8} \sqrt[3]{x}\ dx = \frac{3x^{4/3}}{4}\Big|_{1}^{8} = \frac{3}{4}\,(16 - 1) = \frac{45}{4}$

17. $\displaystyle\int_{1}^{2} \frac{1}{x^2}\ dx = -\frac{1}{x}\Big|_{1}^{2} = -\frac{1}{2} + 1 = \frac{1}{2}$

21. $\displaystyle A = \int_{0}^{1} (x^2 + 1)\,dx$

$\displaystyle\quad = \left(\frac{x^3}{3} + x\right)\Big|_{0}^{1} = \frac{4}{3}$

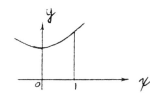

25. $\displaystyle A = \int_{0}^{1} (x-1)^2\,dx = \int_{0}^{1} (x^2 - 2x + 1)\,dx$

$\displaystyle\quad = \left(\frac{x^3}{3} - x^2 + x\right)\Big|_{0}^{1}$

$\displaystyle\quad = \frac{1}{3} - 1 + 1 = \frac{1}{3}$

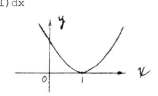

29. $A = \displaystyle\int_0^4 \sqrt{x}\; dx = \dfrac{2}{3}\, x^{3/2} \Big|_0^4$

$= \dfrac{2}{3} \cdot 8 = \dfrac{16}{3}$

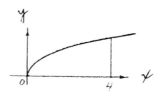

33. (a) $f(x) = \begin{cases} 1 & \text{if } x \leqq 0 \\[2mm] 0 & \text{if } x > 0 \end{cases}$

For $x = 0,\ \displaystyle\lim_{h \to 0} m = 0 \neq f(0)$

(b) $f(x) = \begin{cases} 0 & \text{if } x \neq 0 \text{ and } x \text{ is rational} \\[2mm] 1 & \text{if } x \text{ is irrational} \\[2mm] \dfrac{1}{2} & \text{if } x = 0 \end{cases}$

For $x = 0,\ \displaystyle\lim_{h \to 0} m = 0 \neq f(0)$ and $\displaystyle\lim_{h \to 0} M = 1 \neq f(0)$

Section 7.5, p. 187

1. $\displaystyle\int 2(2x - 3)^2\, dx = \dfrac{1}{3}\,(2x - 3)^3 + C$

5. $\displaystyle\int \sqrt{4x - 3}\; dx = \dfrac{1}{4} \int 4\,\sqrt{4x - 3}\; dx = \dfrac{1}{4}\,\dfrac{(4x - 3)^{3/2}}{3/2} + C$

$= \dfrac{1}{6}\,(4x - 3)^{3/2} + C$

9. $\displaystyle\int (x^2 + 4)^2 x^3\, dx = \int (x^7 + 8x^5 + 16x^3)\, dx = \dfrac{x^8}{8} + \dfrac{4x^6}{3} + 4x^4 + C$

13. $\displaystyle\int \dfrac{(x^{1/3} - 1)^5}{x^{2/3}}\; dx = 3\int \dfrac{1}{3}\, x^{-2/3}(x^{1/3} - 1)^5\, dx$

$= \dfrac{1}{2}\,(x^{1/3} - 1)^6 + C$

17. $\displaystyle\int x \sqrt[3]{3x^2 - 5}\; dx = \dfrac{1}{6} \int 6x \sqrt[3]{3x^2 - 5}\; dx = \dfrac{1}{6} \cdot \dfrac{3}{4}(3x^2 - 5)^{4/3} + C$

$= \dfrac{1}{8}(3x^2 - 5)^{4/3} + C$

21. $\displaystyle\int \frac{x^3 + 8x^2 + 20x + 20}{(x + 2)^2} dx = \int (x + 4 + \frac{4}{(x + 2)^2})\ dx$

$\displaystyle = \frac{x^2}{2} + 4x - \frac{4}{x + 2} + c$

25. $\displaystyle\int_1^2 (x^3 - x)^2 (3x^2 - 1)\, dx = \frac{1}{3}(x^3 - x)^3 \Big|_1^2 = \frac{1}{3}(6^3 - 0) = 72$

29. $\displaystyle\int_1^2 (x^2 - 3)^3 x\ dx = \frac{1}{2}\int_1^2 (x^2 - 3)^3 2x\ dx = \frac{1}{8}(x^2 - 3)^4 \Big|_1^2$

$\displaystyle = \frac{1}{8}(1 - 16) = -\frac{15}{8}$

Section 7.6, p. 191

1. $\displaystyle A = \int_0^2 x\ dy = \int_0^2 (2y - y^2)\, dy$

$\displaystyle = (y^2 - \frac{y^3}{3}) \Big|_0^2 = 4 - \frac{8}{3} = \frac{4}{3}$

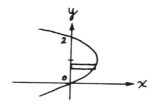

5. $\displaystyle A = \int_2^4 - x\ dy + \int_4^5 x\ dy$

$\displaystyle = \int_2^4 (4y - y^2)\, dy + \int_4^5 (y^2 - 4y)\, dy$

$\displaystyle = (2y^2 - \frac{y^3}{3}) \Big|_2^4 + (\frac{y^3}{3} - 2y^2) \Big|_4^5$

$\displaystyle = (32 - \frac{64}{3}) - (8 - \frac{8}{3}) + (\frac{125}{3} - 50) - (\frac{64}{3} - 32) = \frac{23}{3}$

9. $\displaystyle A = \int_0^3 (y_1 - y_2)\, dx = \int_0^3 (2x - x^2 + x)\, dx$

$\displaystyle = \int_0^3 (3x - x^2)\, dx = (\frac{3x^2}{2} - \frac{x^3}{3}) \Big|_0^3$

$\displaystyle = \frac{27}{2} - 9 = \frac{9}{2}$

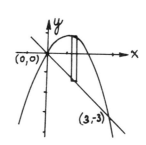

13. $y = \dfrac{4}{x^2}$, $y = 7 - 3x$

$\dfrac{4}{x^2} = 7 - 3x,\quad 3x^3 - 7x^2 + 4 = 0,$

$(x - 1)(3x + 2)(x - 2) = 0$

$x = 1,\quad -2/3,\quad 2$

$A = \displaystyle\int_1^2 (y_1 - y_2)\,dx = \int_1^2 (7 - 3x - \dfrac{4}{x^2})\,dx$

$= (7x - \dfrac{3x^2}{2} + \dfrac{4}{x})\Big|_1^2$

$= (14 - 6 + 2) - (7 - \dfrac{3}{2} + 4) = \dfrac{1}{2}$

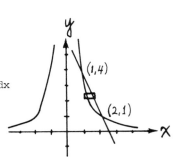

17. $A = \displaystyle\int_{-2}^{1/2} (y_1 - y_2)\,dx = \int_{-2}^{1/2} \big[(2 - x - x^2) - (x^2 + 2x)\big]\,dx$

$= \displaystyle\int_{-2}^{1/2} (2 - 3x - 2x^2)\,dx$

$= (2x - \dfrac{3x^2}{2} - \dfrac{2x^3}{3})\Big|_{-2}^{1/2}$

$= (1 - \dfrac{3}{8} - \dfrac{1}{12}) - (-4 - 6 + \dfrac{16}{3}) = \dfrac{125}{24}$

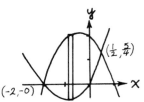

21. $A = \displaystyle\int_0^2 (y_1 - y_2)\,dx = \int_0^2 (2x - x^2)\,dx$

$= (x^2 - \dfrac{x^3}{3})\Big|_0^2 = 4 - \dfrac{8}{3} = \dfrac{4}{3}$

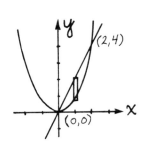

25. $A = \displaystyle\int_1^2 (y_1 - y_2)\,dx + \int_2^5 (y_1 - y_3)\,dx$

$= \displaystyle\int_1^2 (- \dfrac{x}{2} + \dfrac{7}{2} + 5x - 8)\,dx + \int_2^5 (- \dfrac{x}{2} + \dfrac{7}{2} - x + 4)\,dx$

$= \displaystyle\int_1^2 (\dfrac{9x}{2} - \dfrac{9}{2})\,dx + \int_2^5 (- \dfrac{3x}{2} + \dfrac{15}{2})\,dx$

$$= (\frac{9x^2}{4} - \frac{9x}{2}) \Big|_1^2$$

$$+ (- \frac{3x^2}{4} + \frac{15x}{2}) \Big|_2^5$$

$$= (9 - 9) - (\frac{9}{4} - \frac{9}{2})$$

$$+ (- \frac{75}{4} + \frac{75}{2}) - (-3 + 15) = 9$$

$x + 2y_1 - 7 = 0$ $5x + y_2 - 8 = 0$

$x - y_3 - 4 = 0$

$(1, 3)$

$(5, 1)$

$(2, -2)$

Section 7.7, pp. 195-196

1. $F = kx$, $3 = \frac{1}{3} k$, $k = 9$, $F = 9x$

$$W = \int_0^1 9x \ dx = \frac{9x^2}{2} \Big|_0^1 = \frac{9}{2} \text{ ft lb}$$

5. $F = kx$, $5 = \frac{2}{3} k$, $k = \frac{15}{2}$, $F = \frac{15x}{2}$

$$W = \int_1^3 \frac{15x}{2} \ dx = \frac{15x^2}{4} \Big|_1^3 = \frac{15}{4}(9 - 1) = 30 \text{ ft lb}$$

9. $W = \int_{-3}^6 998.4\pi x \ dx = 499.2\pi x^2 \Big|_{-3}^6$

$$= 13,600\pi = 42,330 \text{ ft lb}$$

$$W = \int_0^6 998.4\pi x \ dx = 499.2\pi x^2 \Big|_0^6$$

$$= 17,970\pi = 56,450 \text{ ft lb}$$

— -3

— 0

— 6

13. $W = \int_0^{50} x \ dx = \frac{x^2}{2} \Big|_0^{50} = \frac{2500}{2} = 1250 \text{ ft lb}$

17. $W = \int_0^{15} (\frac{1}{4} \cdot \frac{x}{2} + 100) \, dx = (\frac{x^2}{16} + 100x) \Big|_0^{15}$

$= \frac{225}{16} + 1500 = 1514$ ft lb

Section 7.8, pp. 201-202

1. $x_0 = 1$ $y_0 = 2$

    $x_1 = 5/4$ $y_1 = 2.5625$

    $x_2 = 3/2$ $y_2 = 3.25$

    $x_3 = 7/4$ $y_3 = 4.0625$

    $x_4 = 2$ $y_4 = 5$

    $\int_1^2 (x^2 + 1) \, dx \doteq \frac{1}{8}(2 + 5.1250 + 6.5 + 8.1250 + 5) = 3.344$

    $\int_1^2 (x^2 + 1) \, dx \doteq \frac{1}{12}(2 + 10.25 + 6.5 + 16.25 + 5) = 3.333$

    $\int_1^2 (x^2 + 1) \, dx = (\frac{x^3}{3} + x) \Big|_1^2 = (\frac{8}{3} + 2) - (\frac{1}{3} + 1) = 3.333$

5. $x_0 = 0$ $y_0 = 1$

    $x_1 = 1/4$ $y_1 = 1.118$

    $x_2 = 1/2$ $y_2 = 1.2245$

    $x_3 = 3/4$ $y_3 = 1.323$

    $x_4 = 1$ $y_4 = 1.414$

    $\int_0^1 \sqrt{x+1} \, dx \doteq \frac{1}{8}(1 + 2.236 + 2.449 + 2.646 + 1.414) = 1.218$

    $\int_0^1 \sqrt{x+1} \, dx \doteq \frac{1}{12}(1 + 4.472 + 2.449 + 5.292 + 1.414) = 1.219$

    $\int_0^1 \sqrt{x+1} \, dx = \frac{2}{3}(x+1)^{3/2} \Big|_0^1 = \frac{2}{3}(2\sqrt{2} - 1) = 1.219$

9. $x_0 = 1$ $y_0 = 2$

    $x_1 = 5/4$ $y_1 = 1.64$

    $x_2 = 3/2$ $y_2 = 1.4444$

$$x_3 = 7/4 \qquad y_3 = 1.3265$$

$$x_4 = 2 \qquad y_4 = 1.25$$

$$\int_1^2 \frac{x^2+1}{x^2}\, dx \doteq \frac{1}{8}(2 + 3.28 + 2.8888 + 2.6530 + 1.25) = 1.509$$

$$\int_1^2 \frac{x^2+1}{x^2}\, dx \doteq \frac{1}{12}(2 + 6.56 + 2.8888 + 5.3060 + 1.25) = 1.500$$

$$\int_1^2 \frac{x^2+1}{x^2}\, dx = \int_1^2 (1 + x^{-2})\, dx = \left. \left(x - \frac{1}{x}\right)\right|_1^2$$

$$= \left(2 - \frac{1}{2}\right) - (1 - 1) = 1.500$$

13.
$$x_0 = 1 \qquad y_0 = 2$$

$$x_1 = 1.4 \qquad y_1 = 1.7143$$

$$x_2 = 1.8 \qquad y_2 = 1.5556$$

$$x_3 = 2.2 \qquad y_3 = 1.4545$$

$$x_4 = 2.6 \qquad y_4 = 1.3846$$

$$x_5 = 3 \qquad y_5 = 1.3333$$

$$\int_1^3 \frac{x+1}{x}\, dx \doteq \frac{1}{5}(2 + 3.4286 + 3.1112 + 2.9090 + 2.7692$$

$$+ 1.3333) = 3.110$$

$$f'(x) = -x^{-2}, \quad f''(x) = 2x^{-3}$$

$$E_T = -\frac{h^2}{12}(b-a)f''(c) = -\frac{(2/5)^2}{12}(2)\,\frac{2}{c^3}, \qquad 1 \leqq c \leqq 3$$

$$|E_T| = \frac{4}{75c^3} \leqq \frac{4}{75} = 0.053, \qquad 3.06 \leqq I \leqq 3.11$$

17.
$$x_0 = 0 \qquad y_0 = 1$$

$$x_1 = 1/4 \qquad y_1 = 0.9682$$

$$x_2 = 1/2 \qquad y_2 = 0.8660$$

$$x_3 = 3/4 \qquad y_3 = 0.6614$$

$$x_4 = 1 \qquad y_4 = 0$$

$$\int_0^1 \sqrt{1 - x^2} \; dx \doteq \frac{1}{12}(1 + 3.8728 + 1.7320 + 2.6456 + 0)$$

$$= 0.771$$

$$f'(x) = \frac{-x}{2\sqrt{1 - x^2}} \text{ does not exist when } x = 1$$

21. $x_0 = 0$ $\qquad\qquad$ $y_0 = 0$

$x_1 = 1/4$ $\qquad\qquad$ $y_1 = 0.2$

$x_2 = 1/2$ $\qquad\qquad$ $y_2 = 0.3333$

$x_3 = 3/4$ $\qquad\qquad$ $y_3 = 0.4286$

$x_4 = 1$ $\qquad\qquad$ $y_4 = 0.5$

$$\int_0^1 \frac{x \, dx}{1 + x} \doteq \frac{1}{12}(0 + 0.8 + 0.6666 + 1.7144 + 0.5) = 0.307$$

$$f'(x) = (1 + x)^{-2}, \qquad f''(x) = -2(1 + x)^{-3},$$

$$f'''(x) = 6(1 + x)^{-4}, \qquad f^{(4)}(x) = -\frac{24}{(1 + x)^5}$$

$$E_x = -\frac{h^4}{180}(b - a) f^{(4)}(c)$$

$$= -\frac{(1/4)^4}{180}(1) \frac{-24}{(1 + c)^5}, \qquad 0 \leqq c \leqq 1$$

$$\left| E_s \right| = \frac{1}{1920(1 + c)^5} \leqq \frac{1}{1920} = 0.0005, \quad 0.307 \leqq I \leqq 0.308$$

25. $x_0 = 1$ $\qquad\qquad$ $y_0 = 0$

$x_1 = 5/4$ $\qquad\qquad$ $y_1 = 0.45$

$x_2 = 3/2$ $\qquad\qquad$ $y_2 = 0.8333$

$x_3 = 7/4$ $\qquad\qquad$ $y_3 = 1.1786$

$x_4 = 2$ $\qquad\qquad$ $y_4 = 1.5$

$x_5 = 9/4$ $\qquad\qquad$ $y_5 = 1.8056$

$x_6 = 5/2$ $\qquad\qquad$ $y_6 = 2.1$

$x_7 = 11/4$ $\qquad\qquad$ $y_7 = 2.3864$

$x_8 = 3$ $\qquad\qquad$ $y_8 = 2.6667$

$$\int_1^3 \frac{x^2 - 1}{x}dx \doteq \frac{1}{8}(0 + 0.9 + 1.6666 + 2.3572 + 3 + 3.6112$$

$$+ 4.2 + 4.7728 + 2.6667) = 2.897$$

$$\int_1^3 \frac{x^2 - 1}{x}dx \doteq \frac{1}{12}(0 + 1.8 + 1.6666 + 4.7144 + 3 + 7.2224$$

$$+ 4.2 + 9.5456 + 2.6667) = 2.901$$

29. $f'(x) = \dfrac{x}{\sqrt{x^2 + 4}}$, $\qquad f''(x) = \dfrac{4}{(x^2 + 4)^{3/2}}$

$h = 1/n$

$$E_T = -\frac{h^2}{12}(b - a)f''(c) = -\frac{1/n^2}{12}(1)\frac{4}{(c^2 + 4)^{3/2}}$$

$$0 \leqq c \leqq 1$$

$$|E_T| = \frac{1}{3n^2(c^2 + 4)^{3/2}} \leqq \frac{1}{3n^2 \cdot 4^{3/2}} = \frac{1}{24n^2} < \frac{1}{1000}$$

$$n^2 > \frac{1000}{24} > 41 \qquad n = 7$$

Review 7, pp. 202-203

1. $\displaystyle\sum_{i=1}^{n}\left[\left(\frac{2i}{n}\right)^2 - \frac{6i}{n}\right]\frac{2}{n} = \sum_{i=1}^{n}\left(\frac{8i^2}{n^3} - \frac{12i}{n^2}\right)$

$$= \frac{8}{n^3}\frac{n(n+1)(2n+1)}{6} - \frac{12}{n^2}\frac{n(n+1)}{2}$$

$$= \frac{4(n+1)(2n+1)}{3n^2} - \frac{6(n+1)}{n} = \frac{2(n+1)(4n+2-9n)}{3n^2}$$

$$= \frac{2(n+1)(2-5n)}{3n^2}$$

5. $\displaystyle\int_1^2 \frac{(x^2 - 2)^2}{x^2}dx = \int_1^2 \frac{x^4 - 4x^2 + 4}{x^2} \, dx$

$$= \int_1^2 (x^2 - 4 + 4x^{-2}) \, dx = \left(\frac{x^3}{3} - 4x - \frac{4}{x}\right)\Big|_1^2$$

$$= \left(\frac{8}{3} - 8 - 2\right) - \left(\frac{1}{3} - 4 - 4\right) = \frac{1}{3}$$

9. $A = \int_{-3}^{2}(x^3 - 4x + 15)\,dx$

   $= (\dfrac{x^4}{4} - 2x^2 + 15x)\Big\lfloor_{-3}^{2}$

   $= (4 - 8 + 30) - (\dfrac{81}{4} - 18 - 45)$

   $= \dfrac{275}{4}$

13. $A = \int_{0}^{1}(y_1 - y_2)\,dx$

    $= \int_{0}^{1}(x^3 - x^4)\,dx$

    $= (\dfrac{x^4}{4} - \dfrac{x^5}{5})\Big|_{0}^{1}$

    $= \dfrac{1}{4} - \dfrac{1}{5} = \dfrac{1}{20}$

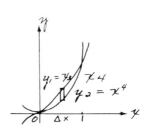

17. $F = kx$, $2 = \dfrac{1}{2}k$, $k = 4$

    $W = \int_{1}^{2} 4x\,dx = 2x^2\Big|_{1}^{2} = 8 - 2 = 6$ ft lb

21. $x_0 = 0 \qquad\qquad y_0 = 1$

    $x_1 = 1 \qquad\qquad y_1 = \sqrt{2} = 1.732$

    $x_2 = 2 \qquad\qquad y_2 = 3$

    $x_3 = 3 \qquad\qquad y_3 = \sqrt{28} = 5.292$

    $x_4 = 4 \qquad\qquad y_4 = \sqrt{65} = 8.062$

    $\int_{0}^{4}\sqrt{x^3 + 1}\,dx \doteq \dfrac{h}{2}(y_0 + 2y_1 + 2y_2 + 2y_3 + y_4)$

    $= \dfrac{1}{2}(1 + 3.464 + 6 + 10.584 + 8.062) = 14.555$

    $\int_{0}^{4}\sqrt{x^3 + 1}\,dx \doteq \dfrac{h}{3}(y_0 + 4y_1 + 2y_2 + 4y_3 + y_4)$

    $= \dfrac{1}{3}(1 + 6.928 + 6 + 21.168 + 8.062) = 14.386$

25. $\displaystyle\int_1^2 (3x^2 - 4x + 1)\,dx = \lim_{n \to +\infty} \sum_{i=1}^{n} f(x_i^*)\triangle x_i$

$\displaystyle = \lim_{n \to +\infty} \sum_{i=1}^{n} f(1 + \frac{i}{n})\frac{1}{n} = \lim_{n \to +\infty} \sum_{i=1}^{n} [3(1 + \frac{i}{n})^2 - 4(1 + \frac{i}{n}) + 1]\frac{1}{n}$

$\displaystyle = \lim_{n \to +\infty} \frac{1}{n} \sum_{i=1}^{n} (\frac{2i}{n} + \frac{3i^2}{n^2})$

$\displaystyle = \lim_{n \to +\infty} \frac{1}{n}(\frac{2}{n}\frac{n(n+1)}{2} + \frac{3}{n^2}\frac{n(n+1)(2n+1)}{6})$

$\displaystyle = \lim_{n \to +\infty} [(1 + \frac{1}{n}) + \frac{(1 + 1/n)(2 + 1/n)}{2}]$

$\displaystyle = 1 + \frac{1 \cdot 2}{2} = 2$

Chapter Eight
Vectors in the Plane

1. $v = (-2-4)i + (1-3)j$

   $= -6i - 2j$

5. $v = (0-1)i + (3+2)j$

   $= -i + 5j$

9. $w = \dfrac{3i-j}{\sqrt{9+1}} = \dfrac{3}{\sqrt{10}}i - \dfrac{1}{\sqrt{10}}j$

13. $w = \dfrac{i+2j}{\sqrt{1+4}} = \dfrac{1}{\sqrt{5}}i + \dfrac{2}{\sqrt{5}}j$

17. $x-1 = 3, \quad x = 4$

   $y-4 = -1, \quad y = 3$

   $B = (4, \; 3)$

21. $x_2 - x_1 = 3$ $\qquad\qquad y_2 - y_1 = 5$

   $\dfrac{x_2 + x_1}{2} = 4$ $\qquad\qquad \dfrac{y_2 + y_1}{2} = 1$

   $x_1 = \dfrac{5}{2}, \quad x_2 = \dfrac{11}{2}$ $\qquad y_1 = -\dfrac{3}{2}, \quad y_2 = \dfrac{7}{2}$

   $A = (\dfrac{5}{2}, \; -\dfrac{3}{2}), \; B = (\dfrac{11}{2}, \; \dfrac{7}{2})$

25. $4-x = 3, \quad x = 1$ $\qquad\qquad 2-y = -1, \quad y = 3$

   $A = (1, \; 3)$

29. $u+v = (3+1)i + (-1+2)j$

   $= 4i + j$

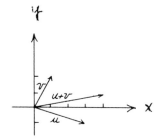

33. $2u + v = (2 \cdot 1 + 2)i + [(2(-3) + 4]j$

$\qquad = 4i - 2j$

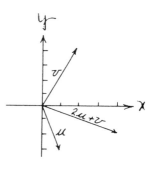

37. $3u + 2v = (3 \cdot 2 + 2 \cdot 4)i + [3(-3) + 2(-1)]j = 14i - 11j$

41. Let $\overrightarrow{AB}$ and $\overrightarrow{A'B'}$ be representations of u and $\overrightarrow{BC}$ and $\overrightarrow{B'C'}$ representations of v. $\overrightarrow{AB}$ and $\overrightarrow{A'B'}$ are the same length and parallel; similarly $\overrightarrow{BC}$ and $\overrightarrow{B'C'}$ are the same length and parallel. Thus $\angle ABC = \angle A'B'C'$ and $\triangle ABC \cong \triangle A'B'C'$. The sides AC and $A'C'$ are the same length and parallel. Thus $\overrightarrow{AC} \equiv \overrightarrow{A'C'}$; both are representative of u + v.

Section 8.2, pp. 213-214

1. $\cos \theta = \dfrac{u \cdot v}{|u| \cdot |v|} = \dfrac{3 \cdot 1 - 1 \cdot 2}{\sqrt{9+1}\sqrt{1+4}} = \dfrac{1}{\sqrt{50}} = \dfrac{\sqrt{2}}{10} = 0.1414$

$\theta = 82°$

5. $\cos \theta = \dfrac{u \cdot v}{|u||v|} = \dfrac{2 \cdot 1 - 1 \cdot 2}{|u||v|} = 0 , \qquad \theta = 90°$

9. $u \cdot v = 1 \cdot 2 - 1 \cdot 1 = 1 \neq 0,$ not orthogonal

13. $u \cdot v = 1 \cdot 3 - 1 \cdot 4 = -1;$ not orthogonal

17. $w = \dfrac{u \cdot v}{|v|^2} v = \dfrac{2 \cdot 1 - 1 \cdot 1}{1 + 1}(i + j) = \dfrac{1}{2}i + \dfrac{1}{2}j$

21. $w = \dfrac{u \cdot v}{|v|^2} v = \dfrac{1 \cdot 2 - 1 \cdot 1}{4 + 1}(2i + j) = \dfrac{2}{5}i + \dfrac{1}{5}j$

25. $u \cdot v = 3 \cdot 1 - 1 \cdot a = 0, \quad a = 3$

29. $\cos \dfrac{\pi}{3} = \dfrac{u \cdot v}{|u| \cdot |v|} = \dfrac{a \cdot 1 + 2(-1)}{\sqrt{a^2 + 4} \sqrt{1 + 1}}$

$\dfrac{1}{2} = \dfrac{a - 2}{\sqrt{2(a^2 + 4)}}$, $\qquad 2a^2 + 8 = 4a^2 - 16a + 16$

$2a^2 - 16a + 8 = 0$, $\qquad a^2 - 8a + 4 = 0$

$a = \dfrac{8 \pm \sqrt{64 - 16}}{2} = 4 \pm 2\sqrt{3}$

$a = 4 + 2\sqrt{3}$ ($a = 4 - 2\sqrt{3}$ is extraneous)

33. $\cos \theta = \dfrac{u \cdot v}{|u| \, |v|} = \dfrac{1 \cdot a + 1(-1)}{\sqrt{1 + 1} \sqrt{a^2 + 1}} = \dfrac{a - 1}{\sqrt{2a^2 + 2}} = \pm 1$

$\dfrac{a^2 - 2a + 1}{2a^2 + 2} = 1 \qquad\qquad\qquad a^2 - 2a + 1 = 2a^2 + 2$

$a^2 + 2a + 1 = 0 \qquad\qquad\qquad\qquad (a + 1)^2 = 0$

$a = -1$

37. $u = i + 4j$, $\qquad v = 2i - j$, $\qquad w = i - 5j$

$p_1(v \text{ on } u) = \dfrac{v \cdot u}{|u|^2} u = \dfrac{2 \cdot 1 - 1 \cdot 4}{1 + 16}(i + 4j) = -\dfrac{2}{17}i - \dfrac{8}{17}j$

$p_2(w \text{ on } u) = \dfrac{w \cdot u}{|u|^2} u = \dfrac{1 \cdot 1 - 5 \cdot 4}{1 + 16}(i + 4j) = -\dfrac{19}{17}i - \dfrac{76}{17}j$

41. Since the zero vector is orthogonal to every other vector and $0 \cdot v = 0$, that case is trivial. Suppose now that neither u nor v is the zero vector. If u and v are orthogonal, then

$$\frac{u \cdot v}{|u| \cdot |v|} = \cos 90° = 0$$

Thus $u \cdot v = 0$. On the other hand, if $u \cdot v = 0$, $\cos \theta = 0$ and $\theta = 90°$.

Section 8.3, pp. 220-221

1. $q = \dfrac{1}{2}u + v + \dfrac{1}{2}w$

$q = -\dfrac{1}{2}u + p - \dfrac{1}{2}w$

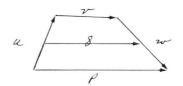

Adding, we have

$$2q = p + v, \quad q = \frac{p + v}{2}$$

5. $v = a + b$

$$|v|^2 = |a + b|^2 = |a|^2 + |b|^2$$

$u = a + d = a - b$

$$|u| = |a - b|^2 = |a|^2 + |-b|^2 = |a|^2 + |b|^2$$

$$|v| = |u|$$

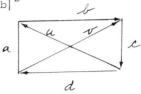

9. $f_1 = 5i, \quad f_2 = 10j, \quad f_1 + f_2 = 5i + 10j$

$$|f_1 + f_2| = \sqrt{25 + 100} = \sqrt{125} = 5\sqrt{5}$$

$$m = b/a = 10/5 = 2, \quad \theta \doteq 63°$$

$5\sqrt{5}$ lb inclined upward to the right and at an angle of 63° with the horizontal

13. $f_1 + f_2 + f_3 = 6i + 2j$

17. $f_1 = 3i, \quad f_2 = \frac{5\sqrt{2}}{2}i + \frac{5\sqrt{2}}{2}j$

$$-(f_1 + f_2) = -(3 + \frac{5\sqrt{2}}{2})i - \frac{5\sqrt{2}}{2}j$$

$$|-(f_1 + f_2)| = \sqrt{9 + 15\sqrt{2} + 25/2 + 25/2} = \sqrt{34 + 15\sqrt{2}}$$

$$m = b/a = \frac{-5\sqrt{2}/2}{-(6 + 5\sqrt{2})/2} = \frac{5\sqrt{2}}{6 + 5\sqrt{2}} = \frac{30\sqrt{2} - 50}{36 - 50} = \frac{25 - 15\sqrt{2}}{7}$$

$$\doteq 0.5414, \quad \theta \doteq 28°$$

$\sqrt{34 + 15\sqrt{2}}$ lb inclined downward to the left and at an angle of 28° with the horizontal

21. $T_o = \dfrac{L^2 w}{8H} = \dfrac{120^2 \cdot 1000}{8 \cdot 12.5} = 144,000$

$y = \dfrac{x^2 w}{2T_o} = \dfrac{1000}{288,000}\, x^2 = \dfrac{x^2}{288}$

$T_{max} = \dfrac{1}{2}\sqrt{4T_o^2 + L^2 w^2} = \dfrac{1}{2}\sqrt{4 \cdot 144,000^2 + (120 \cdot 1000)^2}$

$\qquad = 1000\sqrt{86,544} = 294,000$

25. $T_o = \dfrac{L^2 w}{8H}$, $\qquad 500,000 = \dfrac{400^2 w}{8 \cdot 40}$, $\qquad w = 1000$ lb/ft

Review 8, p. 221

1. $\dfrac{v}{|v|} = \dfrac{2i - 3j}{\sqrt{4+9}} = \dfrac{2}{\sqrt{13}}i - \dfrac{3}{\sqrt{13}}j$

5. $u \cdot v = 2 \cdot 3 - 1(-2) = 6 + 2 = 8$

9. $w = \dfrac{u \cdot v}{|v|^2}\, v = \dfrac{4 \cdot 2 + 1(-2)}{4 + 4}(2i - 2j) = \dfrac{3}{2}i - \dfrac{3}{2}j$

13. $u = a + b,$ $\qquad v = a + d$

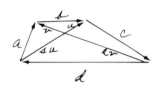

$d = -nb$, $\qquad rv = d + su$

$r(a + d) = -nb + s(a + b)$

$r(a - nb) = -nb + s(a + b)$, $\quad (r - s)a = (rn - n + s)b$

$r - s = 0$, $\qquad rn - n + s = 0$

$rn + r - n = 0$ $\qquad r(1 + n) = n$

$r = \dfrac{n}{1 + n}$ $\qquad s = \dfrac{n}{1 + n}$, $\qquad 1 - r = 1 - s = \dfrac{1}{1 + n}$

$1 - r : r = 1 - s : s = 1 : n$

Chapter Nine
Conic Sections

1. c = 4

    Axis: x axis

    V(0, 0)

    F(4, 0)

    D: x = -4

    ℓr = 4·4 = 16

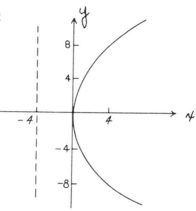

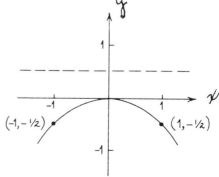

5. c = 5/2

    Axis: x axis

    V(0, 0)

    F(5/2, 0)

    D: x = -5/2

    ℓr 4·5/2 = 10

9. c = -1/2

    Axis: y axis

    V(0, 0)

    F(0, -1/2)

    D: y = 1/2

    ℓr = 4·1/2 = 2

13. $y^2 = 4cx$, $\quad 4|c| = 5$, $\quad |c| = \dfrac{5}{4}$, $\quad c = \pm \dfrac{5}{4}$, $\quad y^2 = \pm 5x$

17. $x^2 = 4cy$, $\quad 4 = 4c \cdot 3$, $\quad c = \dfrac{1}{3}$, $\quad x^2 = \dfrac{4}{3}y$

21. $2x = -5y'$, $\quad y' = -\dfrac{2x}{5} = -2$

$\quad y + 5 = -2(x - 5)$, $\quad 2x + y - 5 = 0$

25. $2x = -8y'$

$\quad y' = -\dfrac{x}{4} = -\dfrac{x_o}{4}$ where $(x_o,\ y_o)$ is the point of

tangency. The equation of the tangent line is

$\quad y = -\dfrac{x_o}{4}(x - 4)$.

Since $(x_o,\ y_o)$ is on the tangent line,

$\quad 4y_o = -x_o^2 + 4x_o$. Since $(x_o,\ y_o)$ is on the parabola,

$\quad x_o^2 = -8y_o$. From the last two equations,

$\quad x_o^2 = 2x_o^2 - 8x_o \qquad\qquad x_o(x_o - 8) = 0$

$\quad x_o = 0 \qquad\qquad\qquad\qquad x_o = 8$

$\quad y = 0 \qquad\qquad\qquad\qquad y = -2(x - 4)$

$\qquad\qquad\qquad\qquad\qquad\qquad 2x + y - 8 = 0$

29. $A_1 = \displaystyle\int_{25,000,000}^{75,000,000} (y_1 - y_2)\, dx$

$\quad = \displaystyle\int_{25,000,000}^{75,000,000} [10,000\sqrt{x} - \sqrt{3}(x - 25,000,000)]\, dx$

$\quad = \dfrac{20,000}{3}\, x^{3/2} - \dfrac{\sqrt{3}}{2}x^2 + 25,000,000\sqrt{3}\, x\ \Big|_{25,000,000}^{75,000,000}$

$\quad = \dfrac{20,000}{3}(75,000,000^{3/2} - 25,000,000^{3/2})$

$\qquad - \dfrac{\sqrt{3}}{2}(75,000,000^2 - 25,000,000^2)$

$\qquad + 25,000,000\sqrt{3}\ (50,000,000) = \dfrac{375\sqrt{3} - 250}{3} \cdot 10^{13}$

$$A_2 = \int_0^{25,000,000} 10,000\sqrt{x}\ dx = \frac{20,000}{3}x^{3/2}\Big|_0^{25,000,000}$$

$$= \frac{20,000}{3}\ 25,000,000^{3/2} = \frac{250}{3}\cdot 10^{13}$$

$$\frac{(375\sqrt{3}-250)/3}{36} = \frac{250/3}{t}$$

$$t = \frac{250\cdot 36}{375\sqrt{3}-250} = \frac{72}{3\sqrt{3}-2} = \frac{72(3\sqrt{3}+2)}{23}\ hr$$

≈ 22 hr 31 min

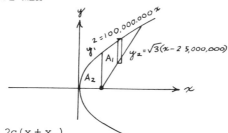

33. $P_1Q: yy_1 = 2c(x+x_1)$

$P_2Q: yy_2 = 2c(x+x_2)$

Solving simultaneously,

we have

$$y(y_1-y_2) = 2c(x_1-x_2)$$

$$y = \frac{2c(x_1-x_2)}{y_1-y_2} = \frac{4cx_1-4cx_2}{2(y_1-y_2)}$$

$$= \frac{y_1^2-y_2^2}{2(y_1-y_2)} = \frac{y_1+y_2}{2}$$

$$\frac{(y_1+y_2)y_1}{2} = 2c(x+x_1)$$

$$x = \frac{y_1^2+y_1y_2}{4c} - \frac{4cx_1}{4c} = \frac{y_1y_2}{4c}$$

$$Q: \left(\frac{y_1y_2}{4c},\ \frac{y_1+y_2}{2}\right)$$

$$m_{P_1F} = \frac{y_1}{x_1-c} = \frac{4cy_1}{4cx_1-4c^2} = \frac{4cy_1}{y_1^2-4c^2}$$

$$m_{P_2F} = \frac{y_2}{x_2 - c} = \frac{4cy_2}{y_2^2 - 4c^2}$$

$$m_{FQ} = \frac{\dfrac{y_1 + y_2}{2}}{\dfrac{y_1 y_2}{4c} - c} = \frac{2c(y_1 + y_2)}{y_1 y_2 - 4c^2}$$

$$\tan\theta_1 = \frac{m_{P_1F} - m_{FQ}}{1 + m_{P_1F}m_{FQ}} = \frac{\dfrac{4cy_1}{y_1^2 - 4c^2} - \dfrac{2c(y_1 + y_2)}{y_1 y_2 - 4c^2}}{1 + \dfrac{8c^2 y_1 (y_1 + y_2)}{(y_1^2 - 4c^2)(y_1 y_2 - 4c^2)}}$$

$$= \frac{4cy_1^2 y_2 - 16c^3 y_1 - 2cy_1^3 - 2cy_1^2 y_2 + 8c^3 y_1 + 8c^3 y_2}{y_1^3 y_2 - 4c^2 y_1^2 - 4c^2 y_1 y_2 + 16c^4 + 8c^2 y_1^2 + 8c^2 y_1 y_2}$$

$$= \frac{2cy_1^2 y_2 - 8c^3 y_1 - 2cy_1^3 + 8c^3 y_2}{y_1^3 y_2 + 4c^2 y_1^2 + 4c^2 y_1 y_2 + 16c^4}$$

$$= \frac{2c(y_1^2 + 4c^2)(y_2 - y_1)}{(y_1^2 + 4c^2)(y_1 y_2 + 4c^2)} = \frac{2c(y_2 - y_1)}{y_1 y_2 + 4c^2}$$

$$\tan\theta_2 = \frac{m_{FQ} - m_{P_2F}}{1 + m_{FQ}m_{P_2F}} = \frac{\dfrac{2c(y_1 + y_2)}{y_1 y_2 - 4c^2} - \dfrac{4cy_2}{y_2^2 - 4c^2}}{1 + \dfrac{8c^2 y_2 (y_1 - y_2)}{(y_2^2 - 4c^2)(y_1 y_2 - 4c^2)}}$$

$$= \frac{2cy_1 y_2^2 - 8c^3 y_1 + 2cy_2^3 - 8c^3 y_2 - 4cy_1 y_2^2 + 16c^3 y_2}{y_1 y_2^3 - 4c^2 y_2^2 - 4c^2 y_1 y_2 + 16c^4 + 8c^2 y_1 y_2 + 8c^2 y_2^2}$$

$$= \frac{-8c^3 y_1 + 2cy_2^3 - 2cy_1 y_2^2 + 8c^3 y_2}{y_1 y_2^3 + 4c^2 y_1 y_2 + 4c^2 y_2^2 + 16c^4}$$

$$= \frac{2c(y_2^2 + 4c^2)(y_2 - y_1)}{(y_2^2 + 4c^2)(y_1 y_2 + 4c^2)} = \frac{2c(y_2 - y_1)}{y_1 y_2 + 4c^2}$$

1. $C(0, 0)$, $V(\pm 13, 0)$, $CV(0, \pm 5)$

$c^2 = a^2 - b^2 = 169 - 25 = 144$

$F(\pm 12, 0)$

$\ell r = \dfrac{2b^2}{a} = \dfrac{2 \cdot 25}{13} = \dfrac{50}{13}$

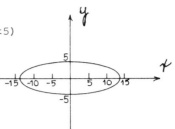

5. $C(0, 0)$, $V(0, \pm 7)$, $CV(\pm 5, 0)$

$c^2 = a^2 - b^2 = 49 - 25 = 24$

$F(0, \pm 2\sqrt{6})$

$\ell r = \dfrac{2b^2}{a} = \dfrac{2 \cdot 25}{7} = \dfrac{50}{7}$

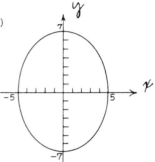

9. $\dfrac{x^2}{9} + \dfrac{y^2}{16} = 1$

$C(0, 0)$, $V(0, \pm 4)$, $CV(3, 0)$

$c^2 = a^2 - b^2 = 16 - 9 = 7$

$F(0, \pm \sqrt{7})$

$\ell r = \dfrac{2b^2}{a} = \dfrac{2 \cdot 9}{4} = \dfrac{9}{2}$

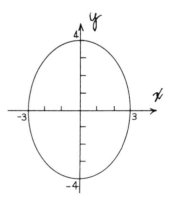

13. $\dfrac{x^2}{a^2} + \dfrac{y^2}{b^2} = 1$, $a = 5$, $\dfrac{x^2}{25} + \dfrac{y^2}{b^2} = 1$

$\dfrac{15}{25} + \dfrac{4}{b^2} = 1$, $\dfrac{4}{b^2} = \dfrac{2}{5}$, $b^2 = 10$, $\dfrac{x^2}{25} + \dfrac{y^2}{10} = 1$

17. $\dfrac{x^2}{a^2} + \dfrac{y^2}{b^2} = 1$, $c = 6$, $e = \dfrac{c}{a}$, $\dfrac{3}{5} = \dfrac{6}{a}$, $a = 10$

$b^2 = a^2 - c^2 = 100 - 36 = 64$, $\dfrac{x^2}{100} + \dfrac{y^2}{64} = 1$

21. $4x + 6yy' = 0$, $y' = \dfrac{2x}{3y} = -\dfrac{4}{3}$

$y - 1 = -\dfrac{4}{3}(x - 2)$, $4x + 3y - 11 = 0$

25. $\dfrac{x^2}{a^2} + \dfrac{y^2}{b^2} = 1$

By Problem 24, the tangent at (x_o, y_o) has slope

$m = -\dfrac{b^2 x_o}{a^2 y_o}$

$m_{PC} = \dfrac{y_o}{x_o - c}$, $m_{PC'} = \dfrac{y_o}{x_o + c}$

$\tan \theta_1 = \dfrac{-\dfrac{b^2 x_o}{a^2 y_o} - \dfrac{y_o}{x_o - c}}{1 - \dfrac{b^2 x_o}{a^2 (x_o - c)}} = \dfrac{-b^2 x_o (x_o - c) - a^2 y_o^2}{a^2 y_o (x_o - c) - b^2 x_o y_o}$

$= \dfrac{b^2 x_o^2 - b^2 c x_o + a^2 y_o^2}{-a^2 x_o y_o + a^2 c y_o + b^2 c_o y_o} = \dfrac{a^2 b^2 - b^2 c x_o}{a^2 c y_o + (b^2 - a^2) x_o y_o}$

$= \dfrac{b^2 (a^2 - c x_o)}{c y_o (a^2 - c x_o)} \quad \dfrac{b^2}{c y_o}$

$\tan \theta_2 = \dfrac{\dfrac{y_o}{x_o + c} + \dfrac{b^2 x_o}{a^2 y_o}}{1 - \dfrac{b^2 x_o}{a^2 (x_o + c)}} = \dfrac{a^2 y_o^2 + b^2 x_o (x_o + c)}{a^2 y_o (x_o + c) - b^2 x_o y_o}$

$= \dfrac{a^2 y_o^2 + b^2 x_o^2 + b^2 c x_o}{a^2 x_o y_o + a^2 c y_o - b^2 x_o y_o} = \dfrac{a^2 b^2 + b^2 c x_o}{a^2 c y_o - (b^2 - a^2) x_o y_o}$

$$= \frac{b^2(a^2 + cx_o)}{cy_o(a^2 + cx_o)} = \frac{b^2}{cy_o}$$

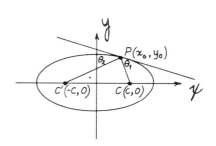

29. $a - c = 91{,}446{,}000$, $a + c = 94{,}560{,}000$

   $2a = 186{,}006{,}000$

   $a = 93{,}003{,}000$

   $2c = 3{,}114{,}000$

   $c = 1{,}557{,}000$

   $e = \dfrac{c}{a} = \dfrac{1{,}557{,}000}{93{,}003{,}000} = 0.0167$

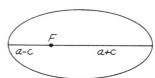

Section 9.4, p. 237

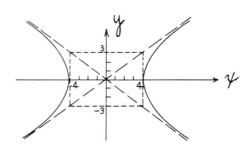

1. $C(0,\ 0)$, $V(\pm 4,\ 0)$

   $c^2 = a^2 + b^2 = 16 + 9 = 25$

   $f(\pm 5,\ 0)$, A: $y = \pm 3x/4$

   $\ell r = \dfrac{2b^2}{a} = \dfrac{2 \cdot 9}{4} = \dfrac{9}{2}$

5. $C(0,\ 0)$, $V(\pm 12,\ 0)$

   $c^2 = a^2 + b^2 = 144 + 25 = 169$

   $F(\pm 13,\ 0)$, A: $y = \pm 5x/12$

   $\ell r = \dfrac{2b^2}{a} = \dfrac{2 \cdot 25}{12} = \dfrac{25}{6}$

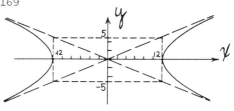

9. $\dfrac{x^2}{1} - \dfrac{y^2}{4} = 1$

C(0, 0), V($\pm$1, 0)

$c^2 = a^2 + b^2 = 1 + 4 = 5$

F($\pm \sqrt{5}$, 0), A: $y = \pm 2x$

$\ell r = \dfrac{2b^2}{a} = \dfrac{2 \cdot 4}{1} = 8$

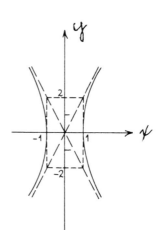

13. $\dfrac{y^2}{25/4} - \dfrac{x^2}{9/4} = 1$

C(0, 0), V(0, $\pm 5/2$)

$c^2 = a^2 + b^2 = \dfrac{25}{4} + \dfrac{9}{4} = \dfrac{34}{4}$

F(0, $\pm\sqrt{34}/2$), A: $y = \pm 5x/3$

$\ell r = \dfrac{2b^2}{a} = \dfrac{2 \cdot 9/4}{5/2} = \dfrac{9}{5}$

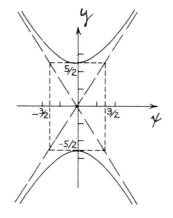

17. $\dfrac{x^2}{a^2} - \dfrac{y^2}{b^2} = 1$, $\quad \dfrac{b}{a} = \dfrac{2}{3}$, $\quad a = 6$, $\quad b = 4$, $\quad \dfrac{x^2}{36} - \dfrac{x^2}{16} = 1$

21. $\dfrac{x^2}{a^2} - \dfrac{y^2}{b^2} = 1$, $\quad a = 5$, $\quad \dfrac{x^2}{25} - \dfrac{y^2}{b^2} = 1$

$\dfrac{81}{625} - \dfrac{16}{b^2} = 1$, $\quad b^2 = -\dfrac{625}{34}$ $\qquad$ No solution

25. $\dfrac{x^2}{a^2} - \dfrac{y^2}{b^2} = 1$, $\quad \dfrac{a^2}{c} = \dfrac{9}{5}$, $\quad e = \dfrac{c}{a} = \dfrac{5}{3}$

$c = 5$, $\quad a = 3$, $\quad b^2 = c^2 - a^2 = 25 - 9 = 16$

$\dfrac{x^2}{9} - \dfrac{y^2}{16} = 1$

29. $2x - 2yy' = 0$

$y' = \dfrac{x}{y} = \dfrac{x_o}{y_o}$ where (x_o, y_o) is the point of tangency

The equation of the tangent line is

$y - 9 = \dfrac{x_o}{y_o}(x - 9)$

Since (x_o, y_o) is on the tangent line,

$y_o - 9 = \dfrac{x_o}{y_o}(x_o - 9)$, $\quad x_o^2 - y_o^2 = 9x_o - 9y_o$

Since (x_o, y_o) is on the hyperbola, $x_o^2 - y_o^2 = 9$

Solving simultaneously, we have

$9x_o - 9y_o = 9$, $\quad y_o = x_o - 1$, $\quad x_o^2 - (x_o - 1)^2 = 9$

$2x_o = 10$, $\quad x_o = 5$, $\quad y_o = 4$

$y - 9 = \dfrac{5}{4}(x - 9)$, $\qquad 5x - 4y - 9 = 0$

There is only one tangent line because the given point is on an asymptote.

33. $\lim\limits_{x \to +\infty}\left[\left(\dfrac{b}{a}\sqrt{x^2 - a^2}\right) - \left(-\dfrac{b}{a}x\right)\right] = \lim\limits_{x \to +\infty}\dfrac{b}{a}\dfrac{(\sqrt{x^2 - a^2} - x)(\sqrt{x^2 - a^2} + x)}{\sqrt{x^2 - a^2} + x}$

$= \lim\limits_{x \to +\infty}\dfrac{b}{a}\dfrac{(x^2 - a^2) - x^2}{\sqrt{x^2 - a^2} + x} = \lim\limits_{x \to +\infty}\dfrac{b}{a}\dfrac{-a^2}{\sqrt{x^2 - a^2} + x} = 0$

37. $\dfrac{\sqrt{(x - 1000)^2 + y^2}}{1100} = 1 + \dfrac{\sqrt{(x + 1000)^2 + y^2}}{1100}$

$\sqrt{(x - 1000)^2 + y^2} = 1100 + \sqrt{(x + 1000)^2 + y^2}$

$x^2 - 2000x + 1,000,000 + y^2$

$$= 1,210,000 + 2200 \sqrt{(x+1000)^2 + y^2} + x^2 + 2000x$$

$$+ 1,000,000 + y^2$$

$$-4000x - 1,210,000 = 2200 \sqrt{(x+1000)^2 + y^2}$$

$$-20x - 6050 = 11 \sqrt{(x+1000)^2 + y^2}$$

$$400x^2 + 242,000x + 36,602,500$$

$$= 121(x^2 + 2000x + 1,000,000 + y^2)$$

$$279x^2 - 121y^2 = 84,397,500$$

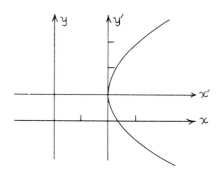

Section 9.5, pp. 244-245

1. $y^2 - 2y + 1 = 4x - 9 + 1$

    $(y - 1)^2 = 4(x - 2)$

    $x' = x - 2, \quad y' = y - 1$

    $y'^2 = 4x'$

5. $9(x^2 + 10x + 25) - 4(y^2 - 8y + 16) = -125 + 225 - 64$

    $9(x + 5)^2 - 4(y - 4)^2 = 36$

    $x' = x + 5, \quad y' = y - 4$

    $\dfrac{x'^2}{4} - \dfrac{y'^2}{9} = 1$

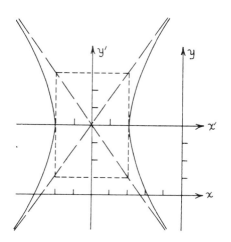

9. $4(x^2 - x + \frac{1}{4}) = 4y + 5 + 1$

$4(x - \frac{1}{2})^2 = 4(y + \frac{3}{2})$

$x' = x - \frac{1}{2}, \quad y' = y + \frac{3}{2}$

$x'^2 = y'$

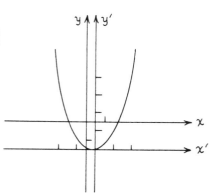

13. $16x^2 - 9(y^2 - 6y + 9) = 225 - 81$

$16x^2 - 9(y - 3)^2 = 144$

$\frac{x^2}{9} - \frac{(y-3)^2}{16} = 1$

$C(0, 3), \quad V(\pm 3, 3)$

$c^2 = a^2 + b^2 = 9 + 16 = 25$

$F(\pm 5, 3), \quad A: y = 3 + 4x/3$

$\ell r = \frac{2b^2}{a} = \frac{2 \cdot 16}{3} = \frac{32}{3}$

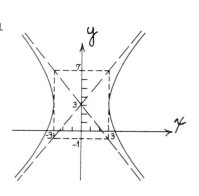

17. $3(x^2 - \frac{2}{3}x + \frac{1}{9}) - 3(y^2 + \frac{4}{3}y + \frac{4}{9}) = 13 + \frac{1}{3} - \frac{4}{3}$

$3(x - \frac{1}{3})^2 - 3(y + \frac{2}{3})^2 = 12$

$\dfrac{(x - 1/3)^2}{4} - \dfrac{(y + 2/3)^2}{4} = 1$

$C(1/3, -2/3)$, $V(1/3 \pm 2, -2/3)$

$c^2 = a^2 + b^2 = 4 + 4 = 8$

$F(1/3 \pm 2\sqrt{2}, -2/3)$,

A: $y = -2/3 \pm (x - 1/3)$

$\ell r = \dfrac{2b^2}{a} = \dfrac{2 \cdot 4}{2} = 4$

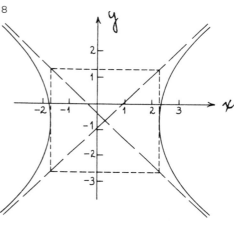

21. $C(-2, 1)$, $c = 4$, $a = 1$, $b^2 = c^2 - a^2 = 16 - 1 = 15$

$\dfrac{(y - 1)^2}{1} - \dfrac{(x + 2)^2}{15} = 1$, $x^2 - 15y^2 + 4x + 30y + 4 = 0$

25. $x = x' + h$, $\qquad y = y' + k$

$x'^2 + 2hx' + h^2 - 2x'y' - 2kx' - 2hy' - 2hk + 4y'^2 + 8ky' + 4k^2$

$\quad + 8x' + 8h - 26y' - 26k + 38 = 0$

$x'^2 - 2x'y' + 4y'^2 + (2h - 2k + 8)x' + (-2h + 8k - 26)y'$

$\quad + (h^2 - 2hk + 4k^2 + 8h - 26k + 38) = 0$

$2h - 2k + 8 = 0$, $\qquad h - k = -4$

$-2h + 8k - 26 = 0$, $\qquad -h + 4k = 13$

$3k = 9$, $\qquad k = 3$, $\qquad h = -1$

$x'^2 - 2x'y' + 4y'^2 - 5 = 0$

29. $x = x' + h$, $\qquad\qquad$ $y = y' + k$

$y' + k = x'^3 + 3hx'^2 + 3h^2x' + h^3 - 3x' - 3h + 6$

$y' = x'^3 + 3hx'^2 + (3h^2 - 3)x' + (h^3 - 3h + 6 - k)$

$3h^2 - 3 = 0$, $\qquad$ $h = \pm 1$, $\qquad$ $h^3 - 3h + 6 - k = 0$

$h = 1$, $\quad k = 4$ $\qquad\qquad$ $h = -1$, $\quad k = 8$

$y' = x'^3 + 3x'^2$ $\qquad\qquad\qquad$ $y' = x'^3 - 3x'^2$

33. $Ax^2 + Bxy + Cy^2 + Dx + Ey + F = 0$

$x = x' + h$, $\qquad\qquad$ $y = y' + k$

$Ax'^2 + 2Ahx' + Ah^2 + Bx'y' + Bkx' + Bhy' + Bhk + Cy'^2 + 2Cky'$

$\quad + Ck^2 + Dx' + Dh + Ey' + Ek + F = 0$

$Ax'^2 + Bx'y' + Cy'^2 + (2Ah + Bk + D)x' + (Bh + 2Ck + E)y'$

$\quad + (Ah^2 + Bhk + Ck^2 + Dh + Ek + F) = 0$

$A' = A$, $B' = B$, $C' = C$ no matter what values of h and k are used

Section 9.6, pp. 250-251

1. $\tan \theta = \dfrac{3}{2}$, $\quad \sin \theta = \dfrac{3}{\sqrt{13}}$, $\cos \theta = \dfrac{2}{\sqrt{13}}$

$x = \dfrac{3x' - 3y'}{\sqrt{13}}$, $\quad y = \dfrac{3x' + 2y'}{\sqrt{13}}$

$\dfrac{4x' - 6y' + 9x' + 6y'}{\sqrt{13}} = 16$

$\dfrac{13x'}{\sqrt{13}} = 6$, $\quad \sqrt{13}\, x' = 6$

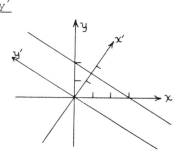

5. $\sin \theta = \dfrac{1}{2}$, $\cos \theta = \dfrac{\sqrt{3}}{2}$

$x = \dfrac{\sqrt{3}x' - y'}{2}$, $y = \dfrac{x' + \sqrt{3}y'}{2}$

$\dfrac{93x'^2 - 62\sqrt{3}x'y' + 31y'^2 + 30x'^2 + 20\sqrt{3}x'y' - 30y'^2 + 21x'^2}{4}$

$\dfrac{+\ 42\sqrt{3}x'y' + 63y'^2}{} = 144$

9. $A = C$, $\theta = 45°$, $\sin \theta = \cos \theta = \dfrac{1}{\sqrt{2}}$

$x = \dfrac{x' - y'}{\sqrt{2}}$, $y = \dfrac{x' + y'}{\sqrt{2}}$

$\dfrac{x'^2 - 2x'y' + y'^2 + x - y'^2 + x'^2 + 2x'y' + y'^2}{2}$

$+\ \dfrac{4\sqrt{2}(x' - y') - 4\sqrt{2}(x' - y')}{\sqrt{2}} = 0$

$\dfrac{3x'^2 + y'^2}{2} - 8y' = 0$

$3x'^2 + y'^2 - 16y'^2 + 64 = 64$

$\dfrac{x'^2}{64/3} + \dfrac{(y' - 8)^2}{64} = 1$

13. $\tan 2\theta = \dfrac{B}{A - C} = \dfrac{-4}{5 - 8} = \dfrac{4}{3}$, $\cos 2\theta = \dfrac{1}{\sqrt{1 + 16/9}} = \dfrac{3}{5}$

$\sin \theta = \sqrt{\dfrac{1 - 3/5}{2}} = \dfrac{1}{\sqrt{5}}$, $\cos \theta = \sqrt{\dfrac{1 + 3/5}{2}} = \dfrac{2}{\sqrt{5}}$,

$\tan \theta = \dfrac{1}{2}$, $x = \dfrac{2x' - y'}{\sqrt{5}}$, $y = \dfrac{x' + 2y'}{\sqrt{5}}$

$\dfrac{5(4x'^2 - 4x'y' + y'^2) - 4(2x'^2 + 3x'y' - 2y'^2) + 8(x'^2 + 4x'y' + 4y'^2)}{5}$

$-\ 36 = 0$

$$4x'^2 + 9y'^2 = 36$$

$$\frac{x'^2}{9} + \frac{y'^2}{4} = 1$$

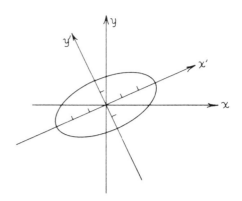

17. $\tan 2\theta = \dfrac{B}{A-C} = \dfrac{-6}{9-1} = -\dfrac{3}{4}$, $\cos 2\theta = \dfrac{-1}{\sqrt{1 + 9/16}} = -\dfrac{4}{5}$

$\sin \theta = \sqrt{\dfrac{1 + 4/5}{2}} = \dfrac{3}{\sqrt{10}}$, $\cos \theta = \sqrt{\dfrac{1 - 4/5}{2}} = \dfrac{1}{\sqrt{10}}$

$\tan \theta = 3$, $x = \dfrac{x' - 3y'}{\sqrt{10}}$, $y = \dfrac{3x' + y'}{\sqrt{10}}$

$\dfrac{9(x'^2 - 6x'y' + 9y'^2) - 6(3x'^2 - 8x'y' - 3y'^2) + (9x'^2 + 6x'y' + y'^2)}{10}$

$+ \dfrac{-12\sqrt{10}(x' - 3y') - 36\sqrt{10}(3x' + y')}{\sqrt{10}} = 0$

$10y'^2 - 120x' = 0$

$y'^2 = 12x'$

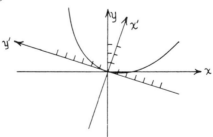

21. $x = x' \cos \theta - y' \sin \theta$, $y = x' \sin \theta + y' \cos \theta$

$Ax'^2\cos^2\theta - 2Ax'y' \sin \theta \cos \theta + Ay'^2\sin^2\theta + Bx'^2\sin \theta \cos \theta$

$+ B(\cos^2\theta - \sin^2\theta)x'y' - By'^2\sin \theta \cos \theta + Cx'^2\sin^2\theta$

$+ 2Cx'y' \sin \theta \cos \theta + Cy'^2\cos^2\theta + Dx' \cos \theta - Dy' \sin \theta$

$$+ Ex' \sin\theta + Ey' \cos\theta + F = 0$$

$$(A \cos^2\theta + B \sin\theta \cos\theta + C \sin^2\theta)x'^2[-2A \sin\theta \cos\theta$$
$$+ B(\cos^2\theta - \sin^2\theta) + 2C \sin\theta \cos\theta]x'y' + (A \sin^2\theta$$
$$- B \sin\theta \cos\theta + C \cos^2\theta)y'^2 + (D \cos\theta + E \sin\theta)x'$$
$$+ (-D \sin\theta + E \cos\theta)y' + F = 0$$

$$B'^2 - 4A'C' = [(C-A)2 \sin\theta \cos\theta + B(\cos^2\theta - \sin^2\theta)]^2$$
$$- 4(A \cos^2\theta + B \sin\theta \cos\theta + C \sin^2\theta)(A \sin^2\theta$$
$$- B \sin\theta + C \cos^2\theta)$$

$$= (C-A)^2 \, 4 \sin^2\theta \cos^2\theta + B(C-A) \, 4 \sin\theta \cos\theta(\cos^2\theta$$
$$- \sin^2\theta) + B^2(\cos^2\theta - \sin^2\theta)^2 - 4A^2 \sin^2\theta \cos^2\theta$$
$$+ 4AB \sin\theta \cos^3\theta - 4AC \cos^4\theta - 4AB \sin^3\theta \cos\theta$$
$$+ 4B^2 \sin^2\theta \cos^2\theta - 4BC \sin\theta \cos^3\theta - 4AC \sin^4\theta$$
$$+ 4BC \sin^3\theta \cos\theta - 4C^2 \sin^2\theta \cos^2\theta$$

$$= B^2(\cos^4\theta + 2 \sin^2\theta \cos^2\theta + \sin^4\theta) - 4AC(\sin^4\theta$$
$$+ 2 \sin^2\theta \cos^2\theta + \cos^4\theta)$$

$$= B^2(\sin^2\theta + \cos^2\theta)^2 - 4AC(\sin^2\theta + \cos^2\theta)^2 = B^2 - 4AC$$

Section 9.7, p. 255

1. $xy - x + y + 3 = 0$

$y(x+1) = x - 3$

$y = \dfrac{x-3}{x+1}$

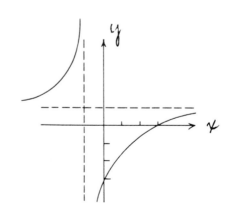

Section 9.7

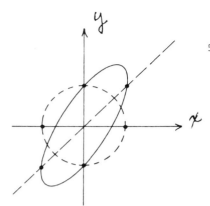

5. $2x^2 - 2xy + y^2 - 1 = 0$

$y^2 - 2xy + (2x^2 - 1) = 0$

$y = \dfrac{2x \pm \sqrt{4x^2 - 8x + 4}}{2}$

$= x \pm \sqrt{1 - x^2}$

$y = x \qquad y = \pm\sqrt{1 - x^2}$

$\qquad\qquad\qquad x^2 + y^2 = 1$

9. $2xy - y^2 - 4 = 0$

$y^2 - 2xy + 4 = 0$

$y = \dfrac{2x \pm \sqrt{4x^2 - 16}}{2} = x \pm \sqrt{x^2 - 4}$

$y = x \qquad y = \pm\sqrt{x^2 - 4}$

$\qquad\qquad x^2 - y^2 = 4$

13. $4x^2 + 4xy + y^2 - 3x + 2y + 1 = 0$

$y^2 + (4x + 2)y + (4x^2 - 3x + 1) = 0$

$y = \dfrac{-4x - 2 \pm \sqrt{16x^2 + 16x + 4 - 16x^2 + 12x - 4}}{2}$

$= -2x - 1 \pm \sqrt{7x}$

$y = -2x - 1 \qquad y = \pm\sqrt{7x}$

$\qquad\qquad\qquad y^2 = 7x$

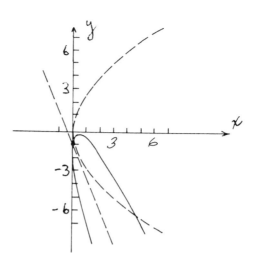

17. $x^2 - xy - x - 2 = 0$

$xy = x^2 - x - 2$

$y = \dfrac{x^2 - x - 2}{x}$

$= \dfrac{(x - 2)(x + 1)}{x}$

$= x - 1 - \dfrac{2}{x}$

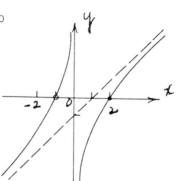

21. Show that $\sqrt{x} + \sqrt{y} = \sqrt{a}$ is a portion of a parabola.

$\sqrt{x} + \sqrt{y} = \sqrt{a}$

$x + 2\sqrt{xy} + y = a$

$2\sqrt{xy} = a - x - y$

$4xy = a^2 + x^2 + y^2 - 2ax - 2ay + 2xy$

$x^2 - 2xy + y^2 - 2ax - 2ay + a^2 = 0$

$B^2 - 4AC = 4 - 4 \cdot 1 \cdot 1 = 0$

Parabola

108   *Section 9.7*

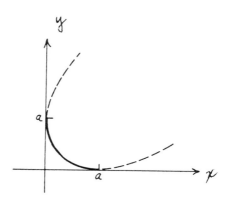

1. $c = 4$

    Axis: x axis

    $V(0, 0)$

    $F(4, 0)$

    D: $x = -4$

    $\ell r = 4 \cdot 4 = 16$

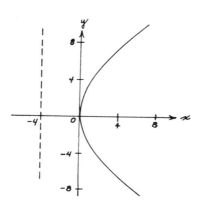

5. $y^2 + 2y + 1 = x - 3$

    $(y + 1)^2 = x - 3$

    Axis: $y = -1$

    $V(3, -1)$, $F(13/4, -1)$

    D: $x = 11/4$

    $\ell r = 1$

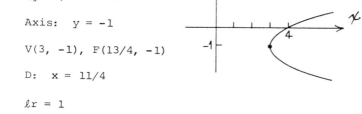

9. $9(x^2 + 2x + 1) - 16(y^2 + y + \frac{1}{4}) = 139 + 9 - 4$

    $9(x + 1)^2 - 16(y + \frac{1}{2})^2 = 144$

$$\frac{(x+1)^2}{16} - \frac{(y+1/2)^2}{9} = 1$$

$C(-1, -1/2), \quad V(-1 \pm 4, -1/2)$

$c^2 = a^2 + b^2 = 16 + 9 = 25$

$F(-1 \pm 5, -1/2),$

A: $\quad y = -1/2 \pm 3(x+1)/4$

$\ell r = \dfrac{2b^2}{a} = \dfrac{2 \cdot 9}{4} = \dfrac{9}{2}$

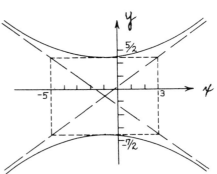

13. $\quad y^2 = 4cx \qquad\qquad\qquad x^2 = 4cy$

$\quad (2,4): \ 16 = 8c, \ c = 2 \qquad (2,4): \ 4 = 16c, \ c = 1/4$

$\quad (8,8): \ 64 = 32c, \ c = 2 \qquad (8,8): \ 64 = 32c, \ c = 2$

$\quad y^2 = 8x \qquad\qquad\qquad\quad$ Impossible

17. $\quad (y-k)^2 = 4c(x-h) \qquad V = (1, 5), \quad c = 2$

$\quad (y-5)^2 = 8(x-1) \qquad\quad y^2 - 8x - 10y + 33 = 0$

21. $\quad y = x^3 + 6x^2 + 3x - 14 \qquad x = x' + h, \quad y = y' + k$

$\quad y' + k = (x'+h)^3 + 6(x'+h)^2 + 3(x'+h) - 14$

$\qquad\qquad = x'^3 + 3hx'^2 + 3h^2x' + h^3 + 6x'^2 + 12hx'$

$\qquad\qquad\quad + 6h^2 + 3x' + 3h - 14$

$\quad y' = x'^3 + (3h+6)x'^2 + (3h^2 + 12h + 3)x'$

$\qquad\qquad + (h^3 + 6h^2 + 3h - 14 - k)$

$\quad 3h + 6 = 0, \quad h = -2, \quad h^3 + 6h^2 + 3h - 14 - k = 0$

$\quad k = -8 + 24 - 6 - 14 = -4$

$$x = x' - 2, \quad y = y' - 4$$

$$y' = x'^3 - 9x' = x'(x' + 3)(x' - 3)$$

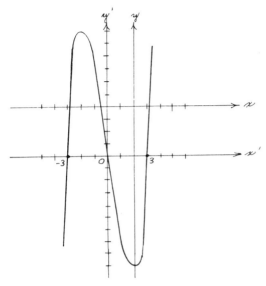

25. $4x^2 - 4xy + y^2 - 5x + 2y + 1 = 0$

$B^2 - 4AC = 16 - 16 = 0$, parabola

$$y^2 + (-4x + 2)y + (4x^2 - 5x + 1) = 0$$

$$y = \frac{4x - 2 \pm \sqrt{16x^2 - 16x + 4 - 16x^2 + 20x - 4}}{2}$$

$$= 2x - 1 \pm \sqrt{x}$$

$y = y_1 + y_2$

$y_1 = 2x - 1$

$y_2 = \pm \sqrt{x}$

$y_2^2 = x$

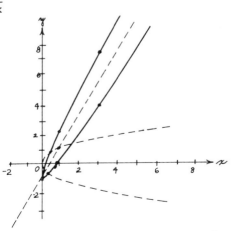

Chapter Ten
Limits and Continuity: A Geometric Approach

Section 10.1, pp. 260-261

1. P = (0,0), A = (0,0)   5. P = (2,4), no A

9. P = (0,0), A = (0,0)   13. Not a limit point
                                   Let h be x = -2 and k be x = 0

17. Limit point   21. Not a limit point
                                   Let h be x = -2 and k be x = 0

25. Limit point

Section 10.2, pp. 263-264

1. Not satisfied
    Let h be x = 2 and k be x = 4

5. Satisfied

9. A = P = (0,0)

    Clearly 0 is a limit
    point of the domain.
    If α and β are two
    horizontal lines
    with P between them,
    let ℓ be any vertical
    line to the left of
    the y axis and m the
    vertical line through
    the point of intersection.

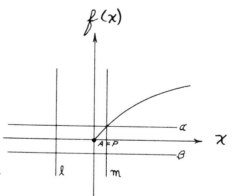

13. A = (1, 2), P = (1, 1)

Clearly 1 is a limit
point of the domain.
If α and β are two
horizontal lines
with P between them,
let ℓ and m be
vertical lines
through the points
of intersection of
α and ℳ. Thus
every point of ℳ-A
between ℓ and m is
also between α and β.

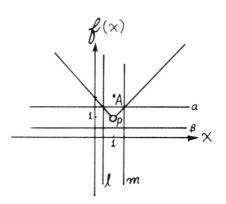

17. A = P = (0, 3)

Clearly 0 is a limit
point of the domain.
If α and β are two
horizontal lines with
P between them, let ℓ
and m be the vertical
lines through the
points of intersection
of α and β with ℳ.
Thus every point of
ℳ-A between ℓ and m is
also between α and β.

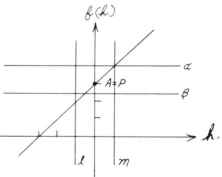

21. A = P = (a, k)

f(x) = k is clearly defined
for all values of x; so the
first part of the definition
is satisfied. Let α and β
be any pair of horizontal
lines with P between them.
Choose ℓ and m to be any
pair of vertical lines with
P between them. Since every
point of ℳ is between α and β, every point of ℳ-A
between ℓ and m is between α and β. Therefore $\lim_{x \to a} k = k$.

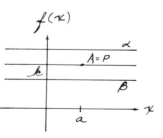

1. A = P = (0,1)

    Choose α to be any
    horizontal line above
    y = 1 and α to be a
    horizontal line
    between y = 1 and the
    x axis. Now, no
    matter how ℓ and m
    are chosen, there are
    points of ℓ -A between
    ℓ and m but not between
    α and β.

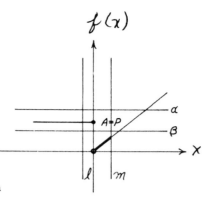

5. No A, P = (0,0)

    Choose α and β to be any
    pair of horizontal lines
    with P between them. No
    matter how ℓ and m are
    chosen, there are points
    of ℓ-A (=ℓ) between ℓ
    and m but not between
    α and β.

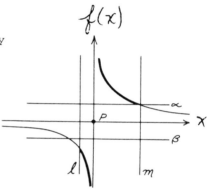

9. A = P = (0,0)

    Clearly 0 is a limit
    point of the domain.
    If α and β are two
    horizontal lines with
    P between them, let ℓ
    and m be the vertical
    lines through the points
    of intersection of α and
    ℓ. Now, every point of
    ℓ-A between ℓ and m is also
    between α and β. The given
    limit statement is true.

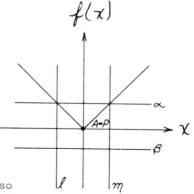

13. No A, P = (0,0)

Note that there are infinitely many waves between the y axis and any vertical line. Choose the vertical lines α and β so that either α is below y = 1 or β is above y = -1. Now no matter how ℓ and m are chosen there are points of $\mathcal{L}$-A (=$\mathcal{L}$) between ℓ and m but not between α and β. The limit statement is false.

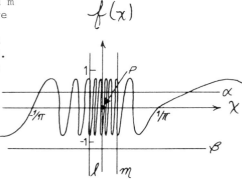

17. Suppose $\lim\limits_{x\to 0} \dfrac{1}{x} = 1$, No A, P = (0, L)

Choose α and β to be any two horizontal lines with P between them. No matter how ℓ and m are chosen there are points of $\mathcal{L}$ -A (=$\mathcal{L}$) between ℓ and m but not between α and β. Thus the given limit does not exist.

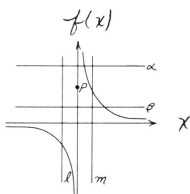

Section 10.4, pp. 270-271

1. This means

$$\lim\limits_{x\to 0} G(x) = 0 \quad \text{where} \quad G(x) = \begin{cases} 1 & \text{if } x = 0 \\ x & \text{if } x > 0 \end{cases}$$

A = (0,1), P = (0,0)

Clearly 0 is a limit point of the domain of G. If α and β are horizontal lines with P between them, let ℓ be any vertical line to the left of the y axis and m the vertical line through the point of intersection of α and ₺. Thus every point of ₺-A between ℓ and m is also between α and β.

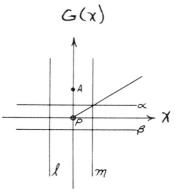

5. This means

    $$\lim_{x \to 0} G(x) = 0 \quad \text{where} \quad G(x) = \begin{cases} x \text{ if } x = 0 \\ 1 \text{ if } x > 0 \end{cases}$$

    A = P = (0,0)

    Choose α to be y = 1/2 and β to be any horizontal line below the x axis. Now, no matter how ℓ and m are chosen, there are points of ₺-A between ℓ and m but not between α and β.

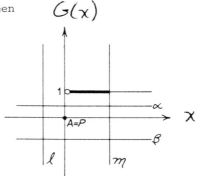

9. Suppose $\lim_{x \to 0^+} \frac{1}{x} = L$. Thus $\lim_{x \to 0} G(x) = L$

    where G(x) = 1/2 if x > 0

    No A,   P = (0,L)

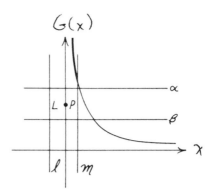

Let α and β be any
two horizontal lines
with P between them.
No matter how ℓ and
m are chosen, there
are points of $\mathfrak{L}$-A
(=$\mathfrak{L}$) between ℓ and
m but not between
α and β.

13. $\lim\limits_{x \to 1^-} f(x) = 2$, $\lim\limits_{x \to 1^+} f(x) = 0$, $\lim\limits_{x \to 1} f(x)$ does not exist

17. $\lim\limits_{x \to 0^-} f(x)$ does not exist, $\lim\limits_{x \to 0^+} f(x) = 0$, $\lim\limits_{x \to 0} f(x)$ does

    not exist

Section 10.5, p. 275

1. No A

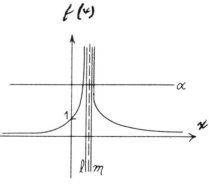

    Clearly 1 is a limit
    point of the domain.
    Let α be a horizontal
    line. If α is on or
    below the x axis, every
    point of $\mathfrak{L}$ is above α
    and the choice of ℓ and
    m is immaterial. If α
    is above the x axis, take
    ℓ and m to be the vertical
    lines through the points
    of intersection of α and $\mathfrak{L}$.
    Now every point of $\mathfrak{L}$-A
    between ℓ and m is above α.

5. Clearly there are points of $\mathcal{L}$ to the right of any vertical line. Suppose α and β are two horizontal lines with y = 0 between them. Let ℓ be the vertical line through the point of intersection of α and $\mathcal{L}$. Now every point of $\mathcal{L}$ to the right of ℓ is between α and β.

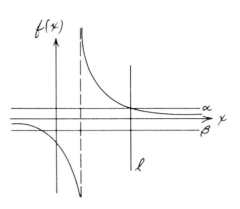

9. lim f(x) = +∞ means that
   x→+∞

   (a) if h is a vertical line, then there is a point of $\mathcal{L}$ to the right of h, and

   (b) if α is a horizontal line, then there is a vertical line ℓ such that every point of $\mathcal{L}$ to the right of ℓ is above α.

13. $\lim\limits_{x\to 0^-} \dfrac{1}{x} = -\infty$ means $\lim\limits_{x\to 0} g(x) = -\infty$ where $g(x) = \dfrac{1}{x}$
    when x < 0

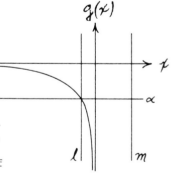

   Clearly 0 is a limit point of the domain of g. Let α be a horizontal line. If α is on or above the x axis, all of $\mathcal{L}$ is below α and the choice of ℓ and m is immaterial. If α is below the x axis, let ℓ be the vertical line through the point of intersection of α and $\mathcal{L}$, and let m be any vertical line to the right of the y axis. Now every point of $\mathcal{L}$ between ℓ and m is below α.

$$\lim_{x \to 0^+} \frac{1}{x} = +\infty \text{ means } \lim_{x \to 0} G(x) = +\infty \text{ where } G(x) = \frac{1}{x}$$

where $x > 0$

Clearly 0 is a limit point of the domain of G.
Let α be a horizontal line. If α is on or below
the x axis, all of $\mathscr{L}$ is below α and the choice
of ℓ and m is immaterial. If α is above the x
axis, let ℓ be any line to
the left of the y axis
and let m be the line
through the point of
intersection of α and
$\mathscr{L}$. Now every point of
$\mathscr{L}$ between ℓ and m is
above α.

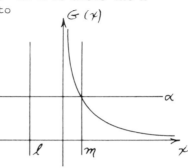

Let α be a horizontal
line. No matter how ℓ
and m are chosen with
$x = 0$ between them,
there are points of $\mathscr{L}$ to the left of the y axis
which are below it, and points of $\mathscr{L}$ to the right
of the y axis which are above it. Thus

$$\lim_{x \to 0} \frac{1}{x} \text{ is neither } +\infty \text{ nor } -\infty .$$

17. Clearly there are
points of $\mathscr{L}$ to the
right of any vertical
line. Suppose α is
a horizontal line.
If α is below the x
axis, all of $\mathscr{L}$ is
above α and the choice
of ℓ is immaterial. If
α is on or above the
x axis, let ℓ be the
vertical line through
the point of intersection
of α and $\mathscr{L}$. Now every point of $\mathscr{L}$ to the right of ℓ
is above α. The limit statement is true.

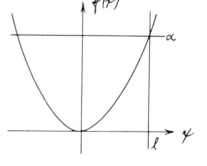

1. Continuous. Given α and β, the vertical lines ℓ
   and m are determined in the same way as in Problem
   15, Section 10.2.

5. Continuous. Suppose α and β are two horizontal lines
   with A between them. Let ℓ be any vertical line to
   the left of A; let m be any vertical line between
   A and the y axis. Now every point of 𝓍 between ℓ and
   m is also between
   α and β.

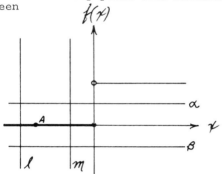

9. Continuous. Suppose α and β are two horizontal
   lines with A between them. Let ℓ be the vertical
   line through the point of intersection of β and
   𝓍; let m be the vertical line through the point
   of intersection of α and 𝓍. Now every point of
   𝓍 between ℓ and m is also between α and β.

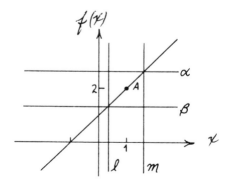

13. Continuous. Given α and β, the vertical lines ℓ and
    m are determined in the same way as in Problem 15,
    Section 10.3.

17. Continuous. Given the
    horizontal lines α and
    β with A between them,
    let ℓ be any vertical
    line to the left of A
    and m any vertical
    line between A and
    x = 2. Now A is the
    only point of ℓ between
    ℓ and m and it is
    between α and β.

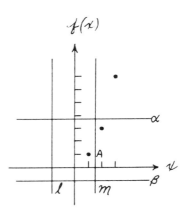

21. Discontinuous at
    (1, 3)

25. Discontinuous everywhere

Section 10.7, p. 287

1. $\lim\limits_{x \to 0} \dfrac{x^2 + x}{x^2 - x} = \lim\limits_{x \to 0} \dfrac{x + 1}{x - 1}$ (by Theorem 10.4)

    $= \dfrac{1}{-1} = -1$ (by Theorem 10.3)

5. $\lim\limits_{h \to 0} \dfrac{(2 + h)^2 - 4}{h} = \lim\limits_{h \to 0} \dfrac{4h + h^2}{h}$

    $= \lim\limits_{h \to 0} (4 + h)$ (by Theorem 10.4)

    $= 4$ (by Theorem 10.3)

9. $\lim\limits_{h \to 0} \dfrac{\dfrac{x + h}{x + h + 1} - \dfrac{x}{x + 1}}{h} = \lim\limits_{h \to 0} \dfrac{\dfrac{h}{(x + h + 1)(x + 1)}}{h}$

    $= \lim\limits_{h \to 0} \dfrac{1}{(x+h+1)(x+1)}$ (by Theorem 10.4)

    $= \dfrac{1}{(x + 1)^2}$ (by Theorem 10.3)

13. $y = (x + 2)^{1/3}(x - 1)^{2/3}$ is continuous everywhere

$y' = (x + 2)^{1/3}\frac{2}{3}(x - 1)^{-1/3} + (x - 1)^{2/3}\frac{1}{3}(x + 2)^{-2/3}$

$= \dfrac{2(x + 2) + (x - 1)}{3(x-1)^{1/3}(x+2)^{2/3}} = \dfrac{x + 1}{(x-1)^{1/3}(x+2)^{2/3}}$

There is no derivative at $x = 1$ or $x = -2$.

17. First we show that $\lim\limits_{x \to a} [-g(x)] = -c$.

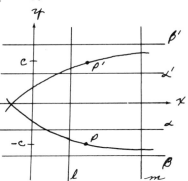

Let G be the graph of $-g(x)$, $P = (a, -c)$, and $A = (a, -g(a))$. Let G' be the graph of $g(x)$, $P' = (a, c)$, and $A' = (a, g(a))$. Suppose α and β are two horizontal lines with P between them and having equations $y = k_1$ and $y = k_2$. Then $y = -k_1$ and $y = -k_2$ are two horizontal lines with P' between them (call them α' and β'). Then there are two vertical lines ℓ and m with P' between them such that every point of $G' - A'$ between ℓ and m is also between α' and β'. But then every point of $G - A$ between ℓ and m is between α and β. Therefore

$$\lim_{x \to a} [-g(x)] = -c$$

Thus
$$\lim_{x \to a} [f(x) - g(x)] = \lim_{x \to a} \{f(x) + [-g(x)]\}$$

$$= \lim_{x \to a} f(x) + \lim_{x \to a} [-g(x)] = b - c$$

Section 10.8, p. 290

1. $f'(x) = -1$ if $x < 0$, $f'(x) = 2x$ if $x > 0$

$f'(0) \lim\limits_{h \to 0} \dfrac{f(h) - f(0)}{h} = \lim\limits_{h \to 0} \dfrac{f(h) - 0}{h}$

If $h < 0$,

$$\lim_{h \to 0} \frac{f(h) - 0}{h} = \lim_{h \to 0^-} \frac{-h}{h} = \lim_{h \to 0^-} (-1) = -1$$

If $h > 0$,

$$\lim_{h \to 0} \frac{f(h) - 0}{h} = \lim_{h \to 0^+} \frac{h^2}{h} = \lim_{h \to 0^+} h = 0$$

$f'(0)$ does not exist

5. $f'(x) = 1$ if $x \neq 0$

$$f'(0) = \lim_{h \to 0} \frac{f(h) - f(0)}{h} = \lim_{h \to 0} \frac{h - 1}{h} = -\infty$$

$f'(0)$ does not exist

9. $f'(x) = 0$ if $x < 0$, $f'(x) = 6x$ if $0 < x < 1$

$f'(x) = 0$ if $x > 1$

$$f'(0) = \lim_{h \to 0} \frac{f(h) - f(0)}{h} = \lim_{h \to 0} \frac{f(h) - 0}{h}$$

If $h < 0$,

$$\lim_{h \to 0} \frac{f(h) - 0}{h} = \lim_{h \to 0^-} \frac{0}{h} = \lim_{h \to 0^-} 0 = 0$$

If $h > 0$ $(h < 1)$,

$$\lim_{h \to 0} \frac{f(h) - 0}{h} = \lim_{h \to 0^+} \frac{3h^2}{h} = \lim_{h \to 0^+} 3h = 0$$

$f'(0) = 0$

$$f'(1) = \lim_{h \to 0} \frac{f(1 + h) - f(1)}{h} = \lim_{h \to 0} \frac{f(1 + h) - 3}{h}$$

If $h < 0$ $(h > -1)$,

$$\lim_{h \to 0} \frac{f(1 + h) - 3}{h} = \lim_{h \to 0^-} \frac{3(1 + h)^2 - 3}{h} = \lim_{h \to 0^-} (6 + 3h^2) = 6$$

If $h > 0$,

$$\lim_{h \to 0} \frac{f(1 + h) - 3}{h} = \lim_{h \to 0^+} \frac{0 - 3}{h} = -\infty$$

$f'(1)$ does not exist

13. $f'(0) = \lim_{h\to 0} \dfrac{f(h)-f(0)}{h} = \lim_{h\to 0} \dfrac{h^2 \sin \frac{1}{h}}{h} = \lim_{h\to 0} h \sin \frac{1}{h} = 0$

Review 10, pp. 290–291

1. False.   A = P = $(1, 2)$

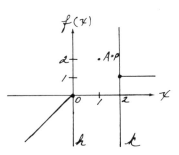

Let h and k be the
vertical lines x = 0
and x = 2. There is
no point of &-A
between h and k. Thus
1 is not a limit point
of the domain of f.

5. True.   A = P = $(2, 5)$

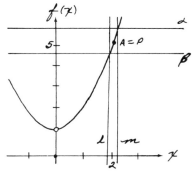

Since f is defined for all
values of x, 2 is a limit
point of the domain of f.
Let α and β be any pair
of horizontal lines with
P between them. Let ℓ
and m be the vertical
lines through the right-
most point of intersection
of & with α and β (if β
does not intersect &, let
m be x = 0). Then any
point of &-A between ℓ and m is also between α and β.

9. True.   G(x) = $1/(x-3)$ if x > 3   No A

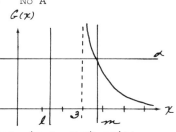

Since G is defined for all
x > 3, 3 is a limit point
of the domain of G.
Suppose α is a horizontal
line. Let ℓ be any
vertical line to the left
of x = 3 and m the vertical
line through the point of
intersection of α and & (if there is no such point,
let m be any vertical line to the right of x = 3).
Then any point of &-A(=&) between ℓ and m is also
above α.

13. False. $P = (0, 0)$; no A

Let α and β be any pair of horizontal lines with P between them. No matter how we choose ℓ and m, there are points of $\mathscr{L}$ $-A(=\mathscr{L})$ between ℓ and m but not between α and β.

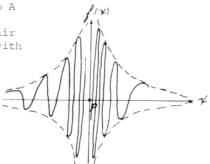

17. Continuous. Let α and β be any pair of horizontal lines with A = (1, 1) between them. Let ℓ be the vertical line through the right-most point of intersection of $\mathscr{L}$ with β that is to the left of A (if there is no point of intersection of $\mathscr{L}$ with β which is to the left of A, let ℓ be x = 0), and let m be the vertical line through the point of intersection of $\mathscr{L}$ with β which is to the right of A. Then every point of $\mathscr{L}$ between ℓ and m is also between α and β.

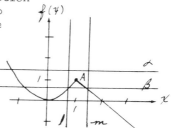

21. $f'(1) = \lim\limits_{h \to 0} \dfrac{f(1 + h) - f(1)}{h}$

$$= \begin{cases} \lim\limits_{h \to 0^-} \dfrac{(1 + h) - 1}{h} = \lim\limits_{h \to 0^-} \dfrac{h}{h} = \lim\limits_{h \to 0^-} 1 = 1 \\[4mm] \lim\limits_{h \to 0^+} \dfrac{[(1 + h)^3 - (1+h)^2 + 1] - 1}{h} = \lim\limits_{h \to 0^+} \dfrac{h^3 + 3h^2 + 3h + 1 - h^2 - 2h - 1}{h} \end{cases}$$

$$= \lim\limits_{h \to 0^+} (h^2 + 2h + 1) = 1$$

Therefore $f'(1) = 1$

Chapter Eleven
Trigonometric Functions

Section 11.1, p. 296

1. $45° = \pi/4$, $-210° = -7\pi/6$, $270° = 3\pi/2$, $30° = \pi/6$,
    $-180° = -\pi$, $-60° = -\pi/3$, $135° = 3\pi/4$, $150° = 5\pi/6$

5.

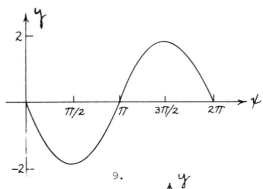

9.

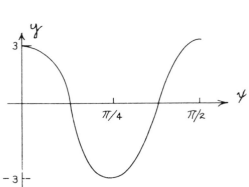

13.

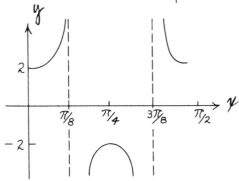

17.

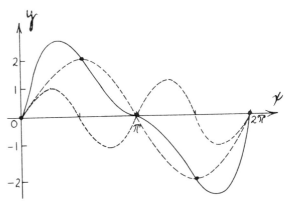

21.

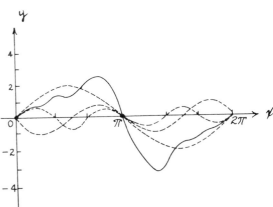

25.

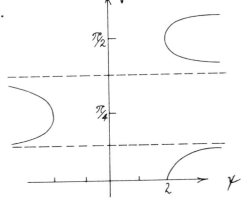

29.

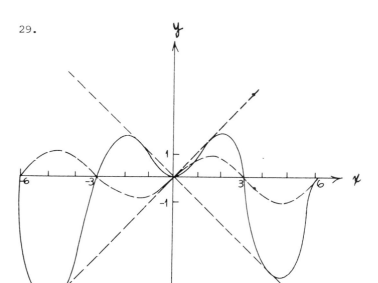

Section 11.2, pp. 303-304

1. $y' = 2 \cos 2x$

5. $y' = 2 \sin 3x \cos 3x \cdot 3 = 6 \sin 3x \cos 3x$

9. $y' = \sec x \cdot \sec^2 x + \tan x \sec x \tan x$

   $= \sec^3 x + \sec x \tan^2 x$

13. $y' = 2 \csc x(-\csc x \cot x) - 2 \cot x(-\csc^2 x)$

   $= -2 \csc^2 x \cot x + 2 \csc^2 x \cot x = 0$ or

   $y = \csc^2 x - (\csc^2 x - 1) = 1, \quad y' = 0$

17. $y' = \dfrac{2x \sec^2 2x \cdot 2 - \tan 2x \cdot 2}{4x^2} = \dfrac{2x \sec^2 2x - \tan 2x}{2x^2}$

21. $y' = \cos \cos x(-\sin x) = -\sin x \cos \cos x$

128   *Section 11.2*

25. $y' = \dfrac{2 \cos 2x}{2\sqrt{\sin 2x}} = \dfrac{1}{0}$ (no value)

29. $y' = 2\sqrt{3} \cos x - 2 \sin 2x = 2\sqrt{3} \cos x - 4 \sin x \cos x$

$\quad = 2 \cos x (\sqrt{3} - 2 \sin x)$

$y'' = -2\sqrt{3} \sin x - 4 \cos 2x = -2\sqrt{3} \sin x - 4(1 - 2 \sin^2 x)$

$\quad = 8 \sin^2 x - 2\sqrt{3} \sin x - 4$

Critical points: $2 \cos x (\sqrt{3} - 2 \sin x) = 0$

$\cos x = 0$ $\qquad\qquad\qquad$ $\sin x = \sqrt{3}/2$

$x = 90°,\ 270°$ $\qquad\qquad$ $x = 60°,\ 120°$

$(90°,\ 2\sqrt{3} - 1),\ (270°,\ -2\sqrt{3} - 1),\ (60°,\ 5/2),\ (120°, 5/2)$

$y'' = -2\sqrt{3}+4 > 0$ $\quad$ $y'' = 2\sqrt{3}+4 > 0$ $\qquad$ $y'' = -1$ $\qquad$ $y'' = -1$

min $\qquad\qquad\qquad$ min $\qquad\qquad\qquad$ max $\qquad\qquad$ max

Points of inflection:

$4 \sin^2 x - \sqrt{3} \sin x - 2 = 0$

$\sin x = \dfrac{\sqrt{3} \pm \sqrt{3 + 32}}{8} = \dfrac{\sqrt{3} \pm \sqrt{35}}{8} = \begin{cases} -0.523 \\ 0.956 \end{cases}$

$x = 73°,\ 107°,\ 211°30',\ 328°30'$

$(73°,\ 2.483),\ (107°,\ 2.483),\ (211°30',\ -1.358)$

$(328°30',\ -1.358)$

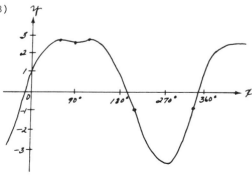

33. $\sec^2(x + y)(1 + y') = y'$

$\sec^2(x + y) = y'[1 - \sec^2(x + y)]$

$y' = \dfrac{\sec^2(x + y)}{1 - \sec^2(x+y)} = \dfrac{\sec^2(x + y)}{-\tan^2(x+y)} = -\dfrac{1}{\sin^2(x+y)}$

$\quad = -\csc^2(x + y)$

37. $y' = \cos x = \dfrac{1}{2}$, $\qquad y - \dfrac{\sqrt{3}}{2} = \dfrac{1}{2}(x - \dfrac{\pi}{3})$

$3x - 6y + 3\sqrt{3} - \pi = 0$

41. $\tan \alpha = \dfrac{x}{2}$

$x = 3.464$, $\tan \alpha = 1.732$, $\alpha = \dfrac{\pi}{3}$

$\dfrac{d\alpha}{dt} = 6\pi$ radians/min

$x = 2 \tan \alpha$

$\dfrac{dx}{dt} = 2 \sec^2\alpha \dfrac{d\alpha}{dt} = 2 \sec^2 \dfrac{\pi}{3} \cdot 6\pi$

$\quad = 48\pi$ mi/min

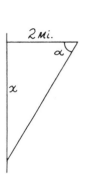

Section 11.3, p. 308

1. $\displaystyle\int \sin 2\theta = \dfrac{1}{2} \int 2 \sin 2\theta \; d\theta = -\dfrac{1}{2} \cos 2\theta + C$

5. $\displaystyle\int x^2 \sec x^3 \tan x^3 dx = \dfrac{1}{3}\int 3x^2 \sec x^3 \tan x^3 dx = \dfrac{1}{3} \sec x^3 + C$

9. $\displaystyle\int (\sin^2 x - \cos^2 x) dx = \int -\cos 2x \; dx = -\dfrac{1}{2}\int 2 \cos 2x \; dx$

$\quad = -\dfrac{1}{2} \sin 2x + C$

13. $\displaystyle\int \dfrac{\tan x}{\sec^3 x} dx = \int \dfrac{\sin x}{\cos x} \cdot \cos^3 x \; dx = \int \cos^2 x \sin x \; dx$

$\quad = -\displaystyle\int \cos^2 x (-\sin x) dx = -\dfrac{1}{3}\cos^3 x + C$

17. $\int \cos^2\theta \, d\theta = \frac{1}{2}\int (1 + \cos 2\theta) \, d\theta = \frac{1}{4}\int (2 + 2\cos 2\theta) \, d\theta$

$\qquad = \frac{1}{4}(2\theta + \sin 2\theta) + C$

21. $\int \sqrt{\cot x} \, \csc^2 x \, dx = -\int \sqrt{\cot x} \, (-\csc^2 x) \, dx = -\frac{2}{3}(\cot x)^{3/2} + C$

25. $\int_{-\pi/4}^{\pi/4} \tan^2 x \, dx = \int_{-\pi/4}^{\pi/4} (\sec^2 x - 1) \, dx = \tan x - x \Big|_{-\pi/4}^{\pi/4}$

$\qquad = (1 - \frac{\pi}{4}) - (-1 + \frac{\pi}{4}) = 2 - \frac{\pi}{2}$

29. From the figure, we have

$\quad A = \int_0^{\pi} \sin x \, dx$

$\qquad = -\cos x \Big|_0^{\pi}$

$\qquad = 2$

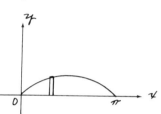

Section 11.4, pp. 314–315

1. $y' = \dfrac{2}{\sqrt{1 - 4x^2}}$

5. $y' = \dfrac{2}{1 + 4x^2}$

9. $y' = 1 - \dfrac{1}{1 + x^2} = \dfrac{x^2}{1 + x^2}$

13. $y' = \dfrac{1/2\sqrt{x}}{\sqrt{1 - x}} = \dfrac{1}{2\sqrt{x - x^2}}$

17. $y' = \dfrac{1}{\sqrt{1-x^2}} - x\dfrac{-x}{\sqrt{1-x^2}} - \sqrt{1-x^2} = \dfrac{1 + x^2 - (1-x^2)}{\sqrt{1-x^2}} = \dfrac{2x^2}{\sqrt{1-x^2}}$

21. $y' = \dfrac{-1/x^2}{\sqrt{1 - 1/x^2}} = \dfrac{-1}{|x|\sqrt{x^2 - 1}}$

25. $y' = \dfrac{1}{\sqrt{1-x^2}} = \dfrac{1}{\sqrt{1 - 1/4}} = \dfrac{1}{\sqrt{3}/2} = \dfrac{2}{\sqrt{3}}$

29. $y' = \dfrac{1}{1 + x^2} = \dfrac{1}{2}$ when $x = -1$

$$y + \frac{\pi}{4} = \frac{1}{2}(x + 1) , \qquad 4y + \pi = 2x + 2 , \qquad 2x - 4y + (2 - \pi) = 0$$

33. $y' = \dfrac{1}{1 + x^2} - 1$

$y'' = \dfrac{-2x}{(1 + x^2)^2}$

$\dfrac{1}{1 + x^2} = 1$

$1 + x^2 = 1$

$x^2 = 0$

$x = 0$

$(0,0)$ neither max. nor min.

37. $y' = \dfrac{1}{1 + x^2} - \dfrac{1}{1 + x^2} = 0, \quad y = C$

Since this shows that y is constant, we can find the constant by determining y for any value of x. When $x = 1$, $y = $ Arctan 1 + Arccot 1 = $\pi/4 + \pi/4 = \pi/2$.

Section 11.5, pp. 318-319

1. $\displaystyle\int \frac{-x \ dx}{\sqrt{1-x^2}} = \frac{1}{2}\int \frac{-2x \, dx}{\sqrt{1-x^2}}$

$= \sqrt{1 - x^2} + C$

5. $\displaystyle\int \frac{dx}{\sqrt{1-9x^2}} = \frac{1}{3}\int \frac{3 \, dx}{\sqrt{1-(3x)^2}}$

$= \frac{1}{3}$ Arcsin 3x + C

9. $\displaystyle\int \frac{dx}{\sqrt{1-(2x+1)^2}} = \frac{1}{2}\int \frac{2 \, dx}{\sqrt{1-(2x+1)^2}} = \frac{1}{2}$ Arcsin$(2x + 1)$ + C

13. $\displaystyle\int \frac{dx}{(x-2)\sqrt{x^2-4x+3}} = \int \frac{dx}{(x-2)\sqrt{(x-2)^2-1}} = $ Arcsec$(x - 2)$ + C

17. $\displaystyle\int \frac{\cos x \ dx}{1 + \sin^2 x} = $ Arctan sin x + C

132　　　*Section 11.5*

21. $\int \dfrac{\text{Arctan } x}{1 + x^2}dx = \dfrac{1}{2} \text{ Arctan}^2 x + C$

25. $\int_0^1 \dfrac{dx}{1 + x^2} = \text{Arctan } x \Big|_0^1 = \dfrac{\pi}{4}$

29. $\theta_1 = \text{Arcsin}(x - 1)$

$\qquad \sin \theta_1 = x - 1$

$\qquad \cos \theta_1 = \sqrt{2x - x^2}$

$\theta_2 = \text{Arcsin } \sqrt{\dfrac{x}{2}}$

$\qquad \sin \theta_2 = \sqrt{\dfrac{x}{2}}$

$\qquad \cos \theta_2 = \sqrt{\dfrac{2 - x}{2}}$

$\sin[\text{Arcsin}(x - 1) - 2 \text{ Arcsin } \sqrt{\dfrac{x}{2}}] = \sin(\theta_1 - 2\theta_2)$

$\qquad = \sin \theta_1 \cos 2\theta_2 - \cos \theta_1 \sin 2\theta_2$

$\qquad = \sin \theta_1 (\cos^2\theta_2 - \sin^2\theta_2) - \cos \theta_1 \cdot 2 \sin \theta_2 \cos \theta_2$

$\qquad = (x - 1)(\dfrac{2-x}{2} - \dfrac{x}{2}) - \sqrt{2x - x^2} \cdot 2\sqrt{\dfrac{x}{2}} \sqrt{\dfrac{2-x}{2}}$

$\qquad = (x-1)(1-x) = (2x-x^2) = -x^2 + 2x - 1 - 2x + x^2 = -1$

33. $A = 2\int_0^1 y \ dx = 2\int_0^1 \dfrac{dx}{1 + x^2}$

$\qquad = 2 \text{ Arctan } x \Big|_0^1$

$\qquad = 2 \cdot \pi/4 - 0 = \pi/2$

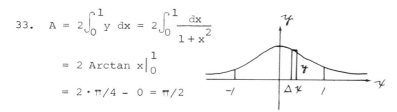

37. $A = \int_0^k y \ dx = \int_0^k \dfrac{dx}{1 + x^2}$

$\qquad = \text{Arctan } x \Big|_0^k$

$\qquad = \text{Arctan } k$

$\lim\limits_{k \to +\infty} A = \dfrac{\pi}{2}$

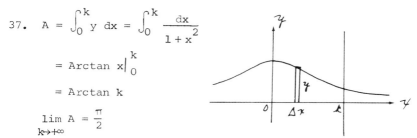

This is the area under the right half of the curve.

1. $y' = \dfrac{\cos x \cdot \cos x - (1 + \sin x)(-\sin x)}{\cos^2 x}$

$= \dfrac{\cos^2 x + \sin x + \sin^2 x}{\cos^2 x} = \dfrac{1 + \sin x}{\cos^2 x}$

5. $y' = 9 - 3 \csc^2 3x \cdot 3 - 3 \cot^2 3x(-\csc^2 3x \cdot 3)$

$= 9 - 9 \csc^2 3x + 9 \csc^2 3x \cot^2 3x$

$= 9 - 9 \csc^2 3x + 9 \csc^2 3x(\csc^2 3x - 1)$

$= 9(1 - 2 \csc^2 3x + \csc^4 3x)$

$= 9(1 - \csc^2 3x) = 9 \cot^4 3x$

9. $y' = 2 \text{ Arccos } 2x \cdot \dfrac{-2}{\sqrt{1 - 4x^2}} = \dfrac{-4 \text{ Arccos } 2x}{\sqrt{1 - 4x^2}}$

13. $\displaystyle\int \cos^{5/2} 3x \, \tan 3x \, dx = \int \cos^{5/2} 3x \, \dfrac{\sin 3x}{\cos 3x} \, dx$

$= -\dfrac{1}{3} \displaystyle\int \cos^{3/2} 3x(-3 \sin 3x) \, dx$

$= -\dfrac{1}{3} \dfrac{\cos^{5/2} 3x}{5/2} + C = -\dfrac{2}{15} \cos^{5/2} 3x + C$

17. $\displaystyle\int \dfrac{dx}{(x + 3)\sqrt{x^2 + 6x + 8}} = \int \dfrac{dx}{(x + 3)\sqrt{(x + 3)^2 - 1}}$

$= \text{Arcsec}(x + 3) + C$

21. $A = \displaystyle\int_0^2 y \, dx = \int_0^2 \dfrac{dx}{1 + x^2}$

$= \text{Arctan } x \Big|_0^2 = \text{Arctan } 2$

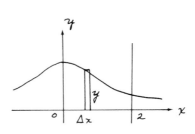

25. $y' = \frac{1}{2} - \sin x$, $y'' = -\cos x$

Critical points: $\sin x = \frac{1}{2}$, $x = \pi/6$, $5\pi/6$

$(\pi/6, (\pi + 6\sqrt{3})/12)$, $\qquad (5\pi/6, (5\pi - 6\sqrt{3})/12)$

$y'' < 0$ $\qquad\qquad\qquad\qquad y'' > 0$

max $\qquad\qquad\qquad\qquad\quad$ min

Points of inflection:

$\cos x = 0$

$x = \pi/2$, $3\pi/2$

$(\pi/2, \pi/4)$, $(3\pi/2, 3\pi/4)$

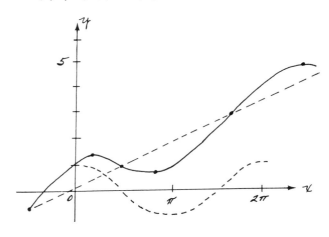

Chapter Twelve
Exponents, Logarithms, and Hyperbolic Functions

1. $y = 2^{x+1} = 2 \cdot 2^x$

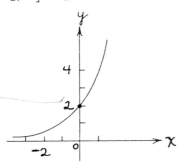

5.

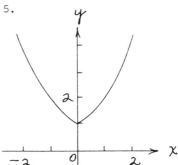

9.

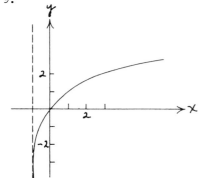

13.

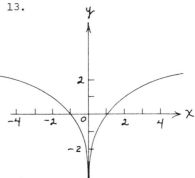

17. $y = e^x$

$x = \ln y$

21. $y = \log_3 3^x$

$x = y$

25. $\log x = 2 \log 3 - \log 4$

$\log x = \log \dfrac{3^2}{4}$

$x = \dfrac{9}{4}$

29. $\ln x^2 - \ln 2x = 4 \ln 3 - \ln 6$

    $\ln \dfrac{x^2}{2x} = \ln \dfrac{3^4}{6}$, $\dfrac{x}{2} = \dfrac{27}{2}$, $x = 27$

33. $\log_5 \dfrac{x^2(x+3)}{x-1} = 2 \log_5 x + \log_5(x+3) - \log_5(x-1)$

37.   I. Let $a^{\log_a n} = x$. Then by the definition of a

        logarithm, $\log_a x = \log_a n$ and $x = n$.

    II. Let $\log_a a^n = y$. Then by the definition of a

        logarithm, $a^n = a^y$ and $y = n$.

41. $F = P(1+\dfrac{r}{n})^n = P[(1+\dfrac{r}{n})^{n/r}]^r$

    $\lim\limits_{n\to+\infty} F = \lim\limits_{r/n\to 0} P[(1+\dfrac{r}{n})^{n/r}]^r = Pe^r$

    (a) $F = P(1+\dfrac{r}{n})^n = 1000(1.06) = \$1060.00$

    (b) $F = P(1+\dfrac{r}{n})^n = 1000(1.03)^2 = \$1060.90$

    (c) $F = Pe^r = 1000\ e^{0.06} = 1000(1.0618) = \$1061.80$

Section 12.2, p. 330

1. $y = \log 4 + \log x$

   $y' = \dfrac{\log e}{x}$

5. $y = -\ln x$, $y' = -\dfrac{1}{x}$

9. $y = \dfrac{2}{3}[\log_5(3x+1) - \log_5(3x-1)]$

   $y' = \dfrac{2}{3}(\dfrac{3}{3x+1} - \dfrac{3}{3x-1})\log_5 e = \dfrac{-4\log_5 e}{9x^2 - 1} = \dfrac{4\log_5 e}{1 - 9x^2}$

13. $y' = \dfrac{x \cdot 1/x - \ln x}{x^2} = \dfrac{1 - \ln x}{x^2}$

17. $y' = \dfrac{\cos \ln x}{x}$

21. $y' = \dfrac{1/x}{\ln x} = \dfrac{1}{x \ln x}$

25. $\dfrac{y'}{y} = \cos x$, $\qquad y' = y \cos x$

29. $2x + 2yy' = \dfrac{1 + y'}{x + y}$, $\qquad 2x^2 + 2xy + 2xyy' + 2y^2y' = 1 + y'$

$\qquad y'(2xy + 2y^2 - 1) = 1 - 2x^2 - 2xy$, $\qquad y' = \dfrac{1 - 2x^2 - 2xy}{2xy + 2y^2 - 1}$

33. $\ln y = \dfrac{2}{3} \ln x + \dfrac{4}{3} \ln(x + 1) - \dfrac{1}{3} \ln(x - 5)$

$\qquad \dfrac{y'}{y} = \dfrac{1}{3}(\dfrac{2}{x} + \dfrac{4}{x + 1} - \dfrac{1}{x - 5})$, $\qquad y' = \dfrac{y}{3}(\dfrac{2}{x} + \dfrac{4}{x + 1} - \dfrac{1}{x - 5})$

37. $y' = x \cdot \dfrac{1}{x} + \ln x = 1 + \ln 1 = 1$

41. $y' = \dfrac{-\sin x}{\cos x} = -1$ when $x = \pi/4$

$\qquad y + \dfrac{\ln 2}{2} = -(x - \dfrac{\pi}{4})$, $\qquad 4y + 2 \ln 2 = -4x + \pi$

$\qquad 4x + 4y + (2 \ln 2 - \pi) = 0$

45. $S = kx^2 \ln 1/x = -kx^2 \ln x$

$\qquad \dfrac{dS}{dx} = -k(x^2 \dfrac{1}{x} + 2x \ln x) = -kx(1 + 2 \ln x)$

$\qquad 1 + 2 \ln x = 0$, $\qquad \ln x = -\dfrac{1}{2}$, $\qquad x = e^{-1/2} = \dfrac{1}{\sqrt{e}}$

Section 12.3, p. 334

1. $y' = 3^x \ln 3$

5. $y' = 2e^{2x+2}$

9. $y' = 1 + e^x$

13. $y' = 3^x \cdot 3x^2 + 3^x(x^3 - 1)\ln 3 = 3^x[3x^2 + (x^3 - 1)\ln 3]$

17. $y' = \sec^2 x \, e^{\tan x}$

21. $\ln y = x \ln \sin x$

$\qquad \dfrac{y'}{y} = x\dfrac{\cos x}{\sin x} + \ln \sin x$

$\qquad y' = (\sin x)^x(x \cot x + \ln \sin x)$

25. $\dfrac{1/x}{\ln x} = e^y \cdot y'$, $\quad y' = \dfrac{1}{xe^y \ln x}$

29. $y' = e^{\sin x} \cos x = e \cdot 0 = 0$

33. $y' = \dfrac{xe^x - e^x}{x^2} = 0$ when $\quad x = 1$

$\quad y - e = 0$

37. $y' = -e^x \sin e^x$

$\quad \sin e^x = 0$, $\quad e^x = n\pi$, $\quad x = \ln(n\pi)$, $\quad y = \pm 1$

$\quad (\ln(n\pi), 1)$ max if n is even

$\quad (\ln(n\pi), -1)$ min if n is odd

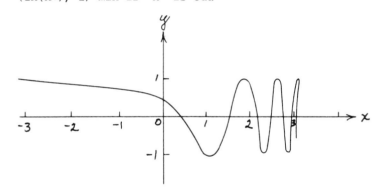

Section 12.4, p. 338

1. $\displaystyle\int \dfrac{dx}{2x+1} = \dfrac{1}{2} \int \dfrac{2dx}{2x+1} = \dfrac{1}{2} \ln \left|2x+1\right| + C$

5. $\displaystyle\int \dfrac{\sin x}{\cos x} dx = - \int \dfrac{-\sin x}{\cos x} dx = -\ln \left|\cos x\right| + C$

9. $\displaystyle\int \dfrac{2x+1}{\sqrt{x^2+x}} dx = \dfrac{\sqrt{x^2+x}}{1/2} + C = 2\sqrt{x^2+x} + C$

13. $\displaystyle\int \sec^2 x \; e^{\tan x} dx = e^{\tan x} + C$

17. $\int \dfrac{x^3 + 3x^2 + 5x + 3}{(x + 1)^2} dx = \int (x + 1 + \dfrac{2x + 2}{x^2 + 2x + 1}) dx$

$$= \dfrac{x^2}{2} + x + \ln(x^2 + 2x + 1) + C = \dfrac{x^2}{2} + x + 2 \ln |x + 1| + C$$

21. $\int (e^x - e^{-x}) dx = e^x + e^{-x} + C$

25. $\int_0^1 \dfrac{dx}{x + 1} = \ln(x + 1) \Big|_0^1 = \ln 2 - \ln 1 = \ln 2$

29. $\int_{\pi/6}^{\pi/2} \dfrac{\cos x}{\sin x} dx = \ln \sin x \Big|_{\pi/6}^{\pi/2} = \ln 1 - \ln \dfrac{1}{2} = \ln 2$

33. $A = \int_0^{\ln 2} (2 - x) dy = \int_0^{\ln 2} (2 - e^y) dy = (2y - e^y) \Big|_0^{\ln 2}$

$$= 2 \ln 2 - 2 + 1 = 2 \ln 2 - 1$$

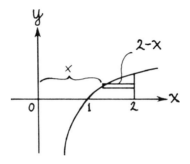

Section 12.5, pp. 344–345

1. $y' = xy + x = x(y + 1)$, $\dfrac{y'}{y + 1} = x$

$\ln |y + 1| = x^2/2 + C$, $y + 1 = ke^{x^2/2}$

$y = ke^{x^2/2} - 1$

5. $y' - e^{x+y} = 0$, $y' = e^x \cdot e^y$

$e^{-y} y' = e^x$, $-e^{-y} = e^x - C$

$e^x + e^y = C$

9. $(1 - x^2)y' = x$

$$y' = \frac{x}{1 - x^2} = -\frac{1}{2}\frac{-2x}{1 - x^2}, \qquad y = -\frac{1}{2}\ln|1 - x^2| + C$$

$$2 = -\frac{1}{2}\ln 1 + C = C, \qquad y = 2 - \frac{1}{2}\ln|1 - x^2|$$

13. $\dfrac{dx}{dt} = kx, \qquad x = x_o e^{kt} = 2,249,000,000\, e^{kt}$

Since $x = 3,008,000,000$ when $t = 20$,

$$3,008,000,000 = 2,249,000,000\, e^{20k}$$

$$e^{20k} = 1.3375, \qquad 20k = \ln 1.3375 = 0.2909$$

$$k = 0.014545, \qquad x = 2,249,000,000\, e^{0.014545t}$$

When $t = 60$,

$$x = 2,249,000,000\, e^{0.014545(60)} = 2,249,000,000\, e^{0.873}$$

$$= 2,249,000,000(2.3948) = 5,386,000,000$$

17. $x = x_o e^{kt} = 0.0001\, e^{kt}$

Since $x = 0.0000947$ when $t = 1$

$$0.0000947 = 0.0001\, e^{k}, \qquad e^{k} = 0.947$$

$$k = \ln 0.947 = -0.0559 \qquad x = 0.0001\, e^{-0.0559t}$$

When $x = 0.00005$, $\qquad\qquad e^{-0.0559t} = 0.5$

$$-0.0559t = \ln 0.5 = -0.6931, \qquad t = 12.4 \text{ yr}$$

21. $\dfrac{dx}{dt} = kx^2, \qquad\qquad \dfrac{x}{x^2} = k$

$$-\frac{1}{x} = kt + C$$

Since $x = 250$ when $t = 0$,

$$C = -\frac{1}{250}, \qquad\qquad -\frac{1}{x} = kt - \frac{1}{250}$$

Since $x = 200$ when $t = 30$,

$$\frac{1}{250} - \frac{1}{200} = 30k, \qquad k = -\frac{1}{30,000}$$

$$\frac{1}{250} - \frac{1}{x} = -\frac{t}{30,000}$$

When $x = 50$

$$\frac{1}{250} - \frac{1}{50} = -\frac{t}{30,000} , \qquad t = 480 \text{ min} = 6 \text{ hr}$$

25. $\dfrac{dx}{dt} = p - kx ,$ $\qquad\qquad\qquad \dfrac{x'}{p - kx} = 1$

$$-\frac{1}{k} \ln(p - kx) = t + C$$

Since $x = 0$ when $t = 0$,

$-\dfrac{1}{k} \ln p = C ,$ $\qquad\qquad -\dfrac{1}{k} \ln(p - kx) = t - \dfrac{1}{k} \ln p$

$$kt = \ln p - \ln(p - kx) = \ln \frac{p}{p - kx} ,$$

$\dfrac{p}{p - kx} = e^{kt} ,$ $\qquad\qquad\qquad p - kx = pe^{-kt}$

$kx = p(1 - e^{-kt}) ,$ $\qquad\qquad\quad x = \dfrac{p}{k}(1 - e^{-kt})$

$$\lim_{t \to +\infty} x = \lim_{t \to +\infty} \frac{p}{k}(1 - e^{-kt}) = \frac{p}{k}$$

Section 12.6, pp. 351-352

1. (a) $\dfrac{15,000,000}{61,900} = 242 \text{ yr},$ $\qquad 1972 + 242 = 2214$

(b) $A = \dfrac{x_0}{k}(e^{kt} - 1)$

$$15,000,000 = \frac{61,900}{0.065}(e^{0.065t} - 1)$$

$e^{0.065t} - 1 = 15.75 ,$ $\qquad\qquad e^{0.065t} = 16.75$

$$0.065t = \ln 16.75 = 2.818$$

$t = 43 ,$ $\qquad\qquad\qquad\qquad 1972 + 43 = 2015$

5. $A = \dfrac{x_0}{k}(e^{kt} - 1)$

$$913,000 = \frac{2,500}{0.067}(e^{0.067t} - 1)$$

$e^{0.067t} - 1 = 24.5 ,$ $\qquad\qquad e^{0.067t} = 25.5$

$$0.067t = \ln 25.5 = 3.2371$$

$t = 48 ,$ $\qquad\qquad\qquad\qquad 1972 + 48 = 2020$

9. $x = 1.4x - 4 \times 10^{-5}x^2 ,$ $\quad 0.04x - 4 \times 10^{-5}x^2 = 0$

$0.04x(1 - 10^{-4}x) = 0 ,$ $\qquad x = 10^4 = 10,000$

$y' = 1.4 - 8 \times 10^{-5} x = 1$, $\quad 8 \times 10^{-5} x = 0.4$

$x = 5,000$, $\qquad\qquad y = 6,000$

$y - x = 1,000$

13. $f'(x) = 1 + i$, $\qquad 1.4 - 8 \times 10^{-5} x = 1.08$

$8 \times 10^{-5} x = 0.32$, $\qquad x = 4000$

17. We want the maximum value of y'.

$y' = 1 + 7.5 \times 10^{-5} x - 3.75 \times 10^{-9} x^2$

$y'' = 7.5 \times 10^{-5} - 7.5 \times 10^{-9} x = 0$

$x = 10,000$, $\qquad\qquad y' = 1.375 = 1 + i$

$i = 0.375 = 37.5\%$

Section 12.7, pp. 356-357

1. $\cosh^2 x - \sinh^2 x = 1$, $\qquad \dfrac{\cosh^2 x}{\cosh^2 x} - \dfrac{\sinh^2 x}{\cosh^2 x} = \dfrac{1}{\cosh^2 x}$

$1 - \tanh^2 x = \operatorname{sech}^2 x$

5. $\sinh x \cosh y - \cosh x \sinh y$

$= \dfrac{e^x - e^{-x}}{2} \dfrac{e^y + e^{-y}}{2} - \dfrac{e^x + e^{-x}}{2} \dfrac{e^y - e^{-y}}{2}$

$= \dfrac{e^{x+y} - e^{-x+y} + e^{x-y} - e^{-x-y} - e^{x+y} + e^{x-y} - e^{-x+y} + e^{-x-y}}{4}$

$= \dfrac{2e^{x-y} - 2e^{-x+y}}{4} = \dfrac{e^{x-y} - e^{-(x-y)}}{2} = \sinh(x - y)$

9. $\cosh 2y = 2 \cosh^2 y - 1$, $\qquad \cosh^2 y = \dfrac{\cosh 2y + 1}{2}$

$\cosh y = \sqrt{\dfrac{\cosh 2y + 1}{2}}$ (cosh y is never negative)

Letting $y = \dfrac{x}{2}$, $\qquad \cosh \dfrac{x}{2} = \sqrt{\dfrac{\cosh x + 1}{2}}$

13. $1 - \tanh^2 x = \operatorname{sech}^2 x$ $\qquad \coth x = \dfrac{1}{\tanh x} = -\dfrac{13}{5}$

$1 - \dfrac{25}{169} = \operatorname{sech}^2 x$ $\qquad \cosh x = \dfrac{1}{\operatorname{sech} x} = \dfrac{13}{12}$

$\dfrac{144}{169} = \operatorname{sech}^2 x$ $\qquad \tanh x = \dfrac{\sinh x}{\cosh x}$

$$\text{sech } x = \frac{12}{13} \qquad\qquad -\frac{5}{13} = \frac{\sinh x}{13/12}$$

$$\sinh x = -\frac{5}{12}$$

$$\text{csch } x = \frac{1}{\sinh x} = -\frac{12}{5}$$

17. $\dfrac{d}{dx}\sinh u = \dfrac{d}{dx}(\dfrac{e^u - e^{-u}}{2}) = \dfrac{e^u \cdot u' + e^{-u} \cdot u'}{2}$

$\qquad = \dfrac{e^u + e^{-u}}{2}u' = \cosh u \cdot u'$

21. $y' = 2x \cosh x^2$

25. $y' = \dfrac{-\text{sech }\sqrt{x}\ \tanh \sqrt{x}}{2\sqrt{x}}$

29. $y' = 2 \sinh x \cosh x + 2 \cosh x \sinh x = 4 \sinh x \cosh x$

33. $y' = -\cosh x \sin \sinh x$

37. $y' = 2 \text{ sech}^2 2x = 2 \quad\text{when}\quad x = 0$

41. $y = \text{gd } x = \text{Arctan } \sinh x$

$\qquad y' = \dfrac{\cosh x}{1 + \sinh^2 x} = \dfrac{\cosh x}{\cosh^2 x} = \dfrac{1}{\cosh x} = \text{sech } x$

$\qquad \lim_{x \to \pm\infty} \sinh x = \pm\infty$

$\qquad \lim_{x \to \pm\infty} \text{gd } x = \lim_{x \to \pm\infty} \text{Arctan } \sinh x = \pm \pi/2$

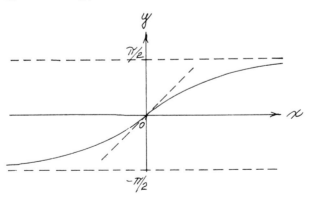

1. $\int \cosh(2x + 1)\,dx = \dfrac{1}{2} \int 2 \cosh(2x + 1)\,dx = \dfrac{1}{2} \sinh(2x + 1) + C$

5. $\int \sinh^2 x \cosh x\,dx = \dfrac{1}{3} \sinh^3 x + C$

9. $\int \dfrac{\sinh x\,dx}{\sqrt{1 - \cosh^2 x}} = \text{Arcsin} \cosh x + C$

13. $A = \displaystyle\int_0^1 y\,dx = \int_0^1 \cosh x\,dx = \sinh x \Big|_0^1 = \sinh 1 - \sinh 0$

    $= \sinh 1$

Section 12.9, p. 363

1. $\sinh^{-1} 3 = \ln(3 + \sqrt{9 + 1}) = \ln(3 + \sqrt{10}) = \ln(3 + 3.162)$

   $= \ln 6.162 = 1.819$

5. $\coth^{-1} 3 = \dfrac{1}{2} \ln \dfrac{4}{2} = \dfrac{1}{2} \ln 2 = \dfrac{1}{2}(0.6931) = 0.347$

9. $y' = \dfrac{1/2\sqrt{x}}{\sqrt{1 + x}} = \dfrac{1}{2\sqrt{x^2 + x}}$

13. $y' = \dfrac{e^x}{1 - e^{2x}}$

17. $y' = 2 \sinh^{-1} x \cdot \dfrac{1}{\sqrt{1 + x^2}} = \dfrac{2 \sinh^{-1} x}{\sqrt{1 + x^2}}$

21. $y = \cosh^{-1} x$, $\qquad\qquad x = \cosh y = \dfrac{e^y + e^{-y}}{2}$

    $e^y - 2x + e^{-y} = 0$, $\qquad\qquad e^{2y} - 2xe^y + 1 = 0$

    $e^y = \dfrac{2x \pm \sqrt{4x^2 - 4}}{2} = x \pm \sqrt{x^2 - 1}$

    $y \geqq 0$, $\quad e^y \geqq 1$; $\quad x - \sqrt{x^2 - 1} \leqq 1$. Thus

    $e^y = x + \sqrt{x^2 - 1}$, $\qquad y = \ln(x + \sqrt{x^2 - 1})$

25. $y = \cosh^{-1} u$, $\qquad\qquad u = \cosh y$

    $u' = \sinh y \cdot y'$

$$y' = \frac{u'}{\sinh y} = \frac{u'}{\sqrt{\cosh^2 y - 1}} = \frac{u'}{\sqrt{u^2 - 1}}$$

or

$$y = \cosh^{-1} u = \ln(u + \sqrt{u^2 - 1})$$

$$y' = \frac{u' + 2uu'/2\sqrt{u^2-1}}{u + \sqrt{u^2-1}} = \frac{\sqrt{u^2-1} + u}{(u + \sqrt{u^2-1})\sqrt{u^2-1}} u' = \frac{u'}{\sqrt{u^2-1}}$$

29. $y = \operatorname{csch}^{-1} u$, $u = \operatorname{csch} y$

$u' = -\operatorname{csch} y \coth y \cdot y'$

$y' = \dfrac{-u'}{\operatorname{csch} y \coth y}$, $\coth^2 y = 1 + \operatorname{csch}^2 y$

If $u > 0$, $y > 0$ and $\coth y = \sqrt{1 + \operatorname{csch}^2 y}$

$$y' = \frac{-u'}{\operatorname{csch} y \sqrt{1 + \operatorname{csch}^2 y}} = \frac{-u'}{u\sqrt{1 + u^2}}$$

If $u < 0$, $y < 0$ and $\coth y = -\sqrt{1 + \operatorname{csch}^2 y}$

$$y' = \frac{u'}{\operatorname{csch} y \sqrt{1 + \operatorname{csch}^2 y}} = \frac{u'}{u\sqrt{1 + u^2}}$$

or

$$y = \operatorname{csch}^{-1} u = \begin{cases} \ln \dfrac{1 + \sqrt{1+u^2}}{u} = \ln(1 + \sqrt{1+u^2}) - \ln u & \text{if } u > 0 \\[3mm] -\ln \dfrac{1 + \sqrt{1+u^2}}{-u} = \ln(-u) - \ln(1 + \sqrt{1+u^2}) \\[3mm] \hspace{6cm} \text{if } u < 0 \end{cases}$$

If $u > 0$, then

$$y' = \frac{uu'/\sqrt{1+u^2}}{1 + \sqrt{1+u^2}} - \frac{u'}{u} = \frac{u^2 u'}{u\sqrt{1+u^2}(1+\sqrt{1+u^2})} - \frac{(\sqrt{1+u^2} + 1 + u^2)u'}{u\sqrt{1+u^2}(1+\sqrt{1+u^2})}$$

$$= \frac{-(1 + \sqrt{1+u^2})u'}{u\sqrt{1+u^2}(1 + \sqrt{1+u^2})} = \frac{-u'}{u\sqrt{1+u^2}}$$

If $u < 0$, then $y' = \dfrac{u'}{u} - \dfrac{uu'/\sqrt{1+u^2}}{1 + \sqrt{1+u^2}} = \dfrac{u'}{u\sqrt{1+u^2}}$

1. $y' = \dfrac{\sin x \cdot e^x - e^x \cdot \cos x}{\sin^2 x} = \dfrac{(\sin x - \cos x)e^x}{\sin^2 x}$

5. $y = \ln \dfrac{x^2(x-1)}{\sqrt{x+1}} = 2 \ln x + \ln(x-1) - \dfrac{1}{2} \ln(x+1)$

$y' = \dfrac{2}{x} + \dfrac{1}{x-1} - \dfrac{1}{2(x+1)} = \dfrac{4(x^2-1) + 2(x^2+x) - (x^2-x)}{2x(x^2-1)}$

$= \dfrac{5x^2 + 3x - 4}{2x(x^2-1)}$

9. $y' = -\text{sech } e^x \text{ tanh } e^x \cdot e^x$

13. $\displaystyle\int \dfrac{(x-2)^2}{x^2+4} dx = \int \dfrac{x^2 - 4x + 4}{x^2+4} dx = \int (1 - 2\dfrac{2x}{x^2+4}) dx$

$= x - 2 \ln(x^2+4) + C$

17. $y' = \dfrac{x^2 \dfrac{1}{x} - 2x \ln x}{x^4} = \dfrac{x - 2x \ln x}{x^4} = \dfrac{1 - 2 \ln x}{x^3}$

$y'' = \dfrac{x^3(-2/x) - (1 - 2 \ln x)3x^2}{x^6} = \dfrac{-2 - 3(1 - 2 \ln x)}{x^4}$

$= \dfrac{6 \ln x - 5}{x^4}$

Critical points:

$1 - 2 \ln x = 0$, $\qquad$ $\ln x = 1/2$

$x = e^{1/2} = \sqrt{e}$, $y = \dfrac{1}{2e}$,

$y'' < 0$, $\qquad\qquad$ $(\sqrt{e}, 1/2e)$ max

Point of inflection:

$6 \ln x - 5 = 0$

$\ln x = 5/6$

$x = e^{5/6}$, $y = 5/6e^{5/3}$

$(e^{5/6}, 5/6e^{5/3})$

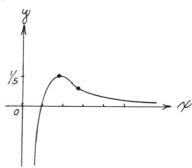

21. $y' = \dfrac{2 \ln x}{x} = \dfrac{2}{e}$ when $x = e$

$y - 1 = \dfrac{2}{e}(x - e)$, $ey - e = 2x - 2e$

$2x - ey - e = 0$

25. $A = \displaystyle\int_0^1 (e^{2x} - e^x)\, dx$

$= \left. (\dfrac{1}{2}e^{2x} - e^x) \right|_0^1$

$= \dfrac{1}{2}e^2 - e - (\dfrac{1}{2} - 1)$

$= \dfrac{(e - 1)^2}{2}$

29. $x = \cosh t$, $y = \sinh t$

$x^2 - y^2 = \cosh^2 t - \sinh^2 t = 1$

33. $\dfrac{dx}{dt} = kx$, $x = x_o\, e^{kt} = 912 \times 10^6 e^{kt}$

Since $x = 1590 \times 10^6$ when $t = 100$,

$1590 \times 10^6 = 912 \times 10^6 e^{100k}$, $e^{100k} = 1.74$

$100k = \ln 1.74 = 0.5535$, $k = 0.005535$

$x = 912 \times 10^6 e^{0.005535t}$

In year 2000: $t = 200$

$x = 912 \times 10^6 e^{1.107} = 912 \times 10^6 (3.0263) = 330,000,000$

In year 2500: $t = 700$

$x = 912 \times 10^6 e^{3.875} = 912 \times 10^6 (48.23) = 44,000,000,000$

37. (a) $\dfrac{340,000}{7100} = 48$ yr, $1972 + 48 = 2020$

(b) $A = \dfrac{x_o}{k}(e^{kt} - 1)$

$$340,000 = \frac{7100}{0.04}(e^{0.04t} - 1)$$

$$e^{0.04t} - 1 = 1.9155, \qquad e^{0.04t} = 2.9155$$

$$0.04t = \ln 2.9155 = 1.0702$$

$$t = 28 \text{ yr}$$

$$1972 + 28 = 2000$$

41. $y' = 1 + 18.75 \times 10^{-5} x - 46.875 \times 10^{-9} x^2 = 1.09$

$$46.875x^2 - 18.75 \times 10^4 x + 9 \times 10^7 = 0$$

$$15.625x^2 - 6.25 \times 10^4 x + 3 \times 10^7 = 0$$

$$x = \frac{6.25 + \sqrt{(6.25)^2 \times 10^8 - 12(15.625) \times 10^8}}{2(15.625)} = 1400$$

$$y - x = 140$$

Chapter Thirteen
Parametric Equations

1. $t = y - 1$

    $x = (y - 1)^2 + 1$

    $x - 1 = (y - 1)^2$

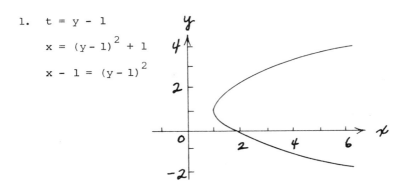

5. $x - y = 2t$, $\quad t = \dfrac{x - y}{2}$

    $x = \dfrac{x^2 - 2xy + y^2}{4} + \dfrac{x - y}{2}$

    $4x = x^2 - 2xy + y^2 + 2x - 2y$

    $x^2 - 2xy + y^2 - 2x - 2y = 0$

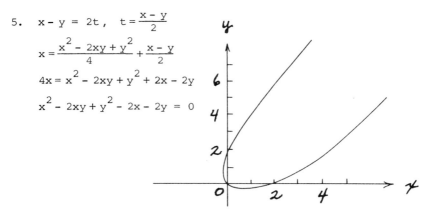

9. $\cos \theta = x - 2$,

    $\sin \theta = y + 1$

    $(x - 2)^2 + (y + 1)^2 = 1$

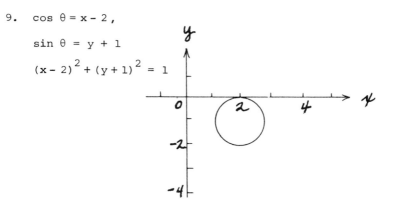

13. $x = t - 1$, $y = t^2$

$t = x + 1$

$y = (x + 1)^2$

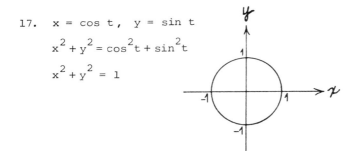

17. $x = \cos t$, $y = \sin t$

$x^2 + y^2 = \cos^2 t + \sin^2 t$

$x^2 + y^2 = 1$

21.

θ	x	y
0	1	0
$\pi/2$	$\pi/2$	1
π	-1	π
$3\pi/2$	$-3\pi/2$	-1
2π	1	-2π
$5\pi/2$	$5\pi/2$	1
3π	-1	3π

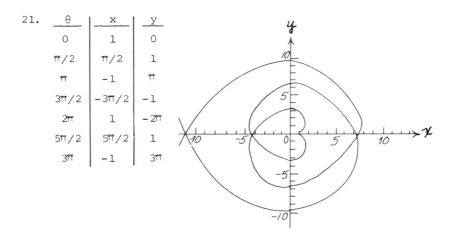

25. $x = x_1 + r(x_2 - x_1) = 1 + 2r$, $y = y_1 + r(y_2 - y_1) = 5 - 4r$

29. $x = x_1 + r(x_2 - x_1) = 2 + 3r$, $y = y_1 + r(y_2 - y_1) = 3$

33. (a) $y^2 = x^2 - 1$

   $x^2 - y^2 = 1$, $y \geqq 0$

(b) $y^2 = x^2 - 1$

   $x^2 - y^2 = 1$, $x \geqq 0$, $y \geqq 0$

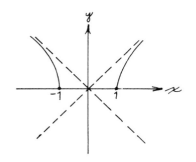

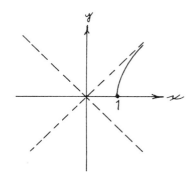

(c) $\sec^2 t = 1 + \tan^2 t$

   $x^2 = 1 + y^2$, $x^2 - y^2 = 1$

(d) $\cosh^2 t - \sinh^2 t = 1$

   $x^2 - y^2 = 1$, $x \geqq 1$

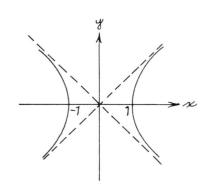

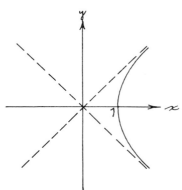

Section 13.2, pp. 375-376

1. $v_x^o = 2000 \cos 60° = 1000$, $v_y^o = 2000 \sin 60° = 1000\sqrt{3}$

$v_x = v_x^o = 1000$ $\qquad a_y = -32$

$x = 1000t$ $\qquad v_y = -32t + v_y^o = -32t + 1000\sqrt{3}$

$\qquad \qquad \qquad \qquad y = -16t^2 + 1000\sqrt{3}t$

5. $x_0 = y_0 = 0$

$v_x^o = v_o \cos \theta$, $v_y^o = v_o \sin \theta$

$v_x = v_o \cos \theta$ $\qquad\qquad a_y = -32$

$x = v_o \cos \theta t + x_0$ $\qquad v_y = -32t + v_y^o$

$\quad = v_o \cos \theta t$ $\qquad\qquad = -32t + v_o \sin \theta$

$\qquad\qquad\qquad\qquad\qquad y = -16t^2 + v_o \sin \theta t + y_0$

$\qquad\qquad\qquad\qquad\qquad\quad = -16t^2 + v_o \sin \theta t$

9. $\overline{OT} = P'T = a\theta$

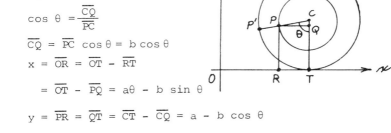

$\sin \theta = \dfrac{\overline{PQ}}{\overline{PC}}$

$\overline{PQ} = \overline{PC} \sin \theta = b \sin \theta$

$\cos \theta = \dfrac{\overline{CQ}}{\overline{PC}}$

$\overline{CQ} = \overline{PC} \cos \theta = b \cos \theta$

$x = \overline{OR} = \overline{OT} - \overline{RT}$

$\quad = \overline{OT} - \overline{PQ} = a\theta - b \sin \theta$

$y = \overline{PR} = \overline{QT} = \overline{CT} - \overline{CQ} = a - b \cos \theta$

13.

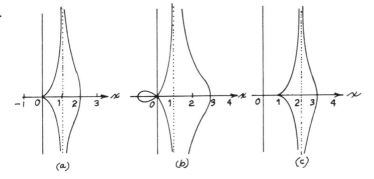

(a)   (b)   (c)

17. $C = (2a \cos \theta, \quad 2a \sin \theta)$

$p = a \cos 2\theta, \qquad q = a \sin 2\theta$

$x = 2a \cos \theta - a \cos 2\theta, \, y = 2a \sin \theta - a \sin 2\theta$

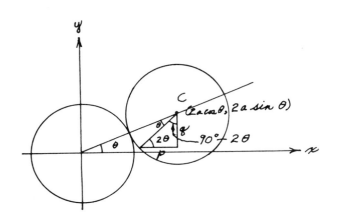

Section 13.3, pp. 378–379

1. $\dfrac{dy}{dx} = \dfrac{dy/dt}{dx/dt} = \dfrac{1}{2t+1}$

$\dfrac{d^2y}{dx^2} = \dfrac{dy'/dt}{dx/dt} = \dfrac{-2/(2t+1)^2}{2t+1} = \dfrac{-2}{(2t+1)^3}$

5. $\dfrac{dy}{dx} = \dfrac{dy/dt}{dx/dt} = \dfrac{2t-1}{2t+1}$

$\dfrac{d^2y}{dx^2} = \dfrac{dy'/dt}{dx/dt} = \dfrac{[(2t+1)2 - (2t-1)2]/(2t+1)^2}{2t+1} = \dfrac{4}{(2t+1)^3}$

9. $\dfrac{dy}{dx} = \dfrac{dy/dt}{dx/dt} = \dfrac{\cosh t}{\sinh t} = \coth t$

$\dfrac{d^2y}{dx^2} = \dfrac{dy'/dt}{dx/dt} = \dfrac{\operatorname{csch}^2 t}{\sinh t} = -\operatorname{csch}^3 t$

13. $\dfrac{dy}{dx} = \dfrac{dy/dt}{dx/dt} = \dfrac{2t}{3t^2} = \dfrac{2}{3t}$. $\qquad \dfrac{dy}{dx} = -\dfrac{2}{9}$ at $t = -3$

$\dfrac{d^2y}{dx^2} = \dfrac{dy'/dt}{dx/dt} = \dfrac{-2/3t^2}{3t^2} = \dfrac{2}{9t^4}$. $\qquad \dfrac{d^2y}{dx^2} = -\dfrac{2}{729}$ at $t = -3$

17. $\dfrac{dy}{dx} = \dfrac{dy/dt}{dx/dt} = \dfrac{8t-2}{2} = 4t - 1 = 3$

$x = 1, \qquad y = 2$

$y - 2 = 3(x - 1), \qquad\qquad 3x - y - 1 = 0$

21. $\dfrac{dy}{dx} = \dfrac{dy/dt}{dx/dt} = \dfrac{3t^2 - 6t}{1} = 3t(t - 2)$

$t = 0 \qquad\quad t = 2$

$(3, 0) \qquad\quad (5, -4)$

25. $\dfrac{dy}{dx} = \dfrac{dy/d\theta}{dx/d\theta} = \dfrac{\sin\theta}{1 - \cos\theta}$

$\sin\theta = 0 \qquad\qquad 1 - \cos\theta = 0$

$\qquad \theta = n\pi \qquad\qquad \cos\theta = 1$

$(2n\pi, 0) \qquad\qquad\qquad \theta = 2n\pi$

$((2n - 1)\pi, 2) \qquad\qquad (2n\pi, 0)$

Although both the numerator and denominator are 0 when $\theta = 2n\pi$, the curve has vertical tangents at these points (see Problem 10 of the previous section).

29. $\dfrac{dy}{dx} = \dfrac{dy/dt}{dx/dt} = \dfrac{-\operatorname{sech} t \tanh t}{1 - \operatorname{sech}^2 t} = - \dfrac{\operatorname{sech} t \tanh t}{\tanh^2 t} = - \dfrac{1}{\sinh t}$

$\sinh t = 0, \qquad t = 0, \qquad (0, 1)$

Section 13.4, pp. 383-384

1. $x = \cos^3\theta \qquad\qquad\qquad y = \sin^3\theta$

$x' = -3\cos^2\theta \sin\theta \qquad\quad y' = 3\sin^2\theta \sin\theta$

$s = \displaystyle\int_0^{\pi/2} \sqrt{9\cos^4\theta \sin^2\theta + 9\sin^4\theta \cos^2\theta}\; d\theta$

$\quad = \displaystyle\int_0^{\pi/2} 3\sin\theta \cos\theta\; d\theta = \left.\dfrac{3\sin^2\theta}{2}\right|_0^{\pi/2} = \dfrac{3}{2}$

5. $y = x^{3/2}, \qquad y' = \dfrac{3}{2} x^{1/2}$

$s = \displaystyle\int_{7/4}^{11/4} \sqrt{1 + \dfrac{9x}{4}}\; dx$

$$= \frac{1}{18} \int_{7/4}^{11/4} 9\sqrt{4 + 9x} \, dx = \frac{1}{27} (4 + 9x)^{3/2} \Big|_{7/4}^{11/4}$$

$$= \frac{1}{27} \frac{115^{3/2} - 79^{3/2}}{8} = \frac{115^{3/2} - 79^{3/2}}{216}$$

9. $y = \dfrac{x^4}{4} + \dfrac{1}{8x^2}$, $\qquad y' = x^3 - \dfrac{1}{4x^3}$

$$s = \int_1^2 \sqrt{1 + (x^3 - 1/4x^3)^2} \, dx = \int_1^2 \sqrt{1 + x^6 - 1/2 + 1/16x^6} \, dx$$

$$= \int_1^2 \sqrt{x^6 + 1/2 + 1/16x^6} \, dx = \int_1^2 (x^3 + 1/4x^3) \, dx$$

$$= \left(\frac{x^4}{4} - \frac{1}{8x^2} \right) \Big|_1^2 = (4 - \frac{1}{32}) - (\frac{1}{4} - \frac{1}{8}) = \frac{123}{32}$$

13. $x = \text{Arcos}(1 - y) + \sqrt{2y - y^2}$

$$x' = \frac{1}{\sqrt{1 - (1 - y)^2}} + \frac{1 - y}{\sqrt{2 - y^2}} = \frac{2 - y}{\sqrt{2y - y^2}}$$

$$s = \int_{1/2}^1 \sqrt{1 + \frac{4 - 4y + y^2}{2y - y^2}} dy = \int_{1/2}^1 \sqrt{\frac{4 - 2y}{2y - y^2}} \, dy$$

$$= \int_{1/2}^1 \sqrt{\frac{2}{y}} \, dy = 2\sqrt{2y} \; \Big|_{1/2}^1 = 2\sqrt{2} - 2$$

17. $d = \sqrt{(-1 - 2)^2 + (5 - 1)^2} = \sqrt{9 + 16} = \sqrt{25} = 5$

$m = -\dfrac{4}{3}$

$$d = \int_{-1}^2 \sqrt{1 + 16/9} \, dx = \frac{5}{3} \int_{-1}^2 dx = \frac{5}{3} x \Big|_{-1}^2 = \frac{5}{3}(2 + 1) = 5$$

Section 13.5, p. 388

1. $T_{min} = T_o = \dfrac{w(s^2 - 4H^2)}{8H} = \dfrac{2(160,000 - 2500)}{200} = 1575 \text{ lb}$

$T_{max} = \sqrt{T_o^2 + w^2 s^2} = \sqrt{1575^2 + 4 \cdot 160,000} = 1767 \text{ lb}$

$y = \dfrac{T_o}{w}(\cosh \dfrac{wx}{T_o} - 1) = \dfrac{1575}{2}(\cosh \dfrac{2x}{1575} - 1)$

5. $T_{max} = \sqrt{T_o^2 + w^2 s^2}$, $\qquad 10,000 = \sqrt{T_o^2 + 4 \cdot 250,000}$

$\qquad 100,000,000 = T_o^2 + 1,000,000$

$\qquad T_o^2 = 99,000,000$, $\qquad T_o = 9950 = \dfrac{w(s^2 - 4H^2)}{8H}$

$\qquad 8H \cdot 9950 = 2(250,000 - 4H^2)$

$\qquad 4H^2 + 39,800H - 250,000 = 0$, $\quad H^2 + 9950H - 62,500 = 0$

$\qquad H = \dfrac{-9950 \pm \sqrt{99,000,000 + 250,000}}{2} = \dfrac{-9950 + 9960}{2} = 5$

Review 13, pp. 388-389

1. $t = x + 3$

$\qquad y = (x + 3)^2 + 4(x + 3) - 2$

$\qquad = x^2 + 10x + 19$

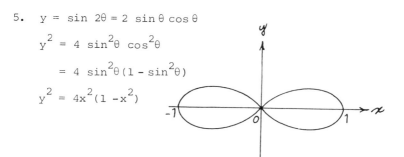

5. $y = \sin 2\theta = 2 \sin \theta \cos \theta$

$\qquad y^2 = 4 \sin^2\theta \cos^2\theta$

$\qquad = 4 \sin^2\theta (1 - \sin^2\theta)$

$\qquad y^2 = 4x^2(1 - x^2)$

9. $x_o = 0$, $y_o = 64$

$v_x^o = 16$, $v_y^o = 16\sqrt{3}$

$v_x = 16$ $\qquad\qquad a_y = -32$

$x = 16t + x_o$ $\qquad v_y = -32t + v_y^o = -32t + 16\sqrt{3}$

$ = 16t$ $\qquad\qquad y = -16t^2 + 16\sqrt{3}t + y_o$

$\qquad\qquad\qquad\qquad = -16t^2 + 16\sqrt{3}t + 64$

If $y = 0$, $-16t^2 + 16\sqrt{3}t + 64 = 0$

$t^2 - \sqrt{3}t - 4 = 0$

$t = \dfrac{\sqrt{3} \pm \sqrt{3 + 16}}{2} = \dfrac{\sqrt{3} + \sqrt{19}}{2}$, $\qquad x = 8(\sqrt{3} + \sqrt{19})$

13. $\dfrac{dy}{dx} = \dfrac{dy/dt}{dx/dt} = \dfrac{3t^2}{4t} = \dfrac{3t}{4}$

$\dfrac{d^2y}{dx^2} = \dfrac{dy'/dt}{dx/dt} = \dfrac{3/4}{4t} = \dfrac{3}{16t}$

17. $\dfrac{dy}{dx} = \dfrac{dy/dt}{dx/dt} = \dfrac{6t^2 + 6t - 12}{2t} = \dfrac{3(t^2 + t - 2)}{t}$

$t^2 + t - 2 = 0$ $\qquad\qquad t = 0$

$(t - 1)(t + 2) = 0$ $\qquad\quad (1, 0)$

$t = 1 \qquad\quad t = -2$

$(2, -7) \qquad\quad (5, 20)$

21. $T_{min} = T_o = \dfrac{w(s^2 - 4H^2)}{8H} = \dfrac{2(160,000 - 6400)}{320} = 960$ lb

$T_{max} = \sqrt{T_o^2 + w^2 s^2} = \sqrt{921,600 + 4 \cdot 160,000} = 1250$ lb

$y = \dfrac{T_o}{w}(\cosh \dfrac{wx}{T_o} - 1) = 480(\cosh \dfrac{x}{480} - 1)$

Chapter Fourteen
Polar Coordinates

1.

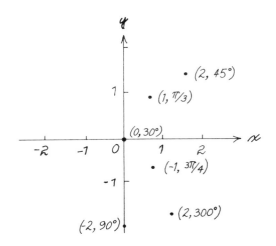

5.

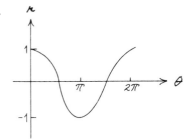

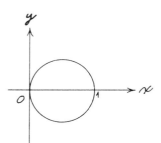

Symmetry about the x axis

9.

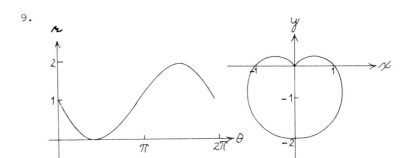

Symmetry about the y axis

13.

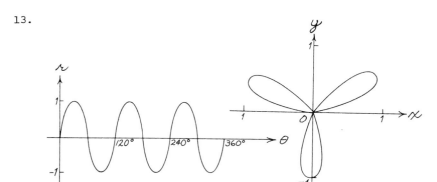

Symmetry about the y axis

17.

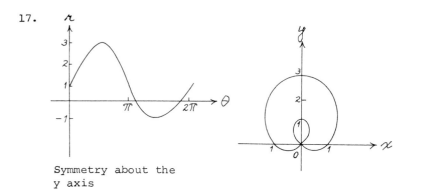

Symmetry about the
y axis

21.

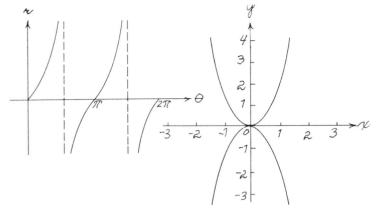

Symmetry about both
axes and the pole

25.

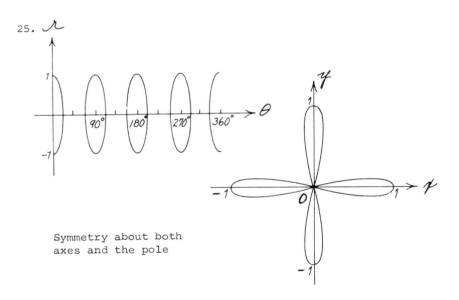

Symmetry about both
axes and the pole

29.

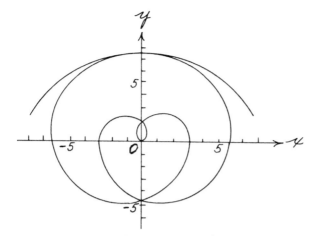

Symmetry about the y axis

33.

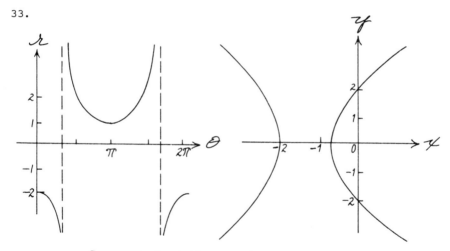

Symmetry about the x axis

Section 14.3, p. 399

1. $r = \sqrt{2}$, $r = 2 \cos \theta$

$2 \cos \theta = \sqrt{2}$

$\cos \theta = \dfrac{1}{\sqrt{2}}$

$\theta = 45°$, $315°$

$(\sqrt{2}, 45°)$, $(\sqrt{2}, 315°)$

5. $r = \cos \theta$, $r = 1 - \cos \theta$

$\cos \theta = 1 - \cos \theta$

$\cos \theta = \dfrac{1}{2}$

$\theta = 60°, 300°$

$(1/2, 60°)$, $(1/2, 300°)$

From the figure:

$(0, 90°) = (0, 0°)$

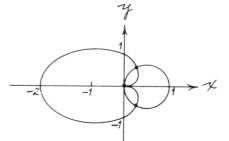

9. $r = \sec \theta$, $r = \csc \theta$

$\sec \theta = \csc \theta$, $\sin \theta = \cos \theta$

$\tan \theta = 1$, $\theta = 45°$, $225°$

$(\sqrt{2}, 45°)$, $(-\sqrt{2}, 225°)$

$= (\sqrt{2}, 45°)$

13. $r = 2(1 + \cos \theta)$, $r(1 - \cos \theta) = 1$

$2(1 - \cos^2\theta) = 1$, $\sin^2\theta = \dfrac{1}{2}$, $\sin \theta = \pm \dfrac{1}{\sqrt{2}}$

$\theta = 45°$, $135°$, $225°$, $315°$

$(2 + \sqrt{2}, 45°)$, $(2 - \sqrt{2}, 135°)$, $(2 - \sqrt{2}, 225°)$, $(2 + \sqrt{2}, 315°)$

17. $r^2 = \cos \theta$, $r^2 = \sec \theta$

$\cos \theta = \sec \theta$

$\cos^2\theta = 1$

$\cos \theta = \pm 1$

$\theta = 0°$, $180°$

$(1, 0°)$, $(-1, 0°)$,

$(-1, 180°) = (1, 0°)$,

$(1, 180°) = (-1, 0°)$

1. $x = r \cos \theta = 1 \cos \pi = -1$

   $y = r \sin \theta = 1 \sin \pi = 0$ $(1, \pi) = (-1, 0)$

   $x = r \cos \theta = \sqrt{3} \cos \frac{\pi}{3} = \frac{\sqrt{3}}{2}$

   $y = r \sin \theta = \sqrt{3} \sin \frac{\pi}{2} = \frac{3}{2}$ $(\sqrt{3}, \pi/3) = (\sqrt{3}/2, 3/2)$

   $x = r \cos \theta = -1 \cos 3\pi = 1$

   $y = r \sin \theta = -1 \sin 3\pi = 0$ $(-1, 3\pi) = (1, 0)$

   $x = r \cos \theta = \sqrt{2} \cos \frac{3\pi}{4} = -1$

   $y = r \sin \theta = \sqrt{2} \sin \frac{3\pi}{4} = 1$ $(\sqrt{2}, 3\pi/4) = (-1, 1)$

   $x = r \cos \theta = 2\sqrt{3} \cos \frac{5\pi}{3} = \sqrt{3}$

   $y = r \sin \theta = 2\sqrt{3} \sin \frac{5\pi}{3} = -3$ $(2\sqrt{3}, 5\pi/3) = (\sqrt{3}, -3)$

   $x = r \cos \theta = -3 \cos \frac{7\pi}{6} = \frac{3\sqrt{3}}{2}$

   $y = r \sin \theta = -3 \sin \frac{7\pi}{6} = \frac{3}{2}$ $(-3, 7\pi/6) = (3\sqrt{3}/2, 3/2)$

   $x = r \cos \theta = 0$

   $y = r \sin \theta = 0$ $(0, 5\pi/4) = (0, 0)$

   $x = r \cos \theta = 4 \cos 0 = 4$

   $y = r \sin \theta = 4 \sin 0 = 0$ $(4, 0) = (4, 0)$

   $x = r \cos \theta = -2 \cos \frac{7\pi}{4} = -\sqrt{2}$

   $y = r \sin \theta = -2 \sin \frac{7\pi}{4} = \sqrt{2}$ $(-2, 7\pi/4) = (-\sqrt{2}, \sqrt{2})$

5. $x^2 + y^2 = 1$

   $r^2 = 1$

   $r = 1$

9. $(x + y)^2 = x - y$

   $(r \cos \theta + r \sin \theta)^2$

        $= r \cos \theta - r \sin \theta$

   $r(1 + 2 \sin \theta \cos \theta)$

        $= \cos \theta - \sin \theta$

   $r = \dfrac{\cos \theta - \sin \theta}{1 + 2 \sin \theta \cos \theta}$

13. $x + 2y - 4 = 0$

$r \cos \theta + 2r \sin \theta - 4 = 0$

$r = \dfrac{4}{\cos \theta + 2 \sin \theta}$

17. $xy = 1$

$r \cos \theta\, r \sin \theta = 1$

$r^2 = \sec \theta \csc \theta$

21. $\theta = \pi/3$

$\tan \theta = \sqrt{3}$

$\dfrac{y}{x} = \sqrt{3}, \quad y = \sqrt{3}\, x$

$\sqrt{3}\, x - y = 0$

25. $r = \cos 2\theta$

$r = \cos^2\theta - \sin^2\theta$

$r^3 = r^2\cos^2\theta - r^2\sin^2\theta$

$(r^2)^3 = (r^2\cos^2\theta - r^2\sin^2\theta)^2$

$(x^2 + y^2)^3 = (x^2 - y^2)^2$

29. $r^2 = 1 + \sin \theta$

$r^2 - 1 = \sin \theta$

$r^2(r^2-1)^2 = r^2\sin^2\theta$

$(x^2+y^2)(x^2+y^2-1)^2 = y^2$

33. $r = 2 \sin \theta + 3 \cos \theta$

$r^2 = 2r \sin \theta + 3r \cos \theta$

$x^2 + y^2 = 2y + 3x$

Section 14.5, pp. 404-406

1. $r = \dfrac{4}{1 + 2 \cos \theta}$

$e = 2, \ p = 2$

hyperbola, $F : (0,0)$,

$D : x = 2$

5. $r = \dfrac{3}{1 + \sin \theta}$

$e = 1, \ p = 3$

parabola, $F : (0,0)$,

$D : y = 3$

9.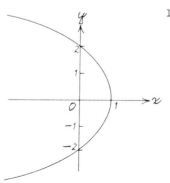

13.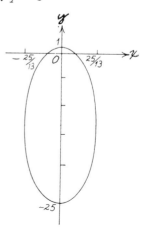

17. $r = \dfrac{ep}{1 + e \sin \theta} = \dfrac{2}{1 + \sin \theta}$

21.

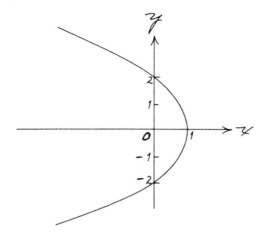

25. $a + c = 4500, \qquad a - c = 4100$

$$2a = 8600 \qquad\qquad 2c = 400$$

$$a = 4300 \qquad\qquad c = 200$$

$$e = \frac{c}{a} = \frac{200}{4300} = \frac{2}{43}, \quad D : x = -\frac{a}{e} = \frac{-4300}{2/43} = -92,450$$

$$r = \frac{ep}{1 - e \cos \theta} = \frac{4300}{1 - \dfrac{2}{43}\cos \theta} = \frac{184,900}{43 - 2 \cos \theta}$$

29. The line through (x_1, y_1) and parallel to the given
line is $Ax + By - (Ax_1 + By_1) = 0$. Both lines can be
put into the normal form by dividing by $\pm\sqrt{A^2 + B^2}$.
Thus we have

$$\frac{A}{\pm\sqrt{A^2 + B^2}}\, x + \frac{B}{\pm\sqrt{A^2 + B^2}}\, y + \frac{C}{\pm\sqrt{A^2 + B^2}} = 0$$

and

$$\frac{A}{\pm\sqrt{A^2 + B^2}}\, x + \frac{B}{\pm\sqrt{A^2 + B^2}}\, y - \frac{Ax_1 + By_1}{\pm\sqrt{A^2 + B^2}} = 0$$

By choosing the same sign in both cases, the polar
coordinates of Q_1 and Q_2 are

$$Q_1 : \left(\frac{C}{\pm\sqrt{A^2 + B^2}}, \alpha \right), \qquad Q_2 : \left(-\frac{Ax_1 + By_1}{\pm\sqrt{A^2 + B^2}}, \alpha \right)$$

Thus the distance between them is

$$d = \left| \frac{C}{\pm\sqrt{A^2 + B^2}} - \left(-\frac{Ax_1 + By_1}{\pm\sqrt{A^2 + B^2}} \right) \right| = \frac{|Ax_1 + By_1 + C|}{\sqrt{A^2 + B^2}}$$

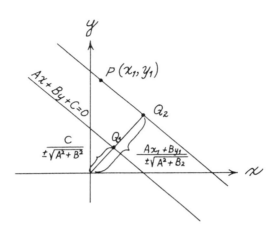

1. $r = 1 + \cos\theta$ $\qquad\qquad r' = -\sin\theta$

$$\frac{dy}{dx} = \frac{r'\sin\theta + r\cos\theta}{r'\cos\theta - r\sin\theta} = \frac{-\sin^2\theta + (1 + \cos\theta)\cos\theta}{-\sin\theta\cos\theta - (1 + \cos\theta)\sin\theta}$$

$$= \frac{\cos\theta + \cos^2\theta - \sin^2\theta}{-\sin\theta - 2\sin\theta\cos\theta} = -\frac{\cos\theta + \cos 2\theta}{\sin\theta + \sin 2\theta}$$

5. $r = \sin 3\theta$ $\qquad\qquad r' = 3\cos 3\theta$

$$\frac{dy}{dx} = \frac{r'\sin\theta + r\cos\theta}{r'\cos\theta - r\sin\theta} = \frac{3\cos 3\theta\sin\theta + \sin 3\theta\cos\theta}{3\cos 3\theta\cos\theta - \sin 3\theta\sin\theta}$$

9. $r = \dfrac{2}{1 - \sin\theta}$ $\qquad\qquad r' = \dfrac{2\cos\theta}{(1 - \sin\theta)^2}$

$$\frac{dy}{dx} = \frac{r'\sin\theta + r\cos\theta}{r'\cos\theta - r\sin\theta} = \frac{\dfrac{2\sin\theta\cos\theta}{(1 - \sin\theta)^2} + \dfrac{2\cos\theta}{1 - \sin\theta}}{\dfrac{2\cos^2\theta}{(1 - \sin\theta)^2} - \dfrac{2\sin\theta}{1 - \sin\theta}}$$

$$= \frac{\sin\theta\cos\theta + \cos\theta\,(1-\sin\theta)}{\cos^2\theta - \sin\theta\,(1-\sin\theta)} = \frac{\cos\theta}{\sin^2\theta + \cos^2\theta - \sin\theta}$$

$$= \frac{\cos\theta}{1-\sin\theta}$$

13. $\displaystyle\frac{dy}{dx} = \frac{r'\sin\theta + r\cos\theta}{r'\cos\theta - r\sin\theta} = \frac{\sin\theta\cos\theta + (1+\sin\theta)\cos\theta}{\cos^2\theta - (1+\sin\theta)\sin\theta}$

$$= \frac{\dfrac{1}{2}' + \left(1 + \dfrac{1}{\sqrt{2}}\right)\dfrac{1}{\sqrt{2}}}{\dfrac{1}{2} - \left(1 + \dfrac{1}{\sqrt{2}}\right)\dfrac{1}{\sqrt{2}}} = \frac{1 + \dfrac{1}{\sqrt{2}}}{-\dfrac{1}{\sqrt{2}}} = (1 + \sqrt{2})$$

17. $r = 1$, $r' = -\sin\theta = -1$, $(1, \pi/2) = (0, 1)$

$$\frac{dy}{dx} = \frac{r'\sin\theta + r\cos\theta}{r'\cos\theta - r\sin\theta} = \frac{-1\cdot 1 + 1\cdot 0}{-1\cdot 0 - 1\cdot 1} = 1$$

$y - 1 = 1\cdot(x - 0)$ $x - y + 1 = 0$

21. $r = 1$, $r' = 8\sin\theta\cos\theta = 8\cdot\dfrac{1}{2}\cdot\dfrac{-\sqrt{3}}{2} = -2\sqrt{3}$,

$(1, 5\pi/6) = (-\sqrt{3}/2, 1/2)$

$$\frac{dy}{dx} = \frac{r'\sin\theta + r\cos\theta}{r'\cos\theta - r\sin\theta} = \frac{-2\sqrt{3}\cdot 1/2 + 1\cdot(-\sqrt{3}/2)}{-2\sqrt{3}(-\sqrt{3}/2) - 1\cdot 1/2}$$

$$= \frac{-\sqrt{3} - \sqrt{3}/2}{3 - 1/2} = \frac{-3\sqrt{3}}{5}$$

$y - \dfrac{1}{2} = -\dfrac{3\sqrt{3}}{5}\left(x + \dfrac{\sqrt{3}}{2}\right)$, $3\sqrt{3}x + 5y + 2 = 0$

25. $r = \sin 2\theta$

$$\frac{dy}{dx} = \frac{r'\sin\theta + r\cos\theta}{r'\cos\theta - r\sin\theta} = \frac{2\sin\theta\cos 2\theta + \cos\theta\sin 2\theta}{2\cos\theta\cos 2\theta - \sin\theta\sin 2\theta}$$

$2\sin\theta\cos 2\theta + \cos\theta\sin 2\theta = 0$

$2\sin\theta(\cos^2\theta - \sin^2\theta) + 2\sin\theta\cos^2\theta = 0$

$2\sin\theta(3\cos^2\theta - 1) = 0$

$\sin\theta = 0$ $3\cos^2\theta - 1 = 0$

$\theta = 0, \pi$ $\cos^2\theta = \dfrac{1}{3}$

$(0,0)$ neither $\cos\theta = \pm\dfrac{1}{\sqrt{3}} = \pm 0.577$

$(0,\pi)$ neither $\theta = 54.8°, 125.2°, 234.8°, 305.2°$

$(2\sqrt{2}/3,\ 54.8°)$ max

$(-2\sqrt{2}/3,\ 125.2°)$ min

$(2\sqrt{2}/3,\ 234.8°)$ min

$(-2\sqrt{2}/3, 305.2°)$ max

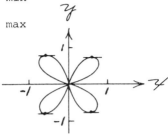

29.

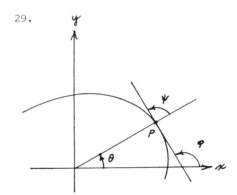

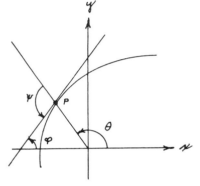

(a) $\psi + \theta + (\pi - \varphi) = \pi$

$\psi = \varphi - \theta$

(b) $(\pi - \psi) + \varphi + (\pi - \theta) = \pi$

$\psi = \pi + \varphi - \theta$

$$\tan \psi = \tan(\varphi - \theta) = \frac{\tan \varphi - \tan \theta}{1 + \tan \varphi \tan \theta}$$

$$= \frac{\dfrac{r'\sin \theta + r \cos \theta}{r'\cos \theta - r \sin \theta} - \dfrac{\sin \theta}{\cos \theta}}{1 + \dfrac{r'\sin \theta + r \cos \theta}{r'\cos \theta - r \sin \theta} \dfrac{\sin \theta}{\cos \theta}}$$

$$= \frac{r'\sin \theta \cos \theta + r \cos^2\theta - r'\sin \theta \cos \theta + r \sin^2\theta}{r'\cos^2\theta - r \sin \theta \cos \theta + r'\sin^2\theta + r \sin \theta \cos \theta}$$

$$= \frac{r(\cos^2\theta + \sin^2\theta)}{r'(\cos^2\theta + \sin^2\theta)} = \frac{r}{r'}$$

33. $l = 2 \sin \theta$

$\sin \theta = 1/2$

$\theta = \pi/6, \ 5\pi/6$

At $(1, \pi/6)$

$r = 1 \qquad\qquad r = 2 \sin \theta = 1$

$r' = 0 \qquad\qquad r' = 2 \cos \theta = \sqrt{3}$

$\tan \psi_1 = \dfrac{r}{r'} = \dfrac{1}{0} \quad \tan \psi_2 = \dfrac{r}{r'} = \dfrac{1}{\sqrt{3}}$

$\psi_1 = \pi/2, \quad \tan(\pi/2 - \psi_2) = -\dfrac{1}{\tan \psi_2} = -\sqrt{3}$

$\theta_2 - \theta_1 = 120°, \quad$ At $(1, 5\pi/6), \ \theta_2 - \theta_1 = 60°$

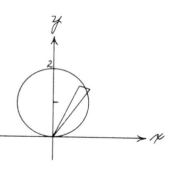

Section 14.7, p. 414

1. $A = 2\displaystyle\int_0^{\pi/2} \dfrac{1}{2} r^2 \, d\theta$

$\quad = \displaystyle\int_0^{\pi/2} 4 \sin^2\theta \, d\theta$

$\quad = 2\displaystyle\int_0^{\pi/2} (1 - \cos 2\theta) \, d\theta$

$\quad = (2\theta - \sin 2\theta) \Big|_0^{\pi/2} = \pi$

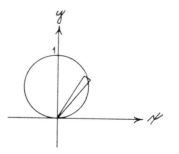

5. $A = 2\displaystyle\int_0^{\pi/2} \dfrac{1}{2} r^2 d\theta$

$\quad = \displaystyle\int_0^{\pi/2} \sin \theta \, d\theta$

$\quad = -\cos \theta \Big|_0^{\pi/2} = 1$

9. $A = 2\int_{-\pi/2}^{\pi/2} \frac{1}{2}\ r^2 d\theta$

$= \int_{-\pi/2}^{\pi/2}(3 - \sin\theta)^2 d\theta$

$= \int_{-\pi/2}^{\pi/2}(9 - 6\sin\theta + \sin^2\theta)\,d\theta$

$= \int_{-\pi/2}^{\pi/2}(9 - 6\sin\theta + \frac{1-\cos 2\theta}{2})\,d\theta$

$= (\frac{19}{2}\ \theta + 6\cos\theta - \frac{1}{4}\sin 2\theta \Big|_{-\pi/2}^{\pi/2} = \frac{19\pi}{2}$

13. $A = 2\left[\int_0^{\pi/2} \frac{1}{2}\ r_1^2 d\theta + \int_{\pi/2}^{\pi} \frac{1}{2}\ r_2^2 d\theta\right]$

$= \int_0^{\pi/2} d\theta + \int_{\pi/2}^{\pi}(1 + \cos\theta)^2 d\theta$

$= \theta\Big|_0^{\pi/2} + \int_{\pi/2}^{\pi}(1 + 2\cos\theta + \cos^2\theta)\,d\theta$

$= \frac{\pi}{2} + \int_{\pi/2}^{\pi}(1 + 2\cos\theta + \frac{1+\cos 2\theta}{2})\,d\theta$

$= \frac{\pi}{2} + (\frac{3}{2}\theta + 2\sin\theta + \frac{1}{4}\sin 2\theta)\Big|_{\pi/2}^{\pi}$

$= \frac{\pi}{2} + \frac{3\pi}{2} - \frac{3\pi}{4} - 2 = \frac{5\pi}{4} - 2$

17. $A = 2\int_0^{\pi/2} \frac{1}{2}(r_1^2 - r_2^2)\,d\theta$

$= \int_0^{\pi/2}[(1 + \sin\theta)^2 - 1]\,d\theta$

$= \int_0^{\pi/2}(2\sin\theta + \sin^2\theta)\,d\theta$

$= \int_0^{\pi/2}(2\sin\theta + \frac{1-\cos 2\theta}{2})\,d\theta$

$= (\frac{1}{2}\theta - 2\cos\theta - \frac{1}{4}\sin 2\theta)\Big|_0^{\pi/2}$

$= \frac{\pi}{4} + 2 = \frac{\pi + 8}{4}$

21. $\dfrac{dy}{dx} = \dfrac{r\cos\theta + r'\sin\theta}{-r\sin\theta + r'\cos\theta}$

$$s = \int_{x_1}^{x_2} \sqrt{1 + (y')^2}\; dx$$

$$= \int_{\theta_1}^{\theta_2} \sqrt{1 + (\dfrac{r\cos\theta + r'\sin\theta}{-r\sin\theta + r'\cos\theta})^2} \;(-r\sin\theta + r'\cos\theta)\,d\theta$$

$$= \int_{\theta_1}^{\theta_2} \sqrt{(-r\sin\theta + r'\cos\theta)^2 + (r\cos\theta + r'\sin\theta)^2}\; d\theta$$

$$= \int_{\theta_1}^{\theta_2} \sqrt{\begin{array}{l} r^2\sin^2\theta - 2\,rr'\sin\theta\cos\theta + (r')^2\cos^2\theta \\[4pt] + r^2\cos^2\theta + 2rr'\sin\theta\cos\theta + (r')^2\sin^2\theta \end{array}}\; d\theta$$

$$= \int_{\theta_1}^{\theta_2} \sqrt{r^2 + (r')^2}\; d\theta$$

25. $r = 2\sin^3\dfrac{\theta}{3}$ $\qquad\qquad$ $r' = 2\sin^2\dfrac{\theta}{3}\cos\dfrac{\theta}{3}$

$$s = \int_0^{\pi/2} \sqrt{4\sin^6\dfrac{\theta}{3} + 4\sin^4\dfrac{\theta}{3}\cos^2\dfrac{\theta}{3}}\; d\theta$$

$$= \int_0^{\pi/2} 2\sin^2\dfrac{\theta}{3}\sqrt{\sin^2\dfrac{\theta}{3} + \cos^2\dfrac{\theta}{3}}\; d\theta = \int_0^{\pi/2} 2\sin^2\dfrac{\theta}{3}\; d\theta$$

$$= \int_0^{\pi/2} (1 - \cos\dfrac{2\theta}{3})\, d\theta = (\theta - \dfrac{3}{2}\sin\dfrac{2\theta}{3})\Big|_0^{\pi/2} = \dfrac{\pi}{2} - \dfrac{3}{2}\dfrac{\sqrt{3}}{2}$$

$$= \dfrac{2\pi - 3\sqrt{3}}{4}$$

<u>Review 14, pp. 414–415</u>

1.

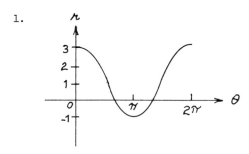

Symmetry about the x axis

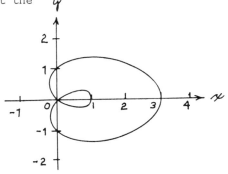

5.

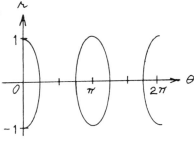

Symmetric about both axes and the pole

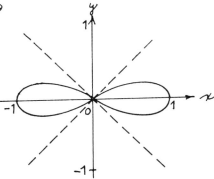

9.

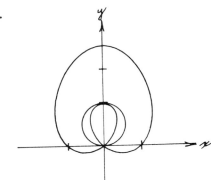

$r = \sin \theta, \quad r = 1 + 2 \sin \theta$

$\sin \theta = 1 + 2 \sin \theta$

$\sin \theta = -1$

$\theta = 3\pi/2$

$(-1, 3\pi/2), (0, 0) = (0, 7\pi/6)$

13. (a) $x = r \cos \theta = -1 \cos \pi/2 = 0$

$y = r \sin \theta = -1 \sin \pi/2 = -1$ $(0, -1)$

$x = r \cos \theta = 2\sqrt{2}(-1/\sqrt{2}) = -2$

$y = r \sin \theta = 2\sqrt{2}(1/\sqrt{2}) = 2$ $(-2, 2)$

$x = r \cos \theta = 4(-\sqrt{3}/2) = -2\sqrt{3}$

$y = r \sin \theta = 4(-1/2) = -2$ $(-2\sqrt{3}, -2)$

$x = r \cos \theta = -\sqrt{3}(-1/2) = \sqrt{3}/2$

$y = r \sin \theta = -\sqrt{3}(\sqrt{3}/2) = -3/2$ $(\sqrt{3}/2, -3/2)$

(b) $r^2 = x^2 + y^2 = 4 + 4 = 8$

$\tan \theta = \dfrac{y}{x} = \dfrac{2}{-2} = -1$ $(2\sqrt{2}, 3\pi/4)$

$r^2 = x^2 + y^2 = 9$

$\tan \theta = \dfrac{y}{x} = \dfrac{0}{-3} = 0$ $(-3, 0)$

$r^2 = x^2 + y^2 = 1 + 3 = 4$

$\tan \theta = \dfrac{y}{x} = -\sqrt{3}$ $(2, -\pi/3)$

$r^2 = x^2 + y^2 = 25 + 4 = 29$

$\tan \theta = \dfrac{y}{x} = \dfrac{2}{5}$ $(\sqrt{29}, \text{Arctan } 2/5)$

17. $F = (0, 0)$

$D : y = -2$

$e = 1$

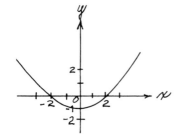

21. $r = \dfrac{1}{1 - \cos\theta}$

   When $\theta = \pi/2$, $r = 1$, $(1, \pi/2) = (0, 1)$

   $r' = \dfrac{-\sin\theta}{(1 - \cos\theta)^2} = \dfrac{-1}{1^2} = -1$

   $\dfrac{dy}{dx} = \dfrac{r\cos\theta + r'\sin\theta}{-r\sin\theta + r'\cos\theta} = \dfrac{1\cdot 0 + (-1)1}{-1\cdot 1 + (-1)0} = 1$

   $y - 1 = 1(x - 0)$, $x - y + 1 = 0$

25. $A = 2\displaystyle\int_{-\pi/2}^{\pi/2} \frac{1}{2}\, r^2 d\theta$

   $= \displaystyle\int_{-\pi/2}^{\pi/2} (2 + \sin\theta)^2 d\theta$

   $= \displaystyle\int_{-\pi/2}^{\pi/2} (4 + 4\sin\theta + \frac{1 - \cos 2\theta}{2})\, d\theta$

   $= \displaystyle\int_{-\pi/2}^{\pi/2} (\frac{9}{2} + 4\sin\theta - \frac{1}{4}\cdot 2\cos 2\theta)\, d\theta$

   $= (\frac{9}{2}\theta - 4\cos\theta - \frac{1}{4}\sin 2\theta) \Big|_{-\pi/2}^{\pi/2}$

   $= \dfrac{9\pi}{4} + \dfrac{9\pi}{4} = \dfrac{9\pi}{2}$

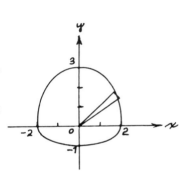

Section 15.1, pp. 420-421

1. $\displaystyle\int \frac{x^2 + x - 3}{x}dx = \int (x + 1 - \frac{3}{x})\,dx = \frac{x^2}{2} + x - 3\,\ell n|x| + C$

5. $\displaystyle\int \frac{u^3 + u}{u - 1}du = \int (u^2 + u + 2 + \frac{2}{u-1})\,du = \frac{u^3}{3} + \frac{u^2}{2} + 2u + 2\,\ell n|u-1| + C$

9. $\displaystyle\int (x + 1)(x^2 + 2x)^{2/3}\,dx = \frac{1}{2}\int (2x + 2)(x^2 + 2x)^{2/3}\,dx$

$\displaystyle = \frac{1}{2}\frac{(x^2 + 2x)^{5/3}}{5/3} + C = \frac{3}{10}(x^2 + 2x)^{5/3} + C$

13. $\displaystyle\int \frac{\ell n\ x}{x}dx = \int \frac{1}{x}\,\ell n\ x\ dx = \frac{\ell n^2 x}{2} + C$

17. $\displaystyle\int \frac{du}{(1 - 2u)^4} = -\frac{1}{2}\int (1 - 2u)^{-4}(-2)\,du = -\frac{1}{2}\frac{(1 - 2u)^{-3}}{-3} + C$

$\displaystyle = \frac{1}{6(1 - 2u)^3} + C$

21. $\displaystyle\int \frac{\sec^2 u}{1 + \tan u}\,du = \ell n|1 + \tan u| + C$

25. $\displaystyle\int \frac{\sinh x}{1 - \cosh x}dx = -\int \frac{-\sinh x}{1 - \cosh x}dx = -\ell n|1 - \cosh x| + C$

29. $\displaystyle\int \frac{x\ dx}{x^4 + 1} = \frac{1}{2}\int \frac{2x\ dx}{1 + (x^2)^2} = \frac{1}{2}\,\text{Arctan}\ x^2 + C$

33. $\displaystyle\int (1 + \sec\ \theta)^2 d\theta = \int (1 + 2\sec\ \theta + \sec^2\theta)\,d\theta$

$\displaystyle = \theta + 2\,\ell n|\sec\ \theta + \tan\ \theta| + \tan\ \theta + C$

37. $\displaystyle\int (\frac{\sec\ \theta}{1 - \tan\ \theta})^2 d\theta = -\int \frac{-\sec^2\theta}{(1 - \tan\ \theta)^2}\,d\theta = \frac{1}{1 - \tan\ \theta} + C$

41. $\displaystyle\int \frac{e^x - 1}{e^x + 1}dx = \int (-1 + \frac{2e^x}{e^x + 1})\,dx = -x + 2\,\ell n(e^x + 1) + C$

or $\displaystyle\int \frac{e^x - 1}{e^x + 1}dx = \int \frac{1 - e^{-x}}{1 + e^{-x}}dx = \int (1 - \frac{2e^{-x}}{1 + e^{-x}})\,dx$

$$= x + 2 \ln(1 + e^{-x}) + C = x + 2 \ln \frac{e^x + 1}{e^x} + C$$

$$= x + 2 \ln(e^x + 1) - 2 \ln e^x + C = -x + 2 \ln(e^x + 1) + C$$

45. $\displaystyle\int \csc u \cdot u' \, dx = \int \frac{(\csc^2 u + \csc u \cot u) u' \, dx}{\csc u + \cot u}$

$$= -\int \frac{(-\csc u \cot u - \csc^2 u) u' \, dx}{\csc u + \cot u} = -\ln \left| \csc u + \cot u \right| + C$$

$$= -\ln \left| \frac{1 + \cos u}{\sin u} \right| + C = \ln \left| \frac{\sin u}{1 + \cos u} \right| + C$$

$$= \ln \left| \frac{\sin u (1 - \cos u)}{1 - \cos^2 u} \right| + C = \ln \left| \frac{\sin u (1 - \cos u)}{\sin^2 u} \right| + C$$

$$= \ln \left| \csc u - \cot u \right| + C$$

Section 15.2, p. 426

1. $u = \sqrt{x + 2}$, $\quad x = u^2 - 2$, $\quad dx = 2u \, du$

$$\int x\sqrt{x + 2} \, dx = \int (u^2 - 2) u \cdot 2u \, du = 2 \int (u^4 - 2u^2) \, du$$

$$= 2 \left(\frac{u^5}{5} - \frac{2u^3}{3} \right) + C = \frac{2}{15} u^3 (3u^2 - 10) + C$$

$$= \frac{2}{15} (x + 2)^{3/2} [3(x + 2) - 10] + C = \frac{2}{15} (x + 2)^{3/2} (3x - 4) + C$$

5. $u = \sqrt{2x + 1}$, $\quad x = \frac{u^2 - 1}{2}$, $\quad dx = u \, du$

$$\int x\sqrt{2x + 1} \, dx = \int \frac{u^2 - 1}{2} u \cdot u \, du = \frac{1}{2} \int (u^4 - u^2) \, du$$

$$= \frac{1}{2} \left(\frac{u^5}{5} - \frac{u^3}{3} \right) + C = \frac{1}{30} u^3 (3u^2 - 5) + C$$

$$= \frac{1}{30} (2x + 1)^{3/2} [3(2x + 1) - 5] = \frac{1}{15} (2x + 1)^{3/2} (3x - 1) + C$$

9. $u = \sqrt{x - 1}$, $\quad x = u^2 + 1$, $\quad dx = 2u \, du$

$$\int \frac{x^2}{\sqrt{x - 1}} dx = \int \frac{(u^2 + 1)^2}{u} \cdot 2u \, du = 2 \int (u^4 + 2u^2 + 1) \, du$$

$$= 2 \left(\frac{u^5}{5} + \frac{2u^3}{3} + u \right) + C = \frac{2}{15} u (3u^4 + 10u^2 + 15) + C$$

$$= \frac{2}{15} \sqrt{x - 1} \, [3(x - 1)^2 + 10(x - 1) + 15] + C$$

$$= \frac{2}{15} \sqrt{x - 1} \, (3x^2 + 4x + 8) + C$$

13. $u = \sqrt[3]{x+1}$, $\quad x = u^3 - 1$, $\quad dx = 3u^2 du$

$\displaystyle \int x\sqrt[3]{x+1}\,dx = \int (u^3 - 1)u \cdot 3u^2 du = 3\int (u^6 - u^3)\,du$

$\displaystyle \quad = 3\left(\frac{u^7}{7} - \frac{u^4}{4}\right) + C = \frac{3}{28}u^4(4u^3 - 7) + C$

$\displaystyle \quad = \frac{3}{28}(x+1)^{4/3}[4(x+1)-7] + C = \frac{3}{28}(x+1)^{4/3}(4x - 3) + C$

17. $u = \sqrt{2x+1}$, $\quad x = \dfrac{u^2 - 1}{2}$, $\quad dx = u\,du$

$\displaystyle \int (2x + 3)\sqrt{2x+1}\,dx = \int (u^2 + 2)u \cdot u\,du = \int (u^4 + 2u^2)\,du$

$\displaystyle \quad = \frac{u^5}{5} + \frac{2u^3}{3} + C = \frac{1}{15}u^3(3u^2 + 10) + C$

$\displaystyle \quad = \frac{1}{15}(2x+1)^{3/2}[3(2x+1) + 10] + C = \frac{1}{15}(2x+1)^{3/2}(6x+13) + C$

21. $u = \sqrt{x-3}$, $\quad x = u^2 + 3$, $\quad dx = 2u\,du$

$\displaystyle \int_3^4 x\sqrt{x-3}\,dx = \int_0^1 (u^2 + 3)u \cdot 2u\,du = 2\int_0^1 (u^4 + 3u^2)\,du$

$\displaystyle \quad = 2\left(\frac{u^5}{5} + u^3\right)\bigg|_0^1 = 2\left(\frac{1}{5} + 1\right) = \frac{12}{5}$

25. $u = \sqrt{2x+1}$, $\quad x = \dfrac{u^2 - 1}{2}$, $\quad dx = u\,du$

$\displaystyle \int_0^4 \frac{x^2}{\sqrt{2x+1}}\,dx = \int_1^3 \frac{(u^2 - 1)^2}{4u}u\,du = \frac{1}{4}\int_1^3 (u^4 - 2u^2 + 1)\,du$

$\displaystyle \quad = \frac{1}{4}\left(\frac{u^5}{5} - \frac{2u^3}{3} + u\right)\bigg|_1^3 = \frac{1}{4}\left(\frac{243}{5} - 18 + 3 - \frac{1}{5} + \frac{2}{3} - 1\right) = \frac{124}{15}$

29. $u = \sqrt[3]{x+1}$, $\quad x = u^3 - 1$, $\quad dx = 3u^2 du$

$\displaystyle \int_0^7 x\sqrt[3]{x+1}\,dx = \int_1^2 (u^3 - 1)u \cdot 3u^2 du$

$\displaystyle \quad = 3\int_1^2 (u^6 - u^3)\,du = 3\left(\frac{u^7}{7} - \frac{u^4}{4}\right)\bigg|_1^2$

$\displaystyle \quad = 3\left(\frac{128}{7} - 2 - \frac{1}{7} + \frac{1}{4}\right) = \frac{153}{28}$

Section 15.3, pp. 429-430

1. $\displaystyle \int \sin^2 x \cos x\,dx = \frac{\sin^3 x}{3} + C$

5. $\int \sin^3 x \cos^2 x \, dx = \int \sin x (1 - \cos^2 x) \cos^2 x \, dx$

$$= \int (-\sin x)(\cos^4 x - \cos^2 x) \, dx = \frac{\cos^5 x}{5} - \frac{\cos^3 x}{3} + C$$

9. $\int \sin^2 \theta \cos^4 \theta \, d\theta = \int \frac{1 - \cos 2\theta}{2} \frac{(1 + \cos 2\theta)^2}{4} \, d\theta$

$$= \frac{1}{8} \int (1 + \cos 2\theta - \cos^2 2\theta - \cos^3 2\theta) \, d\theta$$

$$= \frac{1}{8} \int [1 + \cos 2\theta - \frac{1 + \cos 4\theta}{2} - (1 - \sin^2 2\theta) \cos 2\theta] \, d\theta$$

$$= \frac{1}{8} \int (\frac{1}{2} - \frac{1}{2} \cos 4\theta + \sin^2 2\theta \cos 2\theta) \, d\theta$$

$$= \frac{1}{8} \int (\frac{1}{2} - \frac{1}{8} 4 \cos 4\theta + \frac{1}{2} \sin^2 2\theta \cdot 2 \cos 2\theta) \, d\theta$$

$$= \frac{1}{8} (\frac{\theta}{2} - \frac{1}{8} \sin 4\theta + \frac{1}{2} \frac{\sin^3 2\theta}{3}) + C$$

$$= \frac{\theta}{16} - \frac{\sin 4\theta}{64} + \frac{\sin^3 2\theta}{48} + C$$

13. $\int_{-\pi/4}^{0} \cos^4 2\theta \, d\theta = \int_{-\pi/4}^{0} \frac{(1 + \cos 4\theta)^2}{4} d\theta$

$$= \frac{1}{4} \int_{-\pi/4}^{0} (1 + 2 \cos 4\theta + \cos^2 4\theta) \, d\theta$$

$$= \frac{1}{4} \int_{-\pi/4}^{0} (1 + 2 \cos 4\theta + \frac{1 + \cos 8\theta}{2}) \, d\theta$$

$$= \frac{1}{4} \int_{-\pi/4}^{0} (\frac{3}{2} + \frac{1}{2} \cdot 4 \cos 4\theta + \frac{1}{16} \cdot 8 \cos 8\theta) \, d\theta$$

$$= \frac{1}{4} (\frac{3\theta}{2} \quad \frac{\sin 4\theta}{2} \quad \frac{\sin 8\theta}{16}) \Big|_{-\pi/4}^{0} = -\frac{1}{4}(-\frac{3\pi}{8}) = \frac{3\pi}{32}$$

17. $\int \sin^{3/2} x \cos^3 x \, dx = \int \sin^{3/2} x (1 - \sin^2 x) \cos x \, dx$

$$= \int (\sin^{3/2} x - \sin^{7/2} x) \cos x \, dx = \frac{2}{5} \sin^{5/2} x - \frac{2}{9} \sin^{9/2} x + C$$

21. $\int \frac{dx}{\sec^2 x \tan x} = \int \cos^2 x \frac{\cos x}{\sin x} dx$

$$= \int (1 - \sin^2 x) \frac{\cos x}{\sin x} dx = \int (\frac{\cos x}{\sin x} - \sin x \cos x) \, dx$$

$$= \ln|\sin x| - \frac{\sin^2 x}{2} + C$$

25. $\int \sin^3 t \sqrt{\cos t}\ dt = \int (1 - \cos^2 t)\sqrt{\cos t}\ \sin t\ dt$

$\qquad = \int (\cos^{5/2} t - \cos^{1/2} t)(-\sin t)\,dt$

$\qquad = \frac{2}{7}\cos^{7/2} t - \frac{2}{3}\cos^{3/2} t + C = \frac{2}{21}\cos^{3/2} t(3\cos^2 t - 7) + C$

29. $\int \sin^m u \cdot \cos^n u \cdot u'\,dx = \int \sin^m u \cos^{n-1} u \cdot \cos u \cdot u'\,dx$

$\qquad = \int \sin^m u(1 - \sin^2 u)^{(n-1)/2}\cos u \cdot u'\,dx$

Since n is even, n - 1 is odd and (n - 1)/2 is not a
whole number. Thus $\sin^m u(1 - \sin^2 u)^{(n-1)/2}$ cannot
be written as a polynomial in sin u. A similar
problem arises when we try to express $\sin^m u$ as
$\sin^{m-1} u \cdot \sin u$.

Section 15.4, pp. 433-434

1. $\int \sec^2 x\ \tan^2 x\ dx = \frac{\tan^3 x}{3} + C$

5. $\int \sec\theta\ \tan^3\theta\,d\theta = \int (\sec^2\theta - 1)\sec\theta\ \tan\theta\,d\theta$

$\qquad = \frac{\sec^3\theta}{3} - \sec\theta + C$

9. $\int \sec^6 x\ dx = \int (1 + \tan^2 x)^2 \sec^2 x\ dx$

$\qquad = \int (1 + 2\tan^2 x + \tan^4 x)\sec^2 x\ dx$

$\qquad = \tan x + \frac{2\tan^3 x}{3} + \frac{\tan^5 x}{5} + C$

13. $\int \tan^5 2\theta\,d\theta = \int \tan^3 2\theta(\sec^2 2\theta - 1)\,d\theta$

$\qquad = \int (\tan^3 2\theta\ \sec^2 2\theta - \tan^3 2\theta)\,d\theta$

$\qquad = \int [\tan^3 2\theta\ \sec^2 2\theta - \tan 2\theta(\sec^2 2\theta - 1)]\,d\theta$

$\qquad = \frac{1}{2}\int [\tan^3 2\theta(2\ \sec^2 2\theta) - \tan 2\theta(2\ \sec^2 2\theta) + 2\tan 2\theta]\,d\theta$

$\qquad = \frac{1}{2}(\frac{\tan^4 2\theta}{4} - \frac{\tan^2 2\theta}{2} - \ln|\cos 2\theta|) + C$

$$= \frac{\tan^4 2\theta}{8} - \frac{\tan^2 2\theta}{4} - \frac{1}{2} \ln|\cos 2\theta| + C$$

$$= \frac{(\sec^2 2\theta - 1)^2}{8} - \frac{\sec^2 2\theta - 1}{4} + \frac{1}{2} \ln|\sec 2\theta| + C$$

$$= \frac{\sec^4 2\theta}{8} - \frac{\sec^2 2\theta}{2} + \frac{1}{2} \ln|\sec 2\theta| + C$$

17. $\displaystyle\int \cot^4 3x \; dx = \int (\csc^2 3x - 1)\cot^2 3x \; dx$

$$= \int (\csc^2 3x \cot^2 3x - \cot^2 3x) \, dx$$

$$= \int (\csc^2 3x \cot^2 3x - \csc^2 3x + 1) \, dx$$

$$= \int [-\frac{1}{3} \cot^2 3x(-3 \csc^2 3x) - \frac{1}{3} \cdot 3 \csc^2 3x + 1] dx$$

$$= -\frac{\cot^3 3x}{9} + \frac{\cot 3x}{3} + x + C$$

21. $\displaystyle\int \frac{\tan \theta}{1 - \tan^2 \theta} \; d\theta$

$$= \frac{1}{2}\int \tan 2\theta \; d\theta = -\frac{1}{4} \ln|\cos 2\theta| + C$$

25. $\displaystyle\int \frac{\sin \theta \cos \theta}{\sin^2 \theta - \cos^2 \theta} d\theta = -\frac{1}{2}\int \frac{2 \sin \theta \cos \theta}{\cos^2 \theta - \sin^2 \theta} \; d\theta$

$$= -\frac{1}{2}\int \frac{\sin 2\theta}{\cos 2\theta} \; d\theta = \frac{1}{4} \ln|\cos 2\theta| + C$$

29. $\displaystyle\int \frac{\tan^2 x}{\sec^3 x} dx = \int \sin^2 x \cos x \; dx = \frac{\sin^3 x}{3} + C$

Section 15.5, pp. 439-440

1. $x = \sin \theta, \; dx = \cos \theta \; d\theta$

$$\int \frac{\sqrt{1 - x^2}}{x^2} dx = \int \frac{\sqrt{1 - \sin^2 \theta}}{\sin^2 \theta} \cdot \cos \theta \; d\theta = \int \frac{\cos^2 \theta}{\sin^2 \theta} \; d\theta$$

$$= \int \cot^2 \theta \, d\theta = \int (\csc^2 \theta - 1) d\theta = - \cot \theta - \theta + C$$

$$= -\frac{\sqrt{1 - x^2}}{x} - \text{Arcsin } x + C$$

5. $x = 2 \sin \theta, \; dx = 2 \cos \theta \; d\theta$

$$\int \frac{x^3 \, dx}{\sqrt{4 - x^2}} = \int \frac{8 \sin^3 \theta \cdot 2 \cos \theta \, d\theta}{\sqrt{4 - 4 \sin^2 \theta}} = \int 8 \sin^3 \theta \, d\theta$$

$$= 8 \int (1 - \cos^2\theta)\sin\theta \, d\theta = 8(-\cos\theta + \frac{\cos^3\theta}{3}) + C$$

$$= \frac{8}{3}\cos\theta(\cos^2\theta - 3) + C$$

$$= \frac{8}{3}\frac{\sqrt{4-x^2}}{2}(\frac{4-x^2}{4} - 3) + C = \frac{1}{3}\sqrt{4-x^2}(-x^2 - 8) + C$$

$$= -\frac{1}{3}\sqrt{4-x^2}(x^2 + 8) + C$$

9. $x = \frac{3}{2}\sin\theta$, $dx = \frac{3}{2}\cos\theta \, d\theta$

$$\int \sqrt{9 - 4x^2} \, dx = \int \sqrt{9 - 9\sin^2\theta} \, \frac{3}{2}\cos\theta \, d\theta = \frac{9}{2}\int \cos^2\theta \, d\theta$$

$$= \frac{9}{4}\int (1 + \cos 2\theta) \, d\theta = \frac{9\theta}{4} + \frac{9}{8}\sin 2\theta + C$$

$$= \frac{9\theta}{4} + \frac{9}{4}\sin\theta\cos\theta + C = \frac{9}{4}\text{Arcsin}\frac{2x}{3} + \frac{9}{4}\cdot\frac{2x}{3}\frac{\sqrt{9-4x^2}}{3} + C$$

$$= \frac{9}{4}\text{Arcsin}\frac{2x}{3} + \frac{1}{2}x\sqrt{9-4x^2} + C$$

13. $x = \frac{3}{2}\tan\theta$, $dx = \frac{3}{2}\sec^2\theta \, d\theta$

$$\int \frac{x^3 dx}{\sqrt{4x^2 + 9}} = \int \frac{(27\tan^3\theta)/8}{\sqrt{9\tan^2\theta + 9}} \, \frac{3}{2}\sec^2\theta \, d\theta$$

$$= \frac{27}{16}\int \tan^3\theta \, \sec\theta \, d\theta = \frac{27}{16}\int (\sec^2\theta - 1)\sec\theta\tan\theta \, d\theta$$

$$= \frac{27}{16}(\frac{\sec^3\theta}{3} - \sec\theta) + C = \frac{9}{16}\sec\theta(\sec^2\theta - 3) + C$$

$$= \frac{9}{16}\frac{\sqrt{4x^2+9}}{3}(\frac{4x^2+9}{9} - 3) + C = \frac{1}{24}(2x^2 - 9)\sqrt{4x^2 + 9} + C$$

17. $x = \tan\theta$, $dx = \sec^2\theta \, d\theta$

$$\int_0^1 \frac{dx}{(x^2 + 1)^{3/2}} = \int_0^{\pi/4} \frac{\sec^2\theta \, d\theta}{(\tan^2\theta + 1)^{3/2}} = \int_0^{\pi/4} \frac{\sec^2\theta \, d\theta}{\sec^3\theta}$$

$$= \int_0^{\pi/4} \cos\theta \, d\theta = \sin\theta \Big|_0^{\pi/4} = \frac{1}{\sqrt{2}}$$

21. $x = \text{Arcsin } e^y$ $\qquad x' = \frac{e^y}{\sqrt{1 - e^{2y}}}$

$$s = \int_{-1}^{-1/2} \sqrt{1 + \frac{e^{2y}}{1 - e^{2y}}} \, dy = \int_{-1}^{-1/2} \frac{1}{\sqrt{1 - e^{2y}}} \, dy$$

$$e^y = \sin \theta, \quad y = \ln \sin \theta, \quad dy = \frac{\cos \theta}{\sin \theta} \, d\theta$$

$$= \int_{\text{Arcsin } 1/e}^{\text{Arcsin } 1/\sqrt{e}} \frac{\cos \theta \, d\theta}{\sin \theta \cos \theta} = \int_{\text{Arcsin } 1/e}^{\text{Arcsin } 1/\sqrt{e}} \csc \theta \, d\theta$$

$$= \ln|\csc \theta - \cot \theta| \Big|_{\text{Arcsin } 1/e}^{\text{Arcsin } 1/\sqrt{e}}$$

$$= \ln(\sqrt{e} - \sqrt{e-1}) - \ln(e - \sqrt{e^2 - 1}) = \ln \frac{\sqrt{e} - \sqrt{e-1}}{e - \sqrt{e^2-1}}$$

Section 15.6, p. 443

1. $\displaystyle \int \frac{dx}{x^2 + 2x + 2} = \int \frac{dx}{(x+1)^2 + 1} = \text{Arctan}(x+1) + C$

5. $\displaystyle \int \sqrt{8x - 4x^2} \, dx = \int \sqrt{4 - 4(x-1)^2} \, dx = 2\int \sqrt{1 - (x-1)^2} \, dx$

$$x - 1 = \sin \theta, \quad dx = \cos \theta \, d\theta$$

$$= 2\int \sqrt{1 - \sin^2\theta} \, \cos \theta \, d\theta = 2\int \cos^2\theta \, d\theta = \int (1 + \cos 2\theta) \, d\theta$$

$$= \theta + \frac{1}{2} \sin 2\theta + C = \theta + \sin \theta \cos \theta + C$$

$$= \text{Arcsin}(x-1) + (x-1)\sqrt{2x - x^2} + C$$

9. $\displaystyle \int \frac{\sqrt{-x^2 + 10x - 21}}{x - 5} \, dx = \int \frac{\sqrt{4 - (x-5)^2}}{x - 5} \, dx$

$$x - 5 = 2 \sin \theta, \quad dx = 2 \cos \theta \, d\theta$$

$$= \int \frac{\sqrt{4 - 4\sin^2\theta} \, 2 \cos \theta \, d\theta}{2 \sin \theta} = 2\int \frac{\cos^2\theta}{\sin \theta} \, d\theta$$

$$= 2\int \frac{1 - \sin^2\theta}{\sin \theta} \, d\theta = 2\int (\csc \theta - \sin \theta) \, d\theta$$

$$= 2 \ln|\csc \theta - \cot \theta| + 2 \cos \theta + C$$

$$= 2 \ln\left|\frac{2}{x-5} - \frac{\sqrt{-x^2 + 10x - 21}}{x-5}\right| + 2 \frac{\sqrt{-x^2 + 10x - 21}}{2} + C$$

$$= 2 \ln\left|\frac{2 - \sqrt{-x^2 + 10x - 21}}{x - 5}\right| + \sqrt{-x^2 + 10x - 21} + C$$

13. $\displaystyle \int \frac{x^3 + 1}{x^2 + 2x + 2} \, dx = \int \frac{x^3 + 1}{(x+1)^2 + 1} \, dx$

$$x + 1 = \tan \theta, \quad x = \tan \theta - 1, \quad dx = \sec^2\theta \, d\theta$$

$$= \int \frac{[(\tan \theta - 1)^3 + 1]\sec^2\theta\, d\theta}{\tan^2\theta + 1}$$

$$= \int (\tan^3\theta - 3 \tan^2\theta + 3 \tan \theta)\, d\theta$$

$$= \int [\tan^2\theta(\tan \theta - 3) + 3 \tan \theta]\, d\theta$$

$$= \int [(\sec^2\theta - 1)(\tan \theta - 3) + 3 \tan \theta]\, d\theta$$

$$= \int (\sec^2\theta \tan \theta - 3 \sec^2\theta + 2 \tan \theta + 3)\, d\theta$$

$$= \frac{1}{2} \tan^2\theta - 3 \tan \theta - 2 \ln|\cos \theta| + 3\theta + C$$

$$= \frac{1}{2}(x + 1)^2 - 3(x + 1) - 2 \ln \frac{1}{\sqrt{x^2 + 2x + 2}} + 3 \operatorname{Arctan}(x + 1) + C$$

$$= \frac{1}{x}(x + 1)(x - 5) + \ln(x^2 + 2x + 2) + 3 \operatorname{Arctan}(x + 1) + C$$

17. $\displaystyle \int_{-1}^{2} \frac{dx}{x^2 + 2x + 10} = \int_{-1}^{2} \frac{dx}{(x + 1)^2 + 9}$

$x + 1 = 3 \tan \theta, \quad dx = 3 \sec^2\theta\, d\theta$

$$= \int_{0}^{\pi/4} \frac{3 \sec^2\theta\, d\theta}{9 \tan^2\theta + 9} = \frac{1}{3}\int_{0}^{\pi/4} d\theta = \frac{\pi}{12}$$

Section 15.7, pp. 447-449

1. $u = x \qquad\qquad v' = \cos x$

   $u' = 1 \qquad\qquad v = \sin x$

   $\displaystyle \int x \cos x\, dx = x \sin x - \int \sin x = x \sin x + \cos x + C$

5. $u = \ln x^2 = 2 \ln x \qquad\qquad v' = 1$

   $u' = \dfrac{2}{x} \qquad\qquad\qquad v = x$

   $\displaystyle \int \ln x^2 dx = x \ln x^2 - \int 2\, dx = x \ln x^2 - 2x + C$

9. $u = \operatorname{Arcsin} x \qquad\qquad v' = 1$

   $u' = \dfrac{1}{\sqrt{1 - x^2}} \qquad\qquad v = x$

   $\displaystyle \int \operatorname{Arcsin} x\, dx = x \operatorname{Arcsin} x - \int \frac{x\, dx}{\sqrt{1 - x^2}}$

   $\displaystyle = x \operatorname{Arcsin} x + \frac{1}{2}\int \frac{-2 x\, dx}{\sqrt{1 - x^2}} = x \operatorname{Arcsin} x + \sqrt{1 - x^2} + C$

13. $u = \ln \cos \theta$ $\quad\quad\quad$ $v' = \sin \theta$

$\quad\quad u' = -\dfrac{\sin \theta}{\cos \theta}$ $\quad\quad\quad$ $v = -\cos \theta$

$\quad\quad \displaystyle\int \sin \theta \, \ln \cos \theta \, d\theta = -\cos \theta \, \ln \cos \theta - \int \sin \theta \, d\theta$

$\quad\quad\quad = -\cos \theta \, \ln \cos \theta + \cos \theta + C$

17. $u = x^2$ $\quad\quad\quad$ $v' = x\sqrt{4 - x^2} = -\dfrac{1}{2}(-2x)\sqrt{4 - x^2}$

$\quad\quad u' = 2x$ $\quad\quad\quad$ $v = -\dfrac{1}{3}(4 - x^2)^{3/2}$

$\quad\quad \displaystyle\int x^3\sqrt{4 - x^2} \, dx = -\dfrac{x^2}{3}(4 - x^2)^{3/2} - \int -\dfrac{2x}{3}(4 - x^2)^{3/2} \, dx$

$\quad\quad\quad = -\dfrac{x^2}{3}(4 - x^2)^{3/2} - \dfrac{2}{15}(4 - x^2)^{5/2} + C$

21. $u = \sin \ln x$ $\quad\quad$ $v' = 1$

$\quad\quad u' = \dfrac{\cos \ln x}{x}$ $\quad\quad$ $v = x$

$\quad\quad \displaystyle\int \sin \ln x \, dx = x \sin \ln x - \int \cos \ln x \, dx$

$\quad\quad u = \cos \ln x$ $\quad\quad$ $v' = 1$

$\quad\quad u' = \dfrac{-\sin \ln x}{x}$ $\quad\quad$ $v = x$

$\quad\quad\quad = x \sin \ln x - x \cos \ln x - \displaystyle\int \sin \ln x \, dx$

$\quad\quad 2\displaystyle\int \sin \ln x \, dx = x \sin \ln x - x \cos \ln x$

$\quad\quad \displaystyle\int \sin \ln x \, dx = \dfrac{x}{2}(\sin \ln x - \cos \ln x) + C$

25.

u	v'
x^4	$\cos x$
$4x^3$	$\sin x$
$12x^2$	$-\cos x$
$24x$	$-\sin x$
24	$\cos x$
0	$\sin x$

$\quad \displaystyle\int x^4 \cos x \, dx = x^4 \sin x + 4x^3 \cos x - 12x^2 \sin x - 24x \cos x$

$\quad\quad + 24 \sin x + C$

29. $u = \ln^2 x \qquad\qquad v' = x^3$

$\quad u' = \dfrac{2 \ln x}{x} \qquad\quad v = \dfrac{x^4}{4}$

$\displaystyle\int x^3 \ln^2 x \ dx = \dfrac{x^4}{4} \ln^2 x - \int \dfrac{1}{2} x^3 \ln x \ dx$

$\quad u = \ln x \qquad\qquad v' = \dfrac{x^3}{2}$

$\quad u' = \dfrac{1}{x} \qquad\qquad v = \dfrac{x^4}{8}$

$\qquad = \dfrac{x^4}{4} \ln^2 x - \dfrac{x^4}{8} \ln x + \displaystyle\int \dfrac{x^3}{8} \ dx$

$\qquad = \dfrac{x^4}{4} \ln^2 x - \dfrac{x^4}{8} \ln x + \dfrac{x^4}{32} + C$

$\qquad = \dfrac{x^4}{32} (8 \ln^2 x - 4 \ln x + 1) + C$

33. $u = \sec^3 \theta \qquad\qquad\qquad v' = \sec^2 \theta$

$\quad u' = 3 \sec^3 \theta \ \tan \theta \qquad v = \tan \theta$

$\displaystyle\int \sec^5 \theta \, d\theta = \sec^3 \theta \ \tan \theta - \int 3 \sec^3 \theta \ \tan^2 \theta \, d\theta$

$\qquad = \sec^3 \theta \ \tan \theta - \displaystyle\int 3 \sec^3 \theta \, (\sec^2 \theta - 1) \, d\theta$

$\qquad = \sec^3 \theta \ \tan \theta + 3 \displaystyle\int \sec^3 \theta \, d\theta - 3 \int \sec^5 \theta \, d\theta$

$4 \displaystyle\int \sec^5 \theta \, d\theta = \sec^3 \theta \ \tan \theta + \dfrac{3}{2} (\sec \theta \ \tan \theta + \ln |\sec \theta + \tan \theta |)$

$\displaystyle\int \sec^5 \theta \, d\theta = \dfrac{1}{8} (2 \sec^3 \theta \ \tan \theta + 3 \sec \theta \tan \theta + 3 \ln |\sec \theta$

$\qquad\qquad + \tan \theta |) + C$

37. $u = \sin x \qquad\qquad v' = \sin 3x$

$\quad u' = \cos x \qquad\qquad v = -\dfrac{1}{3} \cos 3x$

$\displaystyle\int \sin x \sin 3x \ dx = -\dfrac{1}{3} \sin x \cos 3x + \int \dfrac{1}{3} \cos x \cos 3x \ dx$

$\quad u = \cos x \qquad\qquad v' = \dfrac{1}{3} \cos 3x$

$\quad u' = -\sin x \qquad\quad v = \dfrac{1}{9} \sin 3x$

$\displaystyle\int \sin x \sin 3x \ dx = -\dfrac{1}{3} \sin x \cos 3x + \dfrac{1}{9} \cos x \sin 3x$

$\qquad + \displaystyle\int \dfrac{1}{9} \sin x \sin 3x \ dx$

$\dfrac{8}{9} \displaystyle\int \sin x \sin 3x \ dx = \dfrac{1}{9} (\cos x \sin 3x - 3 \sin x \cos 3x)$

$\displaystyle\int \sin x \sin 3x \ dx = \dfrac{1}{8} (\cos x \sin 3x - 3 \sin x \cos 3x) + C$

41. $u = \sin x \qquad v' = \sin x$

  $u' = \cos x \qquad v = -\cos x$

  $\int \sin^2 x \, dx = -\sin x \cos x + \int \cos^2 x \, dx = -\sin x \cos x$

  $\quad + \int (1 - \sin^2 x) \, dx$

  $\quad = -\sin x \cos x + x - \int \sin^2 x \, dx$

  $2 \int \sin^2 x \, dx = x - \sin x \cos x$

  $\int \sin^2 x \, dx = \frac{1}{2}(x - \sin x \cos x) + C$

  By the double-angle formula,

  $\int \sin^2 x \, dx = \frac{1}{2}(x - \frac{1}{2} \sin 2x) + C = \frac{1}{2}(x - \sin x \cos x) + C$

Section 15.8, pp. 453-454

1. $\dfrac{1}{x(x+1)} = \dfrac{A}{x} + \dfrac{B}{x+1}$

  $1 = A(x+1) + Bx = (A+B)x + A$

  $A + B = 0, \quad A = 1, \quad B = -1$

  $\int \dfrac{dx}{x(x+1)} = \int (\dfrac{1}{x} - \dfrac{1}{x+1}) \, dx = \ln|x| - \ln|x+1| + C$

  $\quad = \ln\left|\dfrac{x}{x+1}\right| + C$

5. $\int \dfrac{x^2 + 2}{x^2 + 2x} \, dx = \int (1 + \dfrac{-2x + 2}{x(x+2)}) \, dx$

  $\dfrac{-2x + 2}{x(x+2)} = \dfrac{A}{x} + \dfrac{B}{x+2}$

  $-2x + 2 = A(x+2) + Bx = (A+B)x + 2A$

  $A + B = -2, \quad 2A = 2; \quad A = 1, \quad B = -3$

  $\int \dfrac{x^2 + 2}{x^2 + 2x} \, dx = \int (1 + \dfrac{1}{x} - \dfrac{3}{x+2}) \, dx$

  $\quad = x + \ln|x| - 3\ln|x+2| + C = x + \ln\left|\dfrac{x}{(x+2)^3}\right| + C$

9. $\displaystyle\int\frac{(x^3+2)\,dx}{x^3-3x^2+2x} = \int(1+\frac{3x^2-2x+2}{x(x-1)(x-2)})\ dx$

$\dfrac{3x^2-2x+2}{x(x-1)(x-2)} = \dfrac{A}{x}+\dfrac{B}{x-1}+\dfrac{C}{x-2}$

$3x^2-2x+2 = A(x-1)(x-2)+Bx(x-2)+Cx(x-1)$

$\qquad\qquad = (A+B+C)x^2+(-3A-2B-C)x+2A$

$A+B+C = 3, \quad -3A-2B-C = -2, \quad 2A = 2;$

$A = 1, \quad B = -3, \quad C = 5$

$\displaystyle\int\frac{(x^3+2)\,dx}{x^3-3x^2+2x} = \int(1+\frac{1}{x}-\frac{3}{x-1}+\frac{5}{x-2})\ dx$

$\qquad = x+\ln|x|-3\ \ln|x-1|+5\ \ln|x-2|\ +C$

$\qquad = x+\ln\left|\dfrac{x(x-2)^5}{(x-1)^3}\right|\ +C$

13. $\displaystyle\int\frac{x^2\,dx}{(x+1)^2} = \int 1+\left(\frac{-2x-1}{(x+1)^2}\right)dx$

$\dfrac{-2x-1}{(x+1)^2} = \dfrac{A}{x+1}+\dfrac{B}{(x+1)^2}$

$-2x-1 = A(x+1)+B = Ax+(A+B)$

$A = -2, \quad A+B = -1; \quad B = 1$

$\displaystyle\int\frac{x^2\,dx}{(x+1)^2} = \int\left(1-\frac{2}{x+1}+\frac{1}{(x+1)^2}\right)\ dx$

$\qquad = x-2\ \ln|x+1|\ -\dfrac{1}{x+1}+C$

17. $\dfrac{x+2}{x^3(x+3)(x-1)} = \dfrac{A}{x}+\dfrac{B}{x^2}+\dfrac{C}{x+3}+\dfrac{D}{x-1}$

$x+2 = Ax(x+3)(x-1)+B(x+3)(x-1)+Cx^2(x-1)+Dx^2(x+3)$

$\qquad = (A+C+D)x^3+(2A+B-C+3D)x^2+(-3A+2B)x-3B$

$A+C+D = 0, \quad 2A+B-C+3D = 0, \quad -3A+2B = 1, \quad -3B = 2;$

$$A = -7/9, \quad B = -2/3, \quad C = 1/36, \quad D = 3/4$$

$$\int \frac{(x + 2)\,dx}{x^4 + 2x^3 - 3x^2} = \int \left(-\frac{7/9}{x} - \frac{2/3}{x^2} + \frac{1/36}{x + 3} + \frac{3/4}{x - 1}\right)\,dx$$

$$= -\frac{7}{9}\,\ell n|x| + \frac{2}{3x} + \frac{1}{36}\,\ell n|x + 3| + \frac{3}{4}\,\ell n|x - 1| + C$$

21. $u = \sin x, \quad du = \cos x\,dx$

$$\int \frac{\cos x\,dx}{\sin^3 x + \sin^2 x} = \int \frac{du}{u^3 + u^2} = \int \frac{du}{u^2(u + 1)}$$

$$\frac{1}{u^2(u + 1)} = \frac{A}{u} + \frac{B}{u^2} + \frac{C}{u + 1}$$

$$1 = Au(u + 1) + B(u + 1) + Cu^2 = (A + C)u^2 + (A + B)u + B$$

$$A + C = 0, \quad A + B = 0, \quad B = 1; \quad A = -1, \quad B = 1, \quad C = 1$$

$$= \int \left(-\frac{1}{u} + \frac{1}{u^2} + \frac{1}{u + 1}\right)du = -\ell n|u| - \frac{1}{u} + \ell n|u + 1| + C$$

$$= \ell n\left|\frac{u + 1}{u}\right| - \frac{1}{u} + C = \ell n\left|\frac{1 + \sin x}{\sin x}\right| - \frac{1}{\sin x} + C$$

$$= \ell n\left|\frac{1 + \sin x}{\sin x}\right| - \csc x + C$$

Section 15.9, p. 458

1. $\dfrac{1}{x^2(x^2 + 1)} = \dfrac{A}{x} + \dfrac{B}{x^2} + \dfrac{Cx + D}{x^2 + 1}$

$$1 = A(x^3 + x) + B(x^2 + 1) + Cx^3 + Dx^2$$

$$A + C = 0, \quad B + D = 0, \quad A = 0, \quad B = 1$$

$$A = 0, \quad B = 1, \quad C = 0, \quad D = -1$$

$$\int \frac{dx}{x^2(x^2 + 1)} = \int \left(\frac{1}{x^2} - \frac{1}{x^2 + 1}\right)\,dx = -\frac{1}{x} - \text{Arctan } x + C$$

5. $\dfrac{3}{(x + 1)(x^2 + 1)} = \dfrac{A}{x + 1} + \dfrac{Bx + C}{x^2 + 1}$

$$3 = A(x^2 + 1) + (Bx + C)(x + 1) = (A + B)x^2 + (B + C)x + (A + C)$$

$$A + B = 0, \quad B + C = 0, \quad A + C = 3; \quad A = 3/2, \quad B = -3/2, \quad C = 3/2$$

$$\int \frac{3\ dx}{x^3 + x^2 + x + 1} = \frac{3}{2}\int (\frac{1}{x+1} + \frac{1-x}{x^2+1})\ dx$$

$$= \frac{3}{2}\int (\frac{1}{x+1} - \frac{1}{2}\frac{2x}{x^2+1} + \frac{1}{x^2+1})\ dx$$

$$= \frac{3}{2}[\ln|x+1| - \frac{1}{2}\ln(x^2+1) + \text{Arctan } x] + C$$

$$= \frac{3}{4}\left(\ln\frac{(x+1)^2}{x^2+1} + 2\text{ Arctan } x\right) + C$$

9. $\dfrac{x}{(x+1)(x^2+1)} = \dfrac{A}{x+1} + \dfrac{Bx+C}{x^2+1}$

$x = A(x^2+1) + (Bx+C)(x+1) = (A+B)x^2 + (B+C)x + (A+C)$

$A+B = 0$, $B+C = 1$, $A+C = 0$; $A = -1/2$, $B = 1/2$, $C = 1/2$

$$\int\frac{x\ dx}{(x+1)(x^2+1)} = \frac{1}{2}\int(-\frac{1}{x+1} + \frac{x+1}{x^2+1})\ dx$$

$$= \frac{1}{2}\int(-\frac{1}{x+1} + \frac{1}{2}\frac{2x}{x^2+1} + \frac{1}{x^2+1})\ dx$$

$$= \frac{1}{2}[-\ln|x+1| + \frac{1}{2}\ln(x^2+1) + \text{Arctan } x] + C$$

$$= \frac{1}{4}\ln\frac{x^2+1}{(x+1)^2} + \frac{1}{2}\text{Arctan } x + C$$

13. $\dfrac{1}{x(x^2+4)^2} = \dfrac{A}{x} + \dfrac{Bx+C}{x^2+4} + \dfrac{Dx+E}{(x^2+4)^2}$

$1 = A(x^2+4)^2 + (Bx+C)x(x^2+4) + (Dx+E)x$

$\quad = (A+B)x^4 + Cx^3 + (8A+4B+D)x^2 + (4C+E)x + 16A$

$A+B = 0$, $C = 0$, $8A+4B+D = 0$, $4C+E = 0$, $16A = 1$;

$A = 1/16$, $B = -1/16$, $C = 0$, $D = -1/4$, $E = 0$

$$\int\frac{dx}{x(x^2+4)^2} = \int(\frac{1/16}{x} - \frac{1}{16}\frac{x}{x^2+4} - \frac{1}{4}\frac{x}{(x^2+4)^2})\ dx$$

$$= \frac{1}{16}\ln|x| - \frac{1}{32}\ln(x^2+4) + \frac{1}{8(x^2+4)} + C$$

$$= \frac{1}{32}\ln\frac{x^2}{x^2+4} + \frac{1}{8(x^2+4)} + C$$

190   *Section 15.9*

17. $\dfrac{x^2 + 3x + 4}{(x^2 + 1)(x^2 + 4)^2} = \dfrac{Ax + B}{x^2 + 1} + \dfrac{Cx + D}{x^2 + 4} + \dfrac{Ex + F}{(x^2 + 4)^2}$

$x^2 + 3x + 4 = A(x^5 + 8x^3 + 16x) + B(x^4 + 8x^2 + 16)$

$\qquad + C(x^5 + 5x^3 + 4x) + D(x^4 + 5x^2 + 4)$

$\qquad + E(x^3 + x) + F(x^2 + 1)$

$A + C = 0, \quad 8A + 5C + E = 0, \quad 16A + 4C + E = 3$

$B + D = 0, \quad 8B + 5D + F = 1, \quad 16B + 4D + F = 4$

$A = 1/3, \; B = 1/3, \; C = -1/3, \; D = -1/3, \; E = -1, \; F = 0$

$\displaystyle\int \dfrac{(x^2 + 3x + 4)\ dx}{(x^2 + 1)(x^2 + 4)^2} = \int (\dfrac{1}{3} \dfrac{x + 1}{x^2 + 1} - \dfrac{1}{3} \dfrac{x + 1}{x^2 + 4} - \dfrac{x}{(x^2 + 4)^2}) dx$

$\displaystyle = \int (\dfrac{1}{6} \dfrac{2x}{x^2+1} + \dfrac{1}{3}\ \dfrac{1}{x^2+1} - \dfrac{1}{6}\ \dfrac{2x}{x^2+4} - \dfrac{1}{6}\ \dfrac{1/2}{(x/2)^2+1} - \dfrac{1}{2}\ \dfrac{2x}{(x^2+4)^2}) dx$

$= \dfrac{1}{6}\ \ell n(x^2 + 1) + \dfrac{1}{3} \text{ Arctan } x - \dfrac{1}{6}\ \ell n(x^2 + 4) - \dfrac{1}{6} \text{ Arctan } \dfrac{x}{2}$

$\qquad + \dfrac{1}{2}\ \dfrac{1}{x^2 + 4} + C$

$= \dfrac{1}{6}\ \ell n\ \dfrac{x^2 + 1}{x^2 + 4} + \dfrac{1}{3} \text{ Arctan } x - \dfrac{1}{6} \text{ Arctan } \dfrac{x}{2} + \dfrac{1}{2(x^2 + 4)} + C$

Section 15.10, p. 461

1. $u = \sqrt{x + 1}, \quad x = u^2 - 1, \quad dx = 2\ u\ du$

$\displaystyle\int \dfrac{x\ dx}{x + 1 + \sqrt{x + 1}} = \int \dfrac{(u^2 - 1)\,2u\ du}{u^2 + u} = 2\int \dfrac{u^2 - 1}{u + 1}\ du$

$\displaystyle = 2\int (u - 1) du = u^2 - 2u + C = x + 1 - 2\sqrt{x + 1} + C$

5. $u = \sqrt[6]{x}, \quad x = u^6, \quad ex = 6u^5 du$

$\displaystyle\int \dfrac{dx}{\sqrt{x} - \sqrt[3]{x}} = \int \dfrac{6u^5\ du}{u^3 - u^2} = \int \dfrac{6u^3\ du}{u - 1}$

$\displaystyle = \int (6u^2 + 6u + 6 + \dfrac{6}{u - 1}) du = 2u^3 + 3u^2 + 6u + 6\ \ell n|u-1| + C$

$= 2\sqrt{x} + 3\sqrt[3]{x} + 6\sqrt[6]{x} + 6\ \ell n|\ \sqrt[6]{x} - 1| + C$

9. $u = \sqrt{x}$, $x = u^2$, $dx = 2u\ du$

$$\int \frac{\sqrt{x}+1}{\sqrt{x}-1}dx = \int \frac{u+1}{u-1}\cdot 2u\ du = \int \left(2u + 4 + \frac{4}{u-1}\right) du$$

$$= u^2 + 4u + 4\ \ell n|u-1| + C = x + 4\sqrt{x} + 4\ \ell n|\sqrt{x}-1| + C$$

13. $u = \tan\dfrac{\theta}{2}$, $\sin\theta = \dfrac{2u}{1+u^2}$, $\cos\theta = \dfrac{1-u^2}{1+u^2}$, $d\theta = \dfrac{2\ du}{1+u^2}$

$$\int \frac{d\theta}{\sin\theta - \cos\theta + 2} = \int \frac{\dfrac{2\ du}{1+u^2}}{\dfrac{2u}{1+u^2} - \dfrac{1-u^2}{1+u^2} + \dfrac{2+2u^2}{1+u^2}} = \int \frac{2\ du}{3u^2 + 2u + 1}$$

$$= \frac{2}{3}\int \frac{du}{u^2 + 2u/3 + 1/3} = \frac{2}{3}\int \frac{du}{u^2 + 2u/3 + 1/9 + 2/9}$$

$$= \frac{2}{3}\int \frac{du}{(u+1/3)^2 + 2/9}$$

$$= 3\int \frac{du}{[(3u+1)/\sqrt{2}]^2 + 1} = \sqrt{2}\int \frac{3/\sqrt{2}\ du}{[(3u+1)/\sqrt{2}]^2 + 1}$$

$$= \sqrt{2}\ \text{Arctan}\ \frac{3u+1}{\sqrt{2}} + C$$

$$= \sqrt{2}\ \text{Arctan}\ \frac{3\ \tan(\theta/2) + 1}{\sqrt{2}} + C$$

17. $u = \tan\dfrac{\theta}{2}$, $\sin\theta = \dfrac{2u}{1+u^2}$, $\cos\theta = \dfrac{1-u^2}{1+u^2}$, $d\theta = \dfrac{2\ du}{1+u^2}$

$$\int \frac{d\theta}{\sin\theta + \cos\theta} = \int \frac{\dfrac{2\ du}{1+u^2}}{\dfrac{2u}{1+u^2} + \dfrac{1-u^2}{1+u^2}} = \int \frac{2\ du}{1 + 2u - u^2}$$

$$= \int \frac{2\ du}{2 - (u-1)^2} = \int \frac{du}{1 - (u-1)^2/2} = \sqrt{2}\int \frac{1/\sqrt{2}}{1 - (u-1)^2/2}\ du$$

$$= \frac{\sqrt{2}}{2}\ \ell n\left|\frac{1 + (u-1)/\sqrt{2}}{1 - (u-1)/\sqrt{2}}\right| + C = \frac{1}{\sqrt{2}}\ \ell n\left|\frac{\sqrt{2} + u - 1}{\sqrt{2} - u + 1}\right| + C$$

$$= \frac{1}{\sqrt{2}} \ln \frac{(\sqrt{2} + u - 1)^2}{\left| 2 - (u-1)^2 \right|} + C$$

$$= \frac{1}{\sqrt{2}} \ln \frac{(u - 1 + \sqrt{2})^2}{\left| -u^2 + 2u + 1 \right|} + C$$

$$= \frac{1}{\sqrt{2}} \ln \frac{[\tan(\theta/2) - 1 + \sqrt{2}]^2}{\left| -\tan^2(\theta/2) + 2\tan(\theta/2) + 1 \right|} + C$$

21. $u = \sqrt{x^2 - 4}, \quad u^2 = x^2 - 4, \quad x = \sqrt{u^2 + 4}, \quad dx = \dfrac{u \; du}{\sqrt{u^2 + 4}}$

$$\int x^3 \sqrt{x^2 - 4} \; dx = \int (u^2 + 4)^{3/2} u \frac{u \; du}{\sqrt{u^2 + 4}} = \int u^2 (u^2 + 4) \, du$$

$$= \int (u^4 + 4u^2) \, du = \frac{u^5}{5} + \frac{4u^3}{3} + C = \frac{u^3}{15}(3u^2 + 20) + C$$

$$= \frac{1}{15}(x^2 - 4)^{3/2}(3x^2 - 12 + 20) + C$$

$$= \frac{1}{15}(x^2 - 4)^{3/2}(3x^2 + 8) + C$$

Review 15, pp. 462-463

1. $\displaystyle \int (1 + \tan 3\theta)^2 d\theta = \int (1 + 2\tan 3\theta + \tan^2 3\theta) \, d\theta$

$$= \int (2\tan 3\theta + \sec^2 3\theta) \, d\theta = -\frac{2}{3}\ln\left|\cos 3\theta\right| + \frac{1}{3}\tan 3\theta + C$$

5. $\displaystyle \int_0^{\pi/4} (\sec x + 1)(\tan x - 1) \, dx = \int_0^{\pi/4} (\sec x \tan x - \sec x$

$$+ \tan x - 1) \, dx$$

$$= \left. (\sec x - \ln\left|\sec x + \tan x\right| - \ln\left|\cos x\right| - x) \right|_0^{\pi/4}$$

$$= \left. (\sec x - \ln\left|1 + \sin x\right| - x) \right|_0^{\pi/4}$$

$$= \sqrt{2} - \ln\left|1 + 1/\sqrt{2}\right| - \pi/4 - 1 = \sqrt{2} + \ln(2 - \sqrt{2}) - \pi/4 - 1$$

9. $\displaystyle\int x^2 \text{Arctan } x \, dx$ $\qquad\qquad\qquad u = \text{Arctan } x \qquad v' = x^2$

$\displaystyle = \frac{1}{3}x^3 \text{Arctan } x - \frac{1}{3}\int \frac{x^3}{1+x^2} dx \qquad u' = \frac{1}{1+x^2} \qquad v = \frac{x^3}{3}$

$\displaystyle = \frac{1}{3}x^3 \text{Arctan } x - \frac{1}{3}\int \left(x - \frac{x}{1+x^2}\right) dx$

$\displaystyle = \frac{1}{3}x^3 \text{Arctan } x - \frac{x^2}{6} + \frac{1}{6}\ln(1+x^2) + C$

13. $\displaystyle\int_1^2 \frac{dx}{2x^2 - 6x + 5} = \frac{1}{2}\int_1^2 \frac{dx}{x^2 - 3x + 9/4 + 1/4}$

$\displaystyle = \frac{1}{2}\int_1^2 \frac{dx}{(x - 3/2)^2 + 1/4}$

$\displaystyle x - \frac{3}{2} = \frac{1}{2}\tan\theta, \qquad dx = \frac{1}{2}\sec^2\theta \, d\theta$

$\displaystyle = \frac{1}{2}\int_{-\pi/4}^{\pi/4} \frac{\frac{1}{2}\sec^2\theta \, d\theta}{\frac{1}{4}\tan^2\theta + \frac{1}{4}} = \int_{-\pi/4}^{\pi/4} d\theta = \theta\Big|_{-\pi/4}^{\pi/4} = \pi/2$

17. $\displaystyle\int \tan^4 3x \, dx = \int \tan^2 3x(\sec^2 3x - 1) \, dx$

$\displaystyle = \int (\tan^2 3x \, \sec^2 3x - \tan^2 3x) \, dx$

$\displaystyle = \int (\tan^2 3x \, \sec^2 3x - \sec^2 3x + 1) \, dx$

$\displaystyle = \frac{1}{9}\tan^3 3x - \frac{1}{3}\tan 3x + x + C$

21. $\displaystyle\int_1^{1+\sqrt{3}} (2 + 2x - x^2)^{3/2} dx = \int_1^{1+\sqrt{3}} [3 - (x - 1)^2]^{3/2} dx$

$\displaystyle x - 1 = \sqrt{3}\sin\theta, \qquad dx = \sqrt{3}\cos\theta \, d\theta$

$\displaystyle = \int_0^{\pi/2} (3 - 3\sin^2\theta)^{3/2}\sqrt{3}\cos\theta \, d\theta$

$\displaystyle = 9\int_0^{\pi/2} \cos^4\theta \, d\theta = \frac{9}{4}\int_0^{\pi/2} (1 + \cos 2\theta)^2 \, d\theta$

$\displaystyle = \frac{9}{4}\int_0^{\pi/2} (1 + 2\cos 2\theta + \cos^2 2\theta) \, d\theta$

$\displaystyle = \frac{9}{4}\int_0^{\pi/2} \left(1 + 2\cos 2\theta + \frac{1}{2} + \frac{1}{2}\cos 4\theta\right) d\theta$

$\displaystyle = \frac{9}{4}\left(\frac{3}{2}\theta + \sin 2\theta + \frac{1}{8}\sin 4\theta\right)\Big|_0^{\pi/2} = \frac{9}{4}\left(\frac{3\pi}{4}\right) = \frac{27\pi}{16}$

25. $\displaystyle\int \frac{\ln x}{x^3}\, dx \qquad\qquad u = \ln x \qquad v' = \frac{1}{x^3}$

$\displaystyle = -\frac{\ln x}{2x^2} + \int \frac{dx}{2x^3} \qquad\quad u' = \frac{1}{x} \qquad v = -\frac{1}{2x^2}$

$\displaystyle = -\frac{\ln x}{2x^2} - \frac{1}{4x^2} + C = -\frac{2\ln x + 1}{4x^2} + C$

29. $\displaystyle\int_0^1 \frac{x^3\, dx}{\sqrt{9x^2 + 4}} \qquad\qquad x = \frac{2}{3}\tan\theta, \quad dx = \frac{2}{3}\sec^2\theta\, d\theta$

$\displaystyle = \int_0^{\text{Arctan } 3/2} \frac{\frac{8}{27}\tan^3\theta \;\frac{2}{3}\sec^2\theta\, d\theta}{\sqrt{4\tan^2\theta + 4}}$

$\displaystyle = \frac{8}{81}\int_0^{\text{Arctan } 3/2} \frac{\tan^3\theta \;\sec^2\theta}{\sec\theta}\, d\theta$

$\displaystyle = \frac{8}{81}\int_0^{\text{Arctan } 3/2} (\sec^2\theta - 1)\sec\theta \tan\theta\, d\theta$

$\displaystyle = \frac{8}{81}\left(\frac{\sec^3\theta}{3} - \sec\theta\right)\Bigg|_0^{\text{Arctan } 3/2}$

$\displaystyle = \frac{8}{81}\left[\left(\frac{13\sqrt{13}}{24} - \frac{\sqrt{13}}{2}\right) - \left(\frac{1}{3} - 1\right)\right] = \frac{1}{243}(\sqrt{13} + 16)$

33. $\displaystyle\int \sin 2x \cos 3x\, dx \qquad\qquad u = \sin 2x \qquad v' = \cos 3x$

$\qquad\qquad\qquad\qquad\qquad\qquad u' = 2\cos 2x \qquad v = \frac{1}{3}\sin 3x$

$\displaystyle = \frac{1}{3}\sin 2x \sin 3x - \int \frac{2}{3}\cos 2x \sin 3x\, dx$

$\qquad\qquad\qquad\qquad\qquad\qquad u = \frac{2}{3}\cos 2x \qquad v' = \sin 3x$

$\qquad\qquad\qquad\qquad\qquad\qquad u' = \frac{4}{3}\sin 2x \qquad v = -\frac{1}{3}\cos 3x$

$\displaystyle = \frac{1}{3}\sin 2x \sin 3x + \frac{2}{9}\cos 2x \cos 3x + \frac{4}{9}\int \sin 2x \cos 3x\, dx$

$\displaystyle \frac{5}{9}\int \sin 2x \cos 3x\, dx = \frac{3}{9}\sin 2x \sin 3x + \frac{2}{9}\cos 2x \cos 3x$

$\displaystyle \int \sin 2x \cos 3x\, dx = \frac{3}{5}\sin 2x \sin 3x + \frac{2}{5}\cos 2x \cos 3x + C$

37. $\int e^{ax} \sin bx \, dx$ $\qquad u = e^{ax} \qquad v' = \sin bx$

$\qquad\qquad\qquad\qquad\qquad u' = ae^{ax} \qquad v = -\dfrac{1}{b} \cos bx$

$\qquad = -\dfrac{1}{b} e^{ax} \cos bx + \int \dfrac{a}{b} d^{ax} \cos bx \, dx$

$\qquad\qquad\qquad\qquad\qquad u = \dfrac{a}{b} e^{ax} \qquad v' = \cos bx$

$\qquad\qquad\qquad\qquad\qquad u' = \dfrac{a^2}{b} e^{ax} \qquad v = \dfrac{1}{b} \sin bx$

$\qquad = -\dfrac{1}{b} e^{ax} \cos bx + \dfrac{a}{b^2} e^{ax} \sin bx - \dfrac{a^2}{b^2} \int e^{ax} \sin bx \, dx$

$\dfrac{a^2 + b^2}{b^2} \int e^{ax} \sin bx \, dx = \dfrac{a}{b^2} e^{ax} \sin bx - \dfrac{b}{b^2} e^{ax} \cos bx$

$\int e^{ax} \sin bx \, dx = \dfrac{e^{ax}}{a^2 + b^2} (a \sin bx - b \cos bx) + C$

41. $\int \dfrac{\sin \theta \, d\theta}{3 + \sin^2 \theta}$ $\qquad u = \tan \dfrac{\theta}{2}, \quad \sin \theta = \dfrac{2u}{1+u^2}, \quad d\theta = \dfrac{2 \, du}{1+u^2}$

$\qquad = \int \dfrac{\dfrac{2u}{1+u^2} \dfrac{2 \, du}{1+u^2}}{3 + \dfrac{4u^2}{(1+u^2)^2}} = \int \dfrac{4u \, du}{3(1+u^2)^2 + 4u^2} = \int \dfrac{4u \, du}{3u^4 + 10u^2 + 3}$

$\qquad = 4 \int \dfrac{u \, du}{(3u^2 + 1)(u^2 + 3)}$

$\dfrac{u}{(3u^2 + 1)(u^2 + 3)} = \dfrac{Au + B}{3u^2 + 1} + \dfrac{Cu + D}{u^2 + 3}$

$u = A(u^3 + 3u) + B(u^2 + 3) + C(3u^3 + u) + D(3u^2 + 1)$

$A + 3C = 0, \quad B + 3D = 0, \quad 3A + C = 1, \quad 3B + D = 0$

$A = 3/8, \quad B = 0, \quad C = -1/8, \quad D = 0$

$\qquad = 4 \int \left(\dfrac{3}{8} \dfrac{u}{3u^2 + 1} - \dfrac{1}{8} \dfrac{u}{u^2 + 3} \right) du$

$\qquad = \dfrac{1}{4} \ln(3u^2 + 1) - \dfrac{1}{4}(u^2 + 3) + C$

$\qquad = \dfrac{1}{4} \ln \dfrac{3u^2 + 1}{u^2 + 3} + C = \dfrac{1}{4} \ln \dfrac{3 \tan^2 \theta/2 + 1}{\tan^2 \theta/2 + 3} + C$

45. $y = x^2$, $y' = 2x$

$$s = \int_0^2 \sqrt{1 + 4x^2} \, dx \qquad 2x = \tan \theta, \ dx = \frac{1}{2} \sec^2\theta \, d\theta$$

$$= \int_0^{Arctan \, 4} \sqrt{1 + \tan^2\theta} \ \frac{1}{2} \sec^2\theta \, d\theta = \frac{1}{2}\int_0^{Arctan \, 4} \sec^3\theta \, d\theta$$

$$= \frac{1}{4}(\sec \theta \tan \theta + \ell n | \sec \theta + \tan \theta |) \Big|_0^{Arctan \ 4}$$

$$= \frac{1}{4}[\sqrt{17} \cdot 4 + \ell n (\sqrt{17} + 4)] = \sqrt{17} + \frac{1}{4} \ell n (4 + \sqrt{17})$$

Section 16.1, pp. 470-471

1. $\displaystyle\int_2^{+\infty} \frac{dx}{x^3} = \lim_{k\to+\infty} \int_2^k \frac{dx}{x^3} = \lim_{k\to+\infty} (-\frac{1}{2x^2})\Big|_2^k$

   $\displaystyle = \lim_{k\to+\infty} (\frac{1}{8} - \frac{1}{2k^2}) = \frac{1}{8}$

5. $\displaystyle\int_1^{+\infty} \frac{dx}{2x+1} = \lim_{k\to+\infty} \int^k \frac{dx}{2x+1} = \lim_{k\to+\infty} \frac{1}{2} \ln(2x+1)\Big|_1^k$

   $\displaystyle = \lim_{k\to+\infty} \frac{1}{2}[\ln(2k+1) - \ln 3] = +\infty$

9. $\displaystyle\int_1^2 \frac{dx}{\sqrt{x-1}} = \lim_{\epsilon\to0^+} \int_{1+\epsilon}^2 \frac{dx}{\sqrt{x-1}} = \lim_{\epsilon\to0^+} 2\sqrt{x-1}\ \Big|_{1+\epsilon}^2$

   $\displaystyle = \lim_{\epsilon\to0^+} (2 - 2\sqrt{\epsilon}) = 2$

13. $\displaystyle\int_0^2 \frac{dx}{(3x-1)^{2/3}} = \int_0^{1/3} \frac{dx}{(3x-1)^{2/3}} + \int_{1/3}^2 \frac{dx}{(3x-1)^{2/3}}$

   $\displaystyle = \lim_{\epsilon\to0^+} \int_0^{1/3-\epsilon} \frac{dx}{(3x-1)^{2/3}} + \lim_{\delta\to0^+} \int_{1/3+\delta}^2 \frac{dx}{(3x-1)^{2/3}}$

   $\displaystyle = \lim_{\epsilon\to0^+} (3x-1)^{1/3}\Big|_0^{1/3-\epsilon} + \lim_{\delta\to0^+} (3x-1)^{1/3}\Big|_{1/3+\delta}^2$

   $\displaystyle = \lim_{\epsilon\to0^+} (-\sqrt[3]{\epsilon} + 1) + \lim_{\delta\to0^+} (\sqrt[3]{5} - \sqrt[3]{\delta}) = 1 + \sqrt[3]{5}$

17. $\displaystyle\int_0^{+\infty} \frac{dx}{\sqrt{x}} = \int_0^1 \frac{dx}{\sqrt{x}} + \int_1^{+\infty} \frac{dx}{\sqrt{x}} = \lim_{\epsilon\to0^+} \int_\epsilon^1 \frac{dx}{\sqrt{x}} + \lim_{k\to+\infty} \int_1^k \frac{dx}{\sqrt{x}}$

   $\displaystyle = \lim_{\epsilon\to0^+} 2\sqrt{x}\ \Big|_\epsilon^1 + \lim_{k\to+\infty} 2\sqrt{x}\ \Big|_1^k$

   $\displaystyle = \lim_{\epsilon\to0^+} (2 - 2\sqrt{\epsilon}) + \lim_{k\to+\infty} (2\sqrt{k} - 2) = 2 + (+\infty) = +\infty$

21. $\displaystyle\int_{-3}^{6} \frac{dx}{(x+2)^{4/3}} = \int_{-3}^{-2} \frac{dx}{(x+2)^{4/3}} + \int_{-2}^{6} \frac{dx}{(x+2)^{4/3}}$

$\displaystyle = \lim_{\epsilon \to 0^+} \int_{-3}^{-2-\epsilon} \frac{dx}{(x+2)^{4/3}} + \lim_{\delta \to 0^+} \int_{-2+\delta}^{6} \frac{dx}{(x+2)^{4/3}}$

$\displaystyle = \lim_{\epsilon \to 0^+} \frac{-3}{\sqrt[3]{x+2}} \Big|_{-3}^{-2-\epsilon} + \lim_{\delta \to 0^+} \frac{-3}{\sqrt[3]{x+2}} \Big|_{-2+\delta}^{6}$

$\displaystyle = \lim_{\epsilon \to 0^+} \left(\frac{3}{\sqrt[3]{\epsilon}} - 3\right) + \lim_{\delta \to 0^+} \left(-\frac{3}{2} + \frac{3}{\sqrt[3]{\delta}}\right)$　　　　Diverges

　　　　$\underset{\downarrow}{}$　　　　　　　　$\underset{\downarrow}{}$

　　　　$+\infty$　　　　　　　　　$+\infty$

25. $\displaystyle A = \int_{2}^{\infty} \frac{dx}{x^2} = \lim_{k \to +\infty} \int_{2}^{k} \frac{dx}{x^2} = \lim_{k \to +\infty} \left(-\frac{1}{x}\right) \Big|_{2}^{k} = \lim_{k \to +\infty} \left(\frac{1}{2} - \frac{1}{k}\right) = \frac{1}{2}$

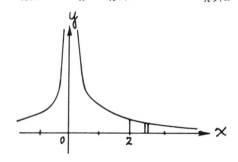

29. $\displaystyle\int_{1}^{+\infty} \frac{dx}{x^n} = \lim_{k \to +\infty} \int_{1}^{k} x^{-n}\, dx = \lim_{k \to +\infty} \frac{x^{1-n}}{1-n} \Big|_{1}^{k}$　　$(n \neq 1)$

$\displaystyle = \lim_{k \to +\infty} \frac{k^{1-n} - 1}{1-n} = \begin{cases} 1/(n-1) & \text{if } n > 1 \\ +\infty & \text{if } n < 1 \end{cases}$

If $\ n = 1, \quad \displaystyle\int_{1}^{+\infty} \frac{dx}{x^n} = \lim_{k \to +\infty} \int_{1}^{k} \frac{dx}{x} = \lim_{k \to +\infty} \ln x \Big|_{1}^{k}$

$\displaystyle = \lim_{k \to +\infty} \ln k = +\infty$

33. Let f be the function whose graph is

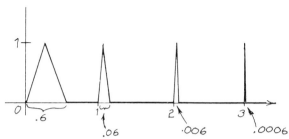

$\int_0^{+\infty} f(x)\,dx$ is the area under the curve, which is

$0.3 + 0.03 + 0.003 + 0.0003 + \cdots = 0.333\cdots = 1/3$

Clearly $\lim\limits_{x \to +\infty} f(x)$ does not exist.

Section 16.2, pp. 474-475

1. $V = \int_0^1 \pi y^2 dx = \int_0^1 \pi x^6 dx = \left.\frac{\pi x^7}{7}\right|_0^1 = \frac{\pi}{7}$

5. $V = \int_0^\pi \pi y^2 dx = \pi \int_0^\pi \sin^2 x\, dx = \frac{\pi}{2} \int_0^\pi (1 - \cos 2x)\,dx$

   $= \frac{\pi}{2}(x - \frac{1}{2} \sin 2x)\big|_0^\pi = \frac{\pi^2}{2}$

9. $V = \int_0^1 \pi (y_1^2 - y_2^2)\,dx$

   $= \pi \int_0^1 (x^2 - x^4)\,dx$

   $= \pi(\frac{x^3}{3} - \frac{x^5}{5})\big|_0^1$

   $= \pi(\frac{1}{3} - \frac{1}{5}) = \frac{2\pi}{15}$

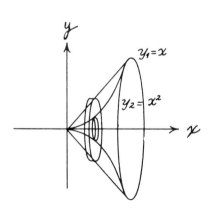

13. $V = \int_0^1 \pi x^2 dy = \pi \int_0^1 \text{Arccos}^2 y \, dy$

$\quad u = \text{Arccos}^2 y \qquad v' = 1$

$\quad u' = \dfrac{-2 \text{ Arccos } y}{\sqrt{1 - y^2}} \qquad v = y$

$\quad = \pi \left[y \text{ Arccos}^2 y \Big|_0^1 + \int_0^1 \dfrac{2y \text{ Arccos } y}{\sqrt{1 - y^2}} \, dy \right]$

$\qquad u = \text{Arccos } y \qquad v' = \dfrac{2y}{\sqrt{1 - y^2}}$

$\qquad u' = \dfrac{1}{\sqrt{1 - y^2}} \qquad v = -2\sqrt{1 - y^2}$

$\quad = \pi \left[(y \text{ Arccos}^2 y - 2\sqrt{1 - y^2} \text{ Arccos } y) \Big|_0^1 - \int_0^1 2 \, dy \right]$

$\quad = \pi (y \text{ Arccos}^2 y - 2\sqrt{1 - y^2} \text{ Arccos } y - 2y) \Big|_0^1$

$\quad = \pi(-2 + \pi) = \pi(\pi - 2)$

17. $V = \int_{-\infty}^0 \pi x^2 dy = \int_{-\infty}^0 \pi \, e^{2y} dy$

$\quad = \dfrac{\pi}{2} \lim_{k \to -\infty} e^{2y} \Big|_k^0 = \dfrac{\pi}{2} \lim_{k \to -\infty} (1 - e^{2k})$

$\quad = \dfrac{\pi}{2}$

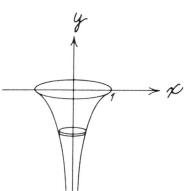

21. $V = \int_0^1 \pi(1-x)^2 dy$

$= \pi \int_0^1 (1 - y^{1/3})^2 dy$

$= \pi \int_0^1 (1 - 2y^{1/3} + y^{2/3}) dy$

25. $V = \int_0^1 \pi[\pi^2 - (\pi - x)^2] dy = \pi \int_0^1 (2\pi x - x^2) dy$

$= \pi \int_0^1 (2\pi \text{ Arccos } y - \text{Arccos}^2 y) dy$

$u = 2\pi \text{ Arccos } y - \text{Arccos}^2 y \qquad v' = 1$

$u' = \dfrac{-2\pi}{\sqrt{1-y^2}} + \dfrac{2 \text{ Arccos } y}{\sqrt{1-y^2}} \qquad v = y$

$= \pi\Big[(2\pi y \text{ Arccos } y - \text{Arccos}^2 y)\Big|_0^1$

$+ \int_0^1 \dfrac{(2\pi - 2 \text{ Arccos } y)y}{\sqrt{1-y^2}} dy \Big]$

$u = 2\pi - 2 \text{ Arccos } y \qquad v' = \dfrac{y}{\sqrt{1-y^2}}$

$u' = \dfrac{2}{\sqrt{1-y^2}} \qquad v = -\sqrt{1-y^2}$

$= \pi\Big[(2\pi y \text{ Arccos } y - y \text{ Arccos}^2 y - 2\pi\sqrt{1-y^2}$

$+ 2\sqrt{1-y^2} \text{ Arccos } y)\Big|_0^1 + \int_0^1 2dy \Big]$

$= \pi(2\pi y \text{ Arccos } y - y \text{ Arccos}^2 y - 2\pi\sqrt{1-y^2}$

$+ 2\sqrt{1-y^2} \text{ Arccos } y + 2y)\Big|_0^1$

$= \pi(2 + 2\pi - \pi) = (2 + \pi)\pi$

29. $V = \int_0^1 \pi[(1-x_1)^2 - (1-x_2)^2]\,dy$

$= \pi\int_0^1 [(1-y)^2 - (1-\sqrt{y})^2]\,dy$

$= \pi\int_0^1 (1 - 2y + y^2 - 1 + 2\sqrt{y} - y)\,dy$

$= \pi\int_0^1 (-3y + y^2 + 2\sqrt{y})\,dy$

$= \pi(-\dfrac{3y^2}{2} + \dfrac{y^3}{3} + \dfrac{4y^{3/2}}{3})\Big|_0^1$

$= \pi(-\dfrac{3}{2} + \dfrac{1}{3} + \dfrac{4}{3})$

$= \pi(\dfrac{-9 + 2 + 8}{6}) = \dfrac{\pi}{6}$

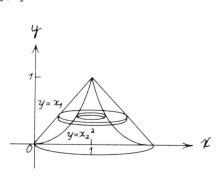

33. $A = 2\int_1^{+\infty} y\,dx$

$= 2\lim_{k\to+\infty} \int_1^k \dfrac{1}{x}\,dx$

$= 2\lim_{k\to+\infty} \ell n\, x\Big|_1^k$

$= 2\lim_{k\to+\infty} \ell n\, k = +\infty$

$V = \int_1^{+\infty} \pi y^2\,dx$

$= \pi\lim_{k\to+\infty} \int_1^k \dfrac{1}{x^2}\,dx = \pi\lim_{k\to+\infty} -\dfrac{1}{x}\Big|_1^k$

$= \lim_{k\to+\infty} (1 - \dfrac{1}{k}) = \pi.$

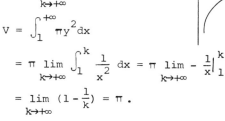

Thus we have a solid with finite volume but having a cross-section with infinite area. This appears to be a contradiction, but area and volume are not comparable.

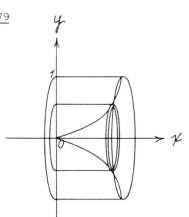

1. $V = \int_0^1 2\pi yx \ dy$

    $= 2\pi \int_0^1 y \cdot y^{1/2} dy$

    $= \frac{4\pi}{5} \ y^{5/2} \Big|_0^1$

    $= \frac{4\pi}{5}$

5. $V = \int_0^1 2\pi (1-x)y \ dx$

    $= 2\pi \int_0^1 (x^2 - x^3) \ dx$

    $= 2\pi (\frac{x^3}{3} - \frac{x^4}{4}) \Big|_0^1$

    $= 2\pi (\frac{1}{3} - \frac{1}{4}) = \frac{\pi}{6}$

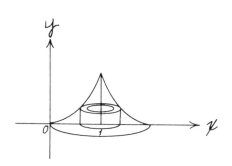

9. $y = x^2 - 4x + 3 = (x-1)(x-3)$

    $V = \int_1^3 2\pi (-y)x \ dx$

    $= 2\pi \int_1^3 (-x^3 + 4x^2 - 3x) \ dx$

    $= 2\pi (-\frac{x^4}{4} + \frac{4x^3}{3} - \frac{3x^2}{2}) \Big|_1^3$

    $= 2\pi (-\frac{81}{4} + 36 - \frac{27}{2} + \frac{1}{4} - \frac{4}{3} + \frac{3}{2})$

    $= \frac{16\pi}{3}$

13. $V = \int_1^{+\infty} 2\pi \ xy \ dx$

$\qquad = 2\pi \int_1^{+\infty} dx = +\infty$

17. $V = \int_0^2 2\pi \ xy \ dx$

$\qquad = 2\pi \int_0^2 x^2 (x-2)^2 dx$

$\qquad = 2\pi \int_0^2 (x^4 - 4x^3 + 4x^2) \, dx$

$\qquad = 2\pi \left(\dfrac{x^5}{5} - x^4 + \dfrac{4x^3}{3} \right) \Big|_0^2$

$\qquad = 2\pi \left(\dfrac{32}{5} - 16 + \dfrac{32}{3} \right)$

$\qquad = \dfrac{32\pi}{15}$

21. $V = \int_0^\pi 2\pi \ xy \ dx$

$\qquad = 2\pi \int_0^\pi x \sin x \ dx$

$\qquad\qquad u = x \quad v' = \sin x$

$\qquad\qquad u' = 1 \quad v = -\cos x$

$\qquad = 2\pi (-x \cos x + \sin x) \Big|_0^\pi$

$\qquad = 2\pi(\pi) = 2\pi^2$

25. $V = \int_0^1 2\pi \ xy \ dy$

$\qquad = \pi \int_0^1 2y(y^2 - 1)^2 dy$

$\qquad = \frac{\pi}{3}(y^2 - 1)^3 \Big|_0^1$

$\qquad = \frac{\pi}{3}$

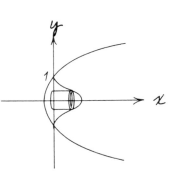

Section 16.4, p. 483

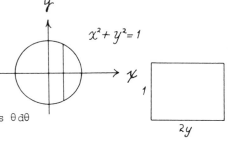

$x^2 + y^2 = 1$

1. $V = 4\int_0^1 y \ dx$

$\qquad = 4\int_0^1 \sqrt{1 - x^2} \ dx$

$\qquad x = \sin \theta, \quad dx = \cos \theta d\theta$

$\qquad = 4\int_0^{\pi/2} \cos^2 \theta \, d\theta$

$\qquad = 2\int_0^{\pi/2} (1 + \cos \ 2\theta) d\theta = 2(\theta + \frac{1}{2} \sin 2\theta) \Big|_0^{\pi/2} = \pi$

5. $V = 2\int_0^1 \frac{1}{2}\pi \ y^2 dx$

$\qquad = \pi \int_0^1 (1 - x^2) dx$

$\qquad = \pi(x - \frac{x^3}{3}) \Big|_0^1$

$\qquad = \pi(1 - \frac{1}{3}) = \frac{2\pi}{3}$

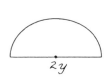

9. $V = \int_0^4 4x \ dy$

$\qquad = 4\int_0^4 \sqrt{4 - y} \ dy = -\frac{8}{3}(4 - y)^{3/2} \Big|_0^4$

$\qquad = \frac{8}{3} \cdot 8 = \frac{64}{3}$

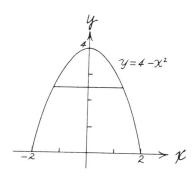

$y = 4 - x^2$

13. $V = \int_0^4 \frac{1}{2} \pi x^2 dy$

$= \frac{1}{2} \pi \int_0^4 (4 - y) \, dy$

$= \frac{1}{2} \pi (4y - \frac{y^2}{2}) \Big|_0^4$

$= \frac{\pi}{2}(16 - 8) = 4\pi$

$\frac{1}{2}\pi x^2$

$2x$

17. $V = 2\int_0^4 8y \, dx$

$= 16\int_0^4 \frac{3}{4} \sqrt{16 - x^2} \, dx$

$x = 4 \sin \theta , \quad dx = 4 \cos \theta \, d\theta$

$= 12\int_0^{\pi/2} 4 \cos \theta \, 4 \cos \theta \, d\theta$

$= 192\int_0^{\pi/2} \cos^2 \theta \, d\theta = 96\int_0^{\pi/2} (1 + \cos 2\theta) \, d\theta$

$= 96(\theta + \frac{1}{2} \sin 2\theta) \Big|_0^{\pi/2} = 48\pi$

4

$2y$

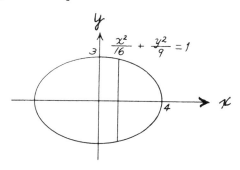

$\frac{x^2}{16} + \frac{y^2}{9} = 1$

21.
$$V = 2\int_0^4 \frac{\pi}{2} y^2 dx$$
$$= \pi \int_0^4 \frac{9}{16}(16 - x^2) dx$$
$$= \frac{9\pi}{16}(16x - \frac{x^3}{3})\Big|_0^4$$
$$= 24\pi$$

25.
$$V = 2\int_0^2 4y^2 dx$$
$$= 8\int_0^2 (4 - x^2) dx$$
$$= 8(4x - \frac{x^3}{3})\Big|_0^2$$
$$= 8 \cdot \frac{2 \cdot 8}{3} = \frac{128}{3}$$

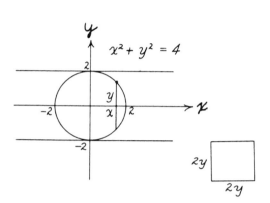

Section 16.5, pp. 487-488

1. $x^2 + y^2 = R^2$, $\qquad y = \sqrt{R^2 - x^2}$, $\qquad y' = \frac{-x}{\sqrt{R^2 - x^2}}$

$$S_x = 2\int_0^R 2\pi y \sqrt{1 + \frac{x^2}{R^2 - x^2}} \ dx$$

$$= 4\pi \int_0^R \sqrt{R^2 - x^2} \sqrt{\frac{R^2}{R^2 - x^2}} \ dx = 4\pi R \int_0^R dx = 4\pi R^2$$

5. $y = \sqrt{4 - x^2}$ $\qquad\qquad y' = \frac{-x}{\sqrt{4 - x^2}}$

$$S_x = \int_{-1/2}^{1/2} 2\pi\sqrt{4 - x^2} \sqrt{1 + \frac{x^2}{4 - x^2}} \ dx = \int_{-1/2}^{1/2} 4\pi \ dx$$

$$= 4\pi \ x\Big|_{-1/2}^{1/2} = 4\pi$$

9. $x = a \cos^3\theta$ $\qquad\qquad\qquad$ $y = a \sin^3\theta$

$\dfrac{dx}{d\theta} = -3a \cos^2\theta \sin\theta$ $\qquad\qquad$ $\dfrac{dy}{d\theta} = 3a \sin^2\theta \cos\theta$

$S_x = \displaystyle\int_0^{\pi/2} 2\pi y \sqrt{9a^2\cos^4\theta \sin^2\theta + 9a^2\sin^4\theta \cos^2\theta}\ d\theta$

$\qquad = \displaystyle\int_0^{\pi/2} 2\pi a \sin^3\theta \cdot 3a \sin\theta \cos\theta\sqrt{\cos^2\theta + \sin^2\theta}\ d\theta$

$\qquad = 6\pi\ a^2 \displaystyle\int_0^{\pi/2} \sin^4\theta \cos\theta\, d\theta = \dfrac{6\pi a^2}{5}\sin^5\theta \Big|_0^{\pi/2} = \dfrac{6\pi a^2}{5}$

13. $y = \cosh x$ $\qquad\qquad\qquad\qquad$ $y' = \sinh x$

$S_x = \displaystyle\int_{-1}^{1} 2\pi \cosh x\sqrt{1 + \sinh^2 x}\ dx = 2\pi \int_{-1}^{1}\cosh^2 x\ dx$

$\qquad = \pi\displaystyle\int_{-1}^{1} (1 + \cosh 2x)\, dx = \pi (x + \tfrac{1}{2} \sinh 2x) \Big|_{-1}^{1}$

$\qquad = \pi(x + \sinh x \cosh x)\Big|_{-1}^{1} = 2\pi(1 + \sinh 1 \cosh 1)$

17. $x = t^3$ $\qquad\qquad\qquad\qquad\qquad$ $y = t^2$

$x' = 3t^2$ $\qquad\qquad\qquad\qquad\qquad$ $y' = 2t$

$S_x = \displaystyle\int_0^2 2\pi t^2\sqrt{9t^4 + 4t^2}\ dt = 2\pi\int_0^2 t^3\sqrt{9t^2 + 4}\ dt$

$\qquad u = t^2$ $\qquad\qquad\qquad\qquad$ $v' = t\sqrt{9t^2 + 4}$

$\qquad u' = 2t$ $\qquad\qquad\qquad\qquad$ $v = \dfrac{1}{27}(9t^2 + 4)^{3/2}$

$\qquad = 2\pi[\dfrac{t^2}{27} (9t^2 + 4)^{3/2}\Big|_0^2 - \displaystyle\int_0^2 \dfrac{2t}{27}(9t^2 + 4)^{3/2}dt]$

$\qquad = 2\pi[\dfrac{t^2}{27} (9t^2 + 4)^{3/2} - \dfrac{2}{1215}(9t^2 + 3)^{5/2}]\Big|_0^2$

$\qquad = 2\pi[\dfrac{4}{27} \cdot 40^{3/2} - \dfrac{2}{1215}\ 40^{5/2} + \dfrac{2}{1215} \cdot 32]$

$\qquad = \dfrac{128\pi(125\sqrt{10} + 1)}{1215}$

21. $y = \dfrac{x^2}{4} - \dfrac{\ln x}{2}$ $\qquad\qquad$ $y' = \dfrac{x}{2} - \dfrac{1}{2x}$

$S_x = \displaystyle\int_1^4 2\pi\,(\dfrac{x^2}{4} - \dfrac{\ln x}{2}) \sqrt{1 + \dfrac{x^2}{4} - \dfrac{1}{2} + \dfrac{1}{4x^2}}\ dx$

$\qquad = 2\pi \displaystyle\int_1^4 (\dfrac{x^2}{4} - \dfrac{\ln x}{2}) \sqrt{\dfrac{x^2}{4} + \dfrac{1}{2} + \dfrac{1}{4x^2}}\ dx$

$\qquad = 2\pi \displaystyle\int_1^4 (\dfrac{x^2}{4} - \dfrac{\ln x}{2})(\dfrac{x^2}{2} + \dfrac{1}{2x})\,dx$

$\qquad = 2\pi \displaystyle\int_1^4 (\dfrac{x^4}{8} - \dfrac{x^2 \ln x}{4} + \dfrac{x}{8} - \dfrac{\ln x}{4x})\ dx$

$\qquad\qquad u = \ln x \qquad\qquad v' = \dfrac{x^2}{4}$

$\qquad\qquad u' = \dfrac{1}{x} \qquad\qquad\ v = \dfrac{x^3}{12}$

$\qquad = 2\pi\,(\dfrac{x^5}{40} + \dfrac{x^2}{16} - \dfrac{\ln^2 x}{8} - \dfrac{x^3 \ln x}{12} + \dfrac{x^3}{36})\,\Big|_1^4$

$\qquad = 2\pi\,(\dfrac{128}{5} + 1 - \dfrac{\ln^2 x}{8} - \dfrac{16\,\ln 4}{3} + \dfrac{16}{9} - \dfrac{1}{40} - \dfrac{1}{16} - \dfrac{1}{36})$

$\qquad = 2\pi\,(\dfrac{2261}{80} - \dfrac{1}{8}\,\ln^2 4 - \dfrac{16}{3}\,\ln 4)$

25. $y = \cosh x$ $\qquad\qquad\qquad$ $y' = \sinh x$

$S_y = \displaystyle\int_0^1 2\pi x \sqrt{1 + \sinh^2 x}\ dx = \displaystyle\int_0^1 2\pi x \cosh x\ dx$

$\qquad\qquad u = x \qquad\qquad\quad v' = \cosh x$

$\qquad\qquad u' = 1 \qquad\qquad\ \ v = \sinh x$

$\qquad = 2\pi\,(x \sinh x - \cosh x)\,\Big|_0^1 = 2\pi\,(\sinh 1 - \cosh 1 + 1)$

29. $x = e^t \cos t$ $\qquad\qquad\qquad$ $y = e^t \sin t$

$\quad x' = e^t(\cos t - \sin t) \qquad\qquad y' = e^t(\cos t + \sin t)$

$\quad S_y = \displaystyle\int_0^\pi 2\pi\,|e^t \cos t| \sqrt{e^{2t}(\cos^2 t - 2\sin t \cos t + \sin^2 t}$

$\qquad\qquad\qquad\overline{+\, \cos^2 t + 2\sin t \cos t + \sin^2 t)}\ dt$

$\qquad = 2\sqrt{2}\,\pi \displaystyle\int_0^\pi e^{2t} \cos t\ dt$

$\qquad\qquad u = e^{2t} \qquad\qquad\qquad v' = \cos t$

$\qquad\qquad u' = 2e^{2t} \qquad\qquad\qquad v = \sin t$

$$\int e^{2t} \cos t \, dt = e^{2t} \sin t - \int 2e^{2t} \sin t \, dt$$

$$u = 2e^{2t} \qquad\qquad v' = -\sin t$$
$$u' = 4e^{2t} \qquad\qquad v = \cos t$$

$$\int e^{2t} \cos t \, dt = e^{2t} \sin t + 2e^{2t} \cos t - \int 4e^{2t} \cos t \, dt$$

$$= \frac{e^{2t}}{5} (\sin t + 2 \cos t)$$

$$= \frac{2\sqrt{2\pi}}{5} [e^{2t}(\sin t + 2 \cos t)\Big|_0^{\pi/2} - e^{2t}(\sin t + 2 \cos t)\Big|_{\pi/2}^{\pi}$$

$$= \frac{2\sqrt{2}\,\pi}{5} (e^{\pi} - 2 + 2e^{2\pi} + e^{\pi}) = \frac{4\sqrt{2}\,\pi}{5} (e^{2\pi} + e^{\pi} - 1)$$

Section 16.6, pp. 491–492

1. $M_y = 2 \cdot 4 + 10 \cdot 2 + 4(-6) = 4$

   $W = 2 + 10 + 4 = 16$, $\qquad\qquad \bar{x} = \dfrac{4}{16} = 1/4$

5. $M_y = 2 \cdot 2 + 5(-1) + 3 \cdot 4 = 11$

   $M_x = 2 \cdot 4 + 5 \cdot 2 + 3(-2) = 12 \qquad W = 2 + 5 + 3 = 10$

   $\bar{x} = \dfrac{11}{10} \qquad\qquad\qquad \bar{y} = \dfrac{12}{10} = \dfrac{6}{5}$

9. $M_y = 10 \cdot 2 + 3 \cdot 5 + 7 \cdot 0 = 35$

   $M_x = 10 \cdot 2 + 3(-1) + 7(-3) = -4$

   $W = 10 + 3 + 7 = 20 \qquad\qquad \bar{x} = \dfrac{35}{20} = \dfrac{7}{4}, \quad y = \dfrac{-4}{20} = -\dfrac{1}{5}$

13. $M_y = 4 \cdot 3 + 2(-1) + 3x = 10 + 3x$

   $W = 4 + 2 + 3 = 9 \qquad\qquad 2 = \dfrac{10 + 3x}{9}, \quad x = \dfrac{8}{3}$

17. $M_y = 4 \cdot 2 + 2(-3) + 2x = 2 + 2x$

   $M_x = 4 \cdot 2 + 2 \cdot 4 + 2y = 16 + 2y$

   $W = 4 + 2 + 2 = 8$, $\qquad\qquad \dfrac{2 + 2x}{8} = -1$

   $x = -5$; $\qquad\qquad \dfrac{16 + 2y}{8} = 1$, $\qquad\qquad y = -4$

21. $M_x = 4 \cdot 1 + \sqrt{3}\ (2 + \frac{\sqrt{3}}{3}) = 5 + 2\sqrt{3}$

$W = 4 + \sqrt{3}$

$\bar{y} = \dfrac{5 + 2\sqrt{3}}{4 + \sqrt{3}}$

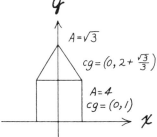

25. $M_x = 4 \cdot 1 + 2\pi \cdot 3 = 4 + 6\pi$

$W = 4 + 2\pi$

$\bar{y} = \dfrac{4 + 6\pi}{4 + 2\pi} = \dfrac{2 + 3\pi}{2 + \pi}$

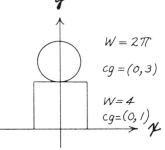

Section 16.7, pp. 495–496

1. $A = \displaystyle\int_0^2 x^2 dx = \dfrac{x^3}{3}\bigg|_0^2 = \dfrac{8}{3}$

$M_x = \displaystyle\int_0^2 \dfrac{x^4}{2} dx = \dfrac{x^5}{10}\bigg|_0^2 = \dfrac{16}{5}$

$M_y = \displaystyle\int_0^2 x^3 dx = \dfrac{x^4}{4}\bigg|_0^2 = 4$

$\bar{x} = \dfrac{4}{8/3} = \dfrac{3}{2}$

$\bar{y} = \dfrac{16/5}{8/3} = \dfrac{6}{5}$

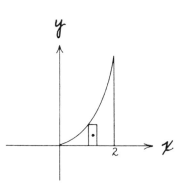

5. $A = \int_0^1 \sqrt{y}\ dy = \frac{2}{3}y^{3/2}\Big|_0^1 = \frac{2}{3}$

$M_x = \int_0^1 y^{3/2}dy = \frac{2}{5}y^{5/2}\Big|_0^1 = \frac{2}{5}$

$M_y = \int_0^1 \frac{y}{2}dy = \frac{y^2}{4}\Big|_0^1 = \frac{1}{4}$

$\bar{x} = \frac{1/4}{2/3} = \frac{3}{8}$

$\bar{y} = \frac{2/5}{2/3} = \frac{3}{5}$

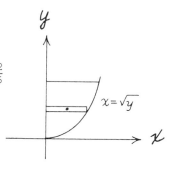

9. $A = \int_0^1 (y^4 + 1)\,dy = (\frac{y^5}{5} + y)\Big|_0^1 = \frac{6}{5}$

$M_x = \int_0^1 (y^5 + y)\,dy = (\frac{y^6}{6} + \frac{y^2}{2})\Big|_0^1 = \frac{2}{3}$

$M_y = \int_0^1 \frac{(y^4 + 1)^2}{2}\ dy$

$\quad = \frac{1}{2}\int_0^1 (y^8 + 2y^4 + 1)\,dy$

$\quad = \frac{1}{2}(\frac{y^9}{9} + \frac{2y^5}{5} + y)\Big|_0^1$

$\quad = \frac{1}{2}(\frac{1}{9} + \frac{2}{5} + 1) = \frac{34}{45}$

$\bar{x} = \frac{34/45}{6/5} = \frac{17}{27}$, $\quad \bar{y} = \frac{2/3}{6/5} = \frac{5}{9}$

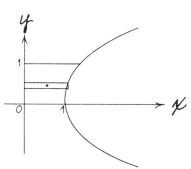

13. $A = 2\int_0^{+\infty} \frac{dx}{x^2 + 1} = \lim_{k\to+\infty} 2\int_0^k \frac{dx}{x^2 + 1}$

$\quad = \lim_{k\to+\infty} 2\ \mathrm{Arctan}\ x\Big|_0^k$

$\quad = \lim_{k\to+\infty} 2\ \mathrm{Arctan}\ k = \pi$

$M_x = \int_0^{+\infty} \frac{dx}{(x^2 + 1)^2}$

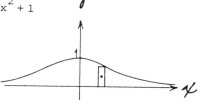

$\quad x = \tan\theta\ .\quad dx = \sec^2\theta\, d\theta$

$$= \int_0^{\pi/2} \frac{\sec^2\theta \, d\theta}{\sec^4\theta} = \int_0^{\pi/2} \cos^2\theta \, d\theta = \frac{1}{2}\int_0^{\pi/2}(1 + \cos 2\theta)\, d\theta$$

$$= \frac{1}{2}(\theta + \frac{1}{2}\sin 2\theta)\Big|_0^{\pi/2} = \frac{\pi}{4}$$

$\bar{x} = 0$ by symmetry, $\quad \bar{y} = \frac{\pi/4}{\pi} = \frac{1}{4}$

17. $A = 2\int_0^2 (4 - x^2 - \frac{x^2}{4} + 1)\, dx$

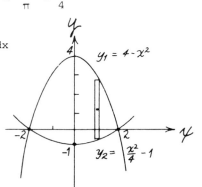

$$= 2\int_0^2 (5 - \frac{5x^2}{4})\, dx$$

$$= 2(5x - \frac{5x^3}{12})\Big|_0^2$$

$$= 2(10 - \frac{10}{3}) = \frac{40}{3}$$

21. $\sqrt{x} + \sqrt{y} = \sqrt{a} \qquad y = (\sqrt{a} - \sqrt{x})^2 = a - 2\sqrt{ax} + x$

$$A = \int_0^a (a - 2\sqrt{ax} + x)\, dx = (ax - \frac{4}{3}\sqrt{a}\, x^{3/2} + \frac{x^2}{2})\Big|_0^a$$

$$= a^2 - \frac{4}{3}a^2 + \frac{1}{2}a^2 = \frac{a^2}{6}$$

$$M_y = \int_0^a (ax - 2\sqrt{a}\, x^{3/2} + x^2)\, dx$$

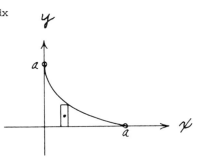

$$= (\frac{ax^2}{2} - \frac{4\sqrt{a}\, x^{5/2}}{5} + \frac{x^3}{3})\Big|_0^a$$

$$= \frac{a^3}{2} - \frac{4a^3}{5} + \frac{a^3}{3}$$

$$= \frac{15 - 24 + 10}{30}a^3 = \frac{a^3}{30}$$

$\bar{x} = \bar{y} = \dfrac{a^3/30}{a^2/6} = \dfrac{a}{5}$

25. $A = \int_0^1 x\, dy = \int_0^1 \sqrt{y}\, dy = \frac{2}{3}y^{3/2}\Big|_0^1 = \frac{2}{3}$

$$M_x = \int_0^1 xy\, dy = \int_0^1 y^{3/2}\, dy = \frac{2}{5}y^{5/2}\Big|_0^2 = \frac{2}{5}$$

$\bar{y} = \dfrac{2/5}{2/3} = \dfrac{3}{5}$

$$V = \frac{2}{3} \cdot 2\pi \cdot \frac{3}{5} = \frac{4\pi}{5}$$

$$V = \int_0^1 2\pi \; xy \; dy$$

$$= 2\pi \int_0^1 y^{3/2} dy$$

$$= 2\pi \cdot \frac{2}{5} \; y^{5/2} \Big|_0^1 = \frac{4\pi}{5}$$

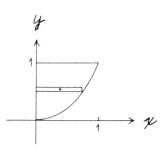

Section 16.8, pp. 498-499

1. $V = \int_0^1 \pi \; y^2 dx = \pi \int_0^1 x^6 dx$

$$= \frac{\pi x^7}{7} \Big|_0^1 = \frac{\pi}{7}$$

$$M_{yz} = \int_0^1 \pi \; y^2 x \; dx = \pi \int_0^1 x^7 dx$$

$$= \frac{\pi x^8}{8} \Big|_0^1 = \frac{\pi}{8}$$

$$\bar{x} = \frac{\pi/8}{\pi/7} = \frac{7}{8}$$

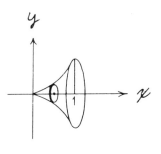

5. $V = \int_0^2 \pi \; x^2 dy = \pi \int_0^2 (1 - \frac{y}{2})^2 dy$

$$= \pi \int_0^2 (1 - y + \frac{y^2}{4}) dy$$

$$= \pi (y - \frac{y^2}{2} + \frac{y^3}{12}) \Big|_0^2$$

$$= \pi (2 - 2 + \frac{2}{3}) = \frac{2\pi}{3}$$

$y = 2 - 2x$

$x = 1 - \frac{y}{2}$

$$M_{xz} = \int_0^2 \pi \; x^2 y \; dy$$

$$= \pi \int_0^2 (y - y^2 + \frac{y^3}{4}) dy$$

$$= \pi (\frac{y^2}{2} - \frac{y^3}{3} + \frac{y^4}{16}) \Big|_0^2 = \pi (2 - \frac{8}{3} + 1) = \frac{\pi}{3}$$

$$\bar{y} = \frac{\pi/3}{2\pi/3} = \frac{1}{2}$$

9. $\dfrac{x^2}{a^2} + \dfrac{y^2}{b^2} = 1$

$V = \displaystyle\int_0^a \pi\, y^2 dx = \pi\int_0^a \dfrac{b^2}{a^2}(a^2 - x^2)\,dx$

$\quad = \dfrac{\pi b^2}{a^2}(a^2 x - \dfrac{x^3}{3})\Big|_0^a = \dfrac{\pi b^2}{a^2}\dfrac{2a^3}{3} = \dfrac{2\pi a b^2}{3}$

$M_{yz} = \displaystyle\int_0^a \pi y^2 x\,dx = \dfrac{\pi b^2}{a^2}\int_0^a (a^2 x - x^3)\,dx$

$\quad = \dfrac{\pi b^2}{a^2}(\dfrac{a^2 x^2}{2} - \dfrac{x^4}{4})\Big|_0^a$

$\quad = \dfrac{\pi b^2}{a^2}\dfrac{a^4}{4} = \dfrac{\pi a^2 b^2}{4}$

$\bar{x} = \dfrac{\pi a^2 b^2/4}{2\pi a b^2/3} = \dfrac{3a}{8}$

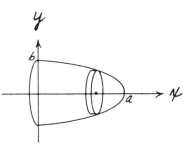

13. $V = \displaystyle\int_0^{+\infty} \pi y^2 dx$

$\quad = \pi\displaystyle\int_0^{+\infty} \dfrac{dx}{(1 + x^2)^2}$

$\quad x = \tan\theta, \ \ dx = \sec^2\theta\,d\theta$

$\quad = \pi\displaystyle\int_0^{\pi/2} \dfrac{\sec^2\theta\,d\theta}{\sec^4\theta}$

$\quad = \pi\displaystyle\int_0^{\pi/2} \cos^2\theta\,d\theta$

$\quad = \dfrac{\pi}{2}\displaystyle\int_0^{\pi/2} (1 + \cos 2\theta)\,d\theta$

$\quad = \dfrac{\pi}{2}(\theta + \dfrac{1}{2}\sin 2\theta)\Big|_0^{\pi/2} = \dfrac{\pi^2}{4}$

$M_{yz} = \displaystyle\int_0^{+\infty} \pi\, y^2 x\,dx = \lim_{k\to+\infty} \pi\int_0^k \dfrac{x}{(1 + x^2)^2}\,dx$

$\quad = \lim_{k\to+\infty} -\dfrac{\pi}{2}\dfrac{1}{1 + x^2}\Big|_0^k = \lim_{k\to+\infty} (\dfrac{\pi}{2} - \dfrac{\pi}{2(1 + k^2)}) = \dfrac{\pi}{2}$

$\bar{x} = \dfrac{\pi/2}{\pi^2/4} = \dfrac{2}{\pi}$

17. $V = \int_0^1 \pi(e^2 - x^2)\,dy$

$\quad = \pi\int_0^1 (e^2 - e^{2y})\,dy$

$\quad = \pi(e^2 y - \frac{1}{2}e^{2y})\Big|_0^1$

$\quad = \pi(e^2 - \frac{1}{2}e^2 + \frac{1}{2})$

$\quad = \frac{\pi}{2}(e^2 + 1)$

$M_{xz} = \int_0^1 \pi y(e^2 - x^2)\,dy$

$\quad = \pi\int_0^1 (ye^2 - ye^{2y})\,dy$

$\qquad u = y \qquad\qquad v' = -e^{2y}$

$\qquad u' = 1 \qquad\qquad v = -\frac{1}{2}e^{2y}$

$\quad = \pi(\frac{e^2 y^2}{2} - \frac{y}{2}e^{2y} + \frac{1}{4}e^{2y})\Big|_0^1$

$\quad = \pi(\frac{e^2}{2} - \frac{e^2}{2} + \frac{e^2}{4} - \frac{1}{4}) = \frac{\pi}{4}(e^2 - 1)$

$\bar{y} = \dfrac{\pi(e^2 - 1)/4}{\pi(e^2 + 1)/2} = \dfrac{e^2 - 1}{2(e^2 + 1)}$

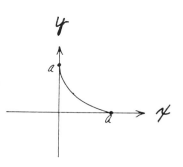

Section 16.9, pp. 502-503

1. $y^{2/3} = a^{2/3} - x^{2/3}$

$\quad y = (a^{2/3} - x^{2/3})^{3/2}$

$\quad y' = \frac{3}{2}(a^{2/3} - x^{2/3})^{1/2}(-\frac{2}{3}x^{-1/3})$

$\qquad = -x^{-1/3}(a^{2/3} - x^{2/3})^{1/2}$

$\quad s = \int_0^a \sqrt{1 + \dfrac{a^{2/3} - x^{2/3}}{x^{2/3}}}\;dx$

$$= \int_0^a \frac{a^{1/3}}{x^{1/3}} \, dx = \frac{3}{2} a^{1/3} x^{2/3} \Big|_0^a = \frac{3}{2} a$$

$$M_y = \int_0^a x \frac{a^{1/3}}{x^{1/3}} \, dx = a^{1/3} \int_0^a x^{2/3} \, dx = \frac{3}{5} a^{1/3} a^{5/3} \Big|_0^a = \frac{3}{5} a^2$$

$$\bar{x} = \bar{y} = \frac{3a^2/5}{3a/2} = \frac{2a}{5}$$

5. $\quad y = \dfrac{x^2}{8} \qquad\qquad\qquad y' = \dfrac{x}{4}$

$$s = \int_{-4}^4 \sqrt{1 + \frac{x^2}{16}} \, dx$$

$$x = 4 \tan \theta, \qquad dx = 4 \sec^2 \theta \, d\theta$$

$$= \int_{-\pi/4}^{\pi/4} \sqrt{1 + \tan^2 \theta} \; 4 \sec^2 \theta \, d\theta = 4 \int_{-\pi/4}^{\pi/4} \sec^3 \theta \, d\theta$$

$$= 2(\sec \theta \tan \theta + \ln|\sec \theta + \tan \theta|) \Big|_{-\pi/4}^{\pi/4}$$

$$= 2[\sqrt{2} + \ln(1 + \sqrt{2}) + \sqrt{2} - \ln|\sqrt{2} - 1|]$$

$$= 2(2\sqrt{2} + \ln \frac{\sqrt{2} + 1}{\sqrt{2} - 1}) = 2[2\sqrt{2} + \ln(\sqrt{2} + 1)^2]$$

$$= 4[\sqrt{2} + \ln(\sqrt{2} + 1)]$$

$$M_x = \int_{-4}^4 \frac{x^2}{8} \sqrt{1 + \frac{x^2}{16}} \, dx$$

$$u = x \qquad\qquad v' = \frac{x}{8} \sqrt{1 + \frac{x^2}{16}}$$

$$u' = 1 \qquad\qquad v = \frac{2}{3} (1 + \frac{x^2}{16})^{3/2}$$

$$= \frac{2}{3} x (1 + \frac{x^2}{16})^{3/2} \Big|_{-4}^4 - \frac{2}{3} \int_{-4}^4 (1 + \frac{x^2}{16})^{3/2} \, dx$$

$$x = 4 \tan \theta, \qquad dx = 4 \sec^2 \theta \, d\theta$$

$$= \frac{32\sqrt{2}}{3} - \frac{2}{3} \int_{-\pi/4}^{\pi/4} (1 + \tan^2\theta)^{3/2} 4 \sec^2\theta \, d\theta$$

$$= \frac{32\sqrt{2}}{3} - \frac{8}{3} \int_{-\pi/4}^{\pi/4} \sec^2\theta \, d\theta$$

$$= \frac{32\sqrt{2}}{3} - \frac{4}{3} \left(\sec\theta \tan\theta + \ln|\sec\theta + \tan\theta| \right) \Big|_{-\pi/4}^{\pi/4}$$

$$= \frac{32\sqrt{2}}{3} - \frac{4}{3} \left(2\sqrt{2} + \ln\frac{\sqrt{2}+1}{\sqrt{2}-1} \right) = \frac{8[3\sqrt{2} - \ln(\sqrt{2}+1)]}{3}$$

$\bar{x} = 0$ by symmetry,

$$\bar{y} = \frac{8[3\sqrt{2} - \ln(\sqrt{2}+1)]/3}{4[\sqrt{2} + \ln(\sqrt{2}+1)]} = \frac{2[3\sqrt{2} - \ln(\sqrt{2}+1)]}{3[\sqrt{2} + \ln(\sqrt{2}+1)]}$$

9. $\quad x = a(\theta - \sin\theta) \qquad\qquad y = a(1 - \cos\theta)$

$\quad x' = a(1 - \cos\theta) \qquad\qquad y' = a\sin\theta$

$$s = \int_0^\pi \sqrt{a^2(1 - 2\cos\theta + \cos^2\theta + \sin^2\theta)} \, d\theta$$

$$= \sqrt{2}a \int_0^\pi \sqrt{1 - \cos\theta} \, d\theta$$

$$u = \sqrt{1 - \cos\theta} \qquad d\theta = \frac{2u}{\sqrt{1 - (1-u^2)^2}} \, du$$

$$u^2 = 1 - \cos\theta$$

$$\cos\theta = 1 - u^2 \qquad\qquad = \frac{2u}{\sqrt{2u^2 - u^4}} \, du$$

$$\theta = \text{Arccos}(1 - u^2) \qquad\qquad = \frac{2}{\sqrt{2 - u^2}} \, du$$

$$= \sqrt{2}a \int_0^{\sqrt{2}} \frac{2u\,du}{\sqrt{2 - u^2}} = -2\sqrt{2}a \sqrt{2 - u^2} \Big|_0^{\sqrt{2}} = 4a$$

$$M_x = \sqrt{2}a^2 \int_0^\pi (1 - \cos\theta)^{3/2} d\theta$$

$$u = \sqrt{1 - \cos\theta}, \qquad d\theta = \frac{2du}{\sqrt{2 - u^2}}$$

$$= \sqrt{2}a^2 \int_0^{\sqrt{2}} \frac{2u^3 du}{\sqrt{2 - u^2}}$$

$$p = u^2 \qquad\qquad q' = \frac{2u}{\sqrt{2 - u^2}}$$

$$p' = 2u \qquad\qquad q = -2\sqrt{2 - u^2}$$

$$= \sqrt{2}a^2[-2u^2\sqrt{2 - u^2}\Big|_0^{\sqrt{2}} + 2\int_0^{\sqrt{2}} 2u\sqrt{2 - u^2}\,du]$$

$$= \sqrt{2}a^2[-2u^2\sqrt{2 - u^2} - \frac{4}{3}(2 - u^2)^{3/2}]\Big|_0^{\sqrt{2}}$$

$$= \sqrt{2}a^2 \cdot \frac{4}{3} \, 2\sqrt{2} = \frac{16a^2}{3}$$

$$M_y = \sqrt{2}a^2\int_0^{\pi} (\theta - \sin\theta)\sqrt{1 - \cos\theta}\,d\theta$$

$$= \sqrt{2}a^2[\int_0^{\pi} \theta\sqrt{1 - \cos\theta}\,d\theta - \frac{2}{3}(1 - \cos\theta)^{3/2}\Big|_0^{\pi}]$$

$$u = \theta \qquad\qquad v' = \sqrt{1 - \cos\theta}$$

$$u' = 1 \qquad\qquad v = -2\sqrt{1 + \cos\theta}$$

$$\text{(by substitution, } x = \sqrt{1 - \cos\theta})$$

$$= \sqrt{2}a^2[-2\theta\sqrt{1 + \cos\theta}\Big|_0^{\pi} + 2\int_0^{\pi}\sqrt{1 + \cos\theta}\,d\theta - \frac{4\sqrt{2}}{3}]$$

$$u = \sqrt{1 + \cos\theta} \qquad d\theta = \frac{-2u\,du}{\sqrt{1 - (u^2 - 1)^2}}$$

$$u^2 = 1 + \cos\theta$$

$$\cos\theta = u^2 - 1 \qquad\qquad = \frac{-2u\,du}{\sqrt{2u^2 - u^4}}$$

$$\theta = \text{Arccos}(u^2 - 1) \qquad = \frac{-2\,du}{\sqrt{2 - u^2}}$$

$$= \sqrt{2}a^2[-\frac{4\sqrt{2}}{3} + 2\int_{\sqrt{2}}^{0} \frac{-2u\,du}{\sqrt{2 - u^2}}]$$

$$= \sqrt{2}a^2[\frac{-4\sqrt{2}}{3} + (4\sqrt{2 - u^2}\Big|_{\sqrt{2}}^{0})] = \sqrt{2}a^2[-\frac{4\sqrt{2}}{3} + 4\sqrt{2}] = \frac{16a^2}{3}$$

$$\bar{x} = \frac{16a^2/3}{4a} = \frac{4a}{3}, \quad \bar{y} = \frac{16a^2/3}{4a} = \frac{4a}{3}$$

13. $y = \cosh x \qquad\qquad y' = \sinh x$

$$S_x = \int_{-1}^{1} 2\pi y\sqrt{1 + \sinh^2 x}\,dx = 2\pi\int_{-1}^{1}\cosh^2 x\,dx$$

$$= \pi\int_{-1}^{1}(1 + \cosh 2x)\,dx = \pi(x + \frac{1}{2}\sinh 2x)\Big|_{-1}^{1}$$

$$= \pi(x + \sinh x \cosh x)\Big|_{-1}^{1} = 2\pi(1 + \sinh 1 \cosh 1)$$

$$M_{yz} = \int_{-1}^{1} 2\pi x \cosh^2 x \, dx$$

$$u = 2\pi x \qquad\qquad v' = \cosh^2 x$$

$$u' = 2\pi \qquad\qquad v = \frac{1}{2}(x + \sinh x \cosh x)$$

$$= \pi x(x + \sinh x \cosh x)\Big|_{-1}^{1} - \pi \int_{-1}^{1}(x + \sinh x \cosh x)\, dx$$

$$= -\pi\left(\frac{x^2}{2} + \frac{\cosh^2 x}{2}\right)\Big|_{-1}^{1} = 0$$

$$\bar{x} = 0$$

17. $x = \sqrt{y}, \qquad\qquad x' = \dfrac{1}{2\sqrt{y}}$

$$S_y = \int_0^4 2\pi\sqrt{y}\sqrt{1 + \frac{1}{4y}}\, dy = \int_0^4 \pi\sqrt{4y + 1}\, dy$$

$$= \frac{\pi}{4}\frac{2}{3}(4y + 1)^{3/2}\Big|_0^4 = \frac{\pi}{6}(17\sqrt{17} - 1)$$

$$M_{xz} = \int_0^4 \pi y\sqrt{4y + 1}\, dy$$

$$u = \pi y \qquad\qquad v' = \sqrt{4y + 1}$$

$$u' = \pi \qquad\qquad v = \frac{1}{6}(4y + 1)^{3/2}$$

$$= \frac{\pi}{6}y(4y + 1)^{3/2}\Big|_0^4 - \frac{\pi}{6}\int_0^4 (4y + 1)^{3/2}\, dy$$

$$= \left[\frac{\pi}{6}y(4y + 1)^{3/2} - \frac{\pi}{60}(4y + 1)^{5/2}\right]\Big|_0^4$$

$$= \frac{2\pi}{3}17\sqrt{17} - \frac{\pi}{60}17^2\sqrt{17} + \frac{\pi}{60} = \frac{\pi(391\sqrt{17} + 1)}{60}$$

$$\bar{y} = \frac{\pi(391\sqrt{17} + 1)/60}{\pi(17\sqrt{17} - 1)/6} = \frac{391\sqrt{17} + 1}{10(17\sqrt{17} - 1)}$$

$$y' = \frac{-x}{\sqrt{1 - x^2}}$$

$$\bar{y} = \frac{2}{\pi} \quad \text{(by Example 1, p. 409)}$$

$$(2\pi \cdot \frac{2}{\pi}) = 4\pi$$

$$x = \int_{-1}^{1} 2\pi\sqrt{1 - x^2}\,\sqrt{1 + \frac{x^2}{1 - x^2}}\ dx = 2\pi\int_{-1}^{1} dx$$

$$= 2\pi x\Big|_{-1}^{1} = 4\pi$$

Section 16.10, pp. 505-506

1. $I_y = 5 \cdot 2^2 + 2(-1)^2 = 22$, $\qquad R = \sqrt{\frac{I}{m}} = \sqrt{\frac{22}{7}}$

5. $m = \int_0^1 3(1 - y)\,dx$

    $\qquad = 3\int_0^1 (1 - x^2)\,dx = 3(x - \frac{x^3}{3})\Big|_0^1$

    $\qquad = 3(1 - \frac{1}{3}) = 2$

    $I_y = \int_0^1 3x^2(1 - y)\,dx$

    $\qquad = 3\int_0^1 (x^2 - x^4)\,dx$

    $\qquad = 3(\frac{x^3}{3} - \frac{x^5}{5})\Big|_0^1 = 3(\frac{1}{3} - \frac{1}{5}) = \frac{2}{5}$

    $R = \sqrt{\frac{I}{m}} = \sqrt{\frac{2}{15}} = \frac{1}{\sqrt{5}}$

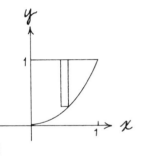

9. $m = \int_0^1 y\,dx = \int_0^1 (x - x^2)\,dx$

    $\qquad = (\frac{x^2}{2} - \frac{x^3}{3})\Big|_0^1 = \frac{1}{2} - \frac{1}{3} = \frac{1}{6}$

    $I_y = \int_0^1 x^2 y\,dx = \int_0^1 (x^3 - x^4)\,dx$

    $\qquad = (\frac{x^4}{4} - \frac{x^5}{5})\Big|_0^1 = \frac{1}{4} - \frac{1}{5} = \frac{1}{20}$

    $R = \sqrt{\frac{I}{m}} = \sqrt{\frac{1/20}{1/6}} = \sqrt{\frac{3}{10}}$

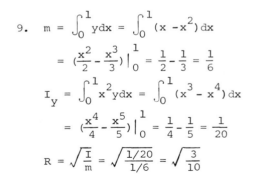

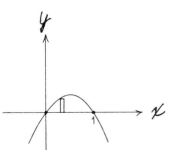

13. $m = \int_0^\pi y\,dx = \int_0^\pi \sin x\,dx = -\cos x\Big|_0^\pi = 2$

$I_y = \int_0^\pi x^2 y\,dx = \int_0^\pi x^2 \sin x\,dx$

u	v'
x^2	$\sin x$
$2x$	$-\cos x$
2	$-\sin x$
0	$\cos x$

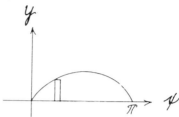

$= (-x^2 \cos x + 2x \sin x + 2 \cos x)\Big|_0^\pi = \pi^2 - 2 - 2 = \pi^2 - 4$

$R = \sqrt{\dfrac{I}{m}} = \sqrt{\dfrac{\pi^2 - 4}{2}} = \sqrt{\dfrac{\pi^2}{2} - 2}$

17. $m = \int_1^2 8y\,dx = 8\int_1^2 \dfrac{dx}{x} = 8\ln x\Big|_1^2$

$= 8\ln 2$

$I_y = \int_1^2 8x^2 y\,dx = 8\int_1^2 x\,dx = 4x^2\Big|_1^2$

$= 4(4 - 1) = 12$

$R = \sqrt{\dfrac{I}{m}} = \sqrt{\dfrac{12}{8\ln 2}} = \sqrt{\dfrac{3}{2\ln 2}}$

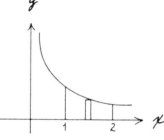

21. $m = 2\pi$ (from Problem 20)

$I_x = 4\int_0^2 y^2\sqrt{1 - \dfrac{y^2}{4}}\,dy$

$u = y \qquad v' = y\sqrt{1 - \dfrac{y^2}{4}}$

$u' = 1 \qquad v = -\dfrac{4}{3}\left(1 - \dfrac{y^2}{4}\right)^{3/2}$

$= 4\left[-\dfrac{4y}{3}\left(1 - \dfrac{y^2}{4}\right)^{3/2}\Big|_0^2 + \dfrac{4}{3}\int_0^2 \left(1 - \dfrac{y^2}{4}\right)^{3/2} dy\right]$

$= \dfrac{16}{3}\int_0^2 \left(1 - \dfrac{y^2}{4}\right)^{3/2} dy$

$$y = 2\sin\theta, \quad dy = 2\cos\theta\,d\theta$$

$$= \frac{16}{3}\int_0^{\pi/2} 2\cos^4\theta\,d\theta$$

$$= \frac{8}{3}\int_0^{\pi/2} (1 + \cos 2\theta)^2\,d\theta$$

$$= \frac{8}{3}\int_0^{\pi/2} (1 + 2\cos 2\theta + \cos^2 2\theta)\,d\theta$$

$$= \frac{8}{3}\int_0^{\pi/2} (1 + 2\cos 2\theta + \frac{1 + \cos 4\theta}{2})\,d\theta$$

$$= \frac{8}{3}\int_0^{\pi/2} (\frac{3}{2} + 2\cos 2\theta + \frac{1}{2}\cos 4\theta)\,d\theta$$

$$= \frac{8}{3}(\frac{3\theta}{2} + \sin 2\theta + \frac{1}{8}\sin 4\theta)\Big|_0^{\pi/2} = \frac{8}{3}\,\frac{3\pi}{4} = 2\pi$$

$$R = \sqrt{\frac{I}{m}} = \sqrt{\frac{2\pi}{2\pi}} = 1$$

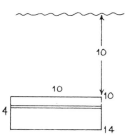

Section 16.11, pp. 509-510

1. $F = \displaystyle\int_{10}^{14} 62.4x \cdot 10\,dx = 624\,\frac{x^2}{2}\Big|_{10}^{14}$

   $= 312(196 - 100) = 29{,}952$ lb

5. $F = 2\displaystyle\int_{-1/2}^{1/2} 62.4(\frac{1}{2} - y)x\,dy = 124.8\int_{-1/2}^{1/2} (\frac{1}{2} - y)\sqrt{\frac{1}{4} - y^2}\,dy$

   $= 62.4\displaystyle\int_{-1/2}^{1/2} \sqrt{\frac{1}{4} - y^2}\,dy + 124.8\int_{-1/2}^{1/2} -y\sqrt{\frac{1}{4} - y^2}\,dy$

   $y = \frac{1}{2}\sin\theta, \quad dy = \frac{1}{2}\cos\theta\,d\theta$

   $= 62.4\displaystyle\int_{-\pi/2}^{\pi/2} \frac{1}{4}\cos^2\theta\,d\theta$

   $\quad + (124.8)\frac{1}{3}(\frac{1}{4} - y^2)^{3/2}\Big|_{-1/2}^{1/2}$

   $= \dfrac{62.4}{8}\displaystyle\int_{-\pi/2}^{\pi/2} (1 + \cos 2\theta)\,d\theta$

   $= 7.8(\theta + \frac{1}{2}\sin 2\theta)\Big|_{-\pi/2}^{\pi/2}$

   $= 7.8\pi = 24.5$ lb

9. $F = 2\int_{-1}^{0} -62.4yx\,dy$

$= 2\int_{-1}^{0} -62.4y(\frac{1+y}{\sqrt{3}})\,dy$

$= -124.8\int_{-1}^{0} (\frac{y+y^2}{\sqrt{3}})\,dy$

$= \frac{-124.8}{\sqrt{3}}(\frac{y^2}{2} + \frac{y^3}{3})\Big|_{-1}^{0}$

$= \frac{124.8}{\sqrt{3}}(\frac{1}{2} - \frac{1}{3}) = \frac{20.8}{3}\sqrt{3}$

$= 12.0$ lb

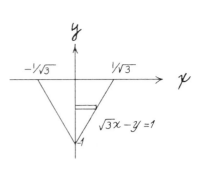

$\sqrt{3}x - y = 1$

13. $F = 2\int_{-3}^{0} -62.4yx\,dy$

$= -124.8\int_{-3}^{0} y\cdot\frac{5}{3}\sqrt{9-y^2}\,dy$

$= 104\int_{-3}^{0} -2y\sqrt{9-y^2}\,dy$

$= \frac{208}{3}(9-y^2)^{3/2}\Big|_{-3}^{0}$

$= \frac{208}{3}\cdot 27 = 1872$ lb

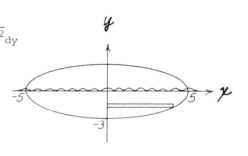

17. $y = kx^2$, $10 = 16k$

$k = \frac{10}{16} = \frac{5}{8}$

$y = \frac{5}{8}x^2$, $x = \frac{2\sqrt{2}}{\sqrt{5}}\sqrt{y}$

$F = 2\int_{0}^{10} 62.4(10-y)x\,dy$

$= 124.8\int_{0}^{10} (10-y)\frac{2\sqrt{2}}{\sqrt{5}}\sqrt{y}\,dy$

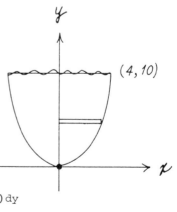

$(4, 10)$

$= \frac{249.6\sqrt{2}}{\sqrt{5}}\int_{0}^{10} (10y^{1/2} - y^{3/2})\,dy$

$= \frac{249.6}{5}\sqrt{10}(\frac{20}{3}y^{3/2} - \frac{2}{5}y^{5/2})\Big|_{0}^{10}$

$= \frac{249.6}{5}\sqrt{10}(\frac{200\sqrt{10}}{3} - \frac{200\sqrt{10}}{5}) = 13,300$ lb

21. $F = \int_0^4 62.4(20\pi)x\,dx$

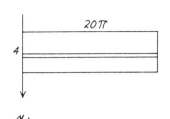

$= 624\pi x^2 \big|_0^4 = 624\pi \cdot 16$

$= 9984\pi = 31,350$ lb

25. $F = \int_{y_1}^{y_2} 62.4y(x_1 - x_2)\,dy$

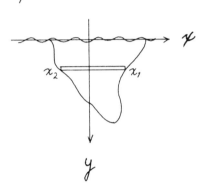

$= 62.4 \int_{y_1}^{y_2} y(x_1 - x_2)\,dy$

$M_x = A\bar{y} = \int_{y_1}^{y_2} y(x_1 - x_2)\,dy$

$F = 62.4A\bar{y}$

Review 16, pp. 510–511

1. $\int_{-\infty}^{+\infty} \frac{x\,dx}{x^4+1} = 0$ provided $\int_0^{+\infty} \frac{x\,dx}{x^4+1}$ converges since

$\frac{x}{x^4+1}$ is an odd function.

$\int_0^{+\infty} \frac{x\,dx}{x^4+1} = \lim_{k\to+\infty} \frac{1}{2}\int_0^k \frac{2x\,dx}{1+x^4} = \lim_{k\to+\infty} \frac{1}{2} \operatorname{Arctan} x^2 \big|_0^k$

$= \lim_{k\to+\infty} \frac{1}{2} \operatorname{Arctan} k^2 = \frac{\pi}{4}$. Hence $\int_{-\infty}^{+\infty} \frac{x\,dx}{x^4+1} = 0$

5. $V = \int_0^2 \pi y^2\,dx = \pi \int_0^2 (2x - x^2)^2\,dx$

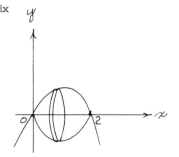

$= \pi \int_0^2 (4x^2 - 4x^3 + x^4)\,dx$

$= \pi\left(\frac{4x^3}{3} - x^4 + \frac{x^5}{5}\right)\big|_0^2$

$= \pi\left(\frac{32}{3} - 16 + \frac{32}{5}\right) = \frac{16\pi}{15}$

9. $V = 2\pi \int_0^2 (3 - x) y \, dx$

$= 2\pi \int_0^2 (3 - x) 3x^2 \, dx$

$= 6\pi \int_0^2 (3x^2 - x^3) \, dx$

$= 6\pi (x^3 - \frac{x^4}{4}) \Big|_0^2$

$= 6\pi (8 - 4) = 24\pi$

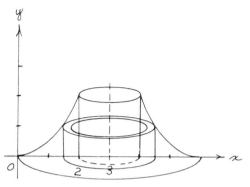

13. $V = \int_0^1 \frac{1}{2}\pi x^2 \, dy$

$= \frac{\pi}{2} \int_0^1 (1 - y)^2 \, dy$

$= -\frac{\pi}{2} \frac{(1 - y)^3}{3} \Big|_0^1$

$= \frac{\pi}{6}$

17. $y = \frac{2}{3}(x^2 - 1)^{3/2}$, $y' = 2x(x^2 - 1)^{1/2}$

$S_y = 2\pi \int_1^2 x \sqrt{1 + 4x^2(x^2 - 1)} \, dx$

$= 2\pi \int_1^2 x(2x^2 - 1) \, dx = 2\pi \int_1^2 (2x^3 - x) \, dx$

$= 2\pi (\frac{x^4}{2} - \frac{x^2}{2}) \Big|_1^2 = \pi(16 - 4) = 12\pi$

21. $m = 12 + 16 + 12 = 40$

$M_x = 12 \cdot 2 + 16 \cdot 2 + 12 \cdot \frac{3}{2} = 74$

$M_y = 12(-1) + 16 \cdot 2 + 12 \cdot 5 = 80$

$\bar{x} = \dfrac{M_y}{m} = \dfrac{80}{40} = 2$, $\qquad \bar{y} = \dfrac{M_x}{m} = \dfrac{74}{40} = \dfrac{37}{20}$

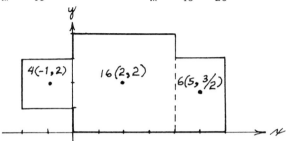

25. $A = \displaystyle\int_0^1 (y_1 - y_2)\,dx$

$= \displaystyle\int_0^1 (x - x^n)\,dx$

$= \left.\left(\dfrac{x^2}{2} - \dfrac{x^{n+1}}{n+1}\right)\right|_0^1$

$= \dfrac{1}{2} - \dfrac{1}{n+1} = \dfrac{n-1}{2(n+1)}$

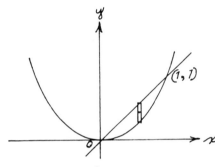

$M_x = \displaystyle\int_0^1 (y_1 - y_2)(y_1 + y_2)/2 \ dx = \dfrac{1}{2}\int_0^1 (y_1^2 - y_2^2)\,dx$

$= \dfrac{1}{2}\displaystyle\int_0^1 (x^2 - x^{2n})\,dx = \dfrac{1}{2}\left.\left(\dfrac{x^3}{3} - \dfrac{x^{2n+1}}{2n+1}\right)\right|_0^1$

$= \dfrac{1}{2}\left(\dfrac{1}{3} - \dfrac{1}{2n+1}\right) = \dfrac{n-1}{3(2n+1)}$

$M_y = \displaystyle\int_0^1 (y_1 - y_2)x \ dx = \int_0^1 (x^2 - x^{n+1})\,dx$

$= \left.\left(\dfrac{x^3}{3} - \dfrac{x^{n+2}}{n+2}\right)\right|_0^1 = \dfrac{1}{3} - \dfrac{1}{n+2} = \dfrac{n-1}{3(n+2)}$

$\bar{x} = \dfrac{M_y}{A} = \dfrac{\dfrac{n-1}{3(n+2)}}{\dfrac{n-1}{2(n+1)}} = \dfrac{2(n+1)}{3(n+2)}$

$$\bar{y} = \frac{M_x}{A} = \frac{\dfrac{n-1}{3(2n+1)}}{\dfrac{n-1}{2(n+1)}} = \frac{2(n+1)}{3(2n+1)}$$

$$\lim_{n \to +\infty} \bar{x} = \frac{2}{3}, \qquad\qquad \lim_{n \to +\infty} \bar{y} = \frac{1}{3}$$

Note that this is the centroid of the triangle bounded by $y = x$, $y = 0$, and $x = 1$.

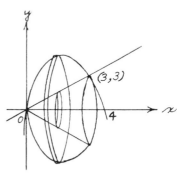

29. $$V = \pi \int_0^3 (y_1^2 - y_2^2)\,dx$$

$$= \pi \int_0^3 [(4x - x^2)^2 - x^2]\,dx$$

$$= \pi \int_0^3 (15x^2 - 8x^3 + x^4)\,dx$$

$$= \pi (5x^3 - 2x^4 + \frac{x^5}{5})\Big|_0^3$$

$$= \pi (135 - 162 + \frac{243}{5}) = \frac{108\pi}{5}$$

$$M_{yz} = \pi \int_0^3 x(y_1^2 - y_2^2)\,dx = \pi \int_0^3 (15x^3 - 8x^4 + x^5)\,dx$$

$$= \pi (\frac{15x^4}{4} - \frac{8x^5}{5} + \frac{x^6}{6})\Big|_0^3 = 81\pi (\frac{15}{4} - \frac{24}{5} + \frac{3}{2}) = \frac{729\pi}{20}$$

$$\bar{x} = \frac{M_{yz}}{V} = \frac{729\pi/20}{108\pi/5} = \frac{27}{16}$$

33. $$x = \cos^3\theta \qquad\qquad y = \sin^3\theta$$

$$x' = -3\cos^2\theta\,\sin\theta \qquad y' = 3\sin^2\theta\,\cos\theta$$

$$(x')^2 + (y')^2 = 9\sin^2\theta\,\cos^4\theta + 9\sin^4\theta\,\cos^2\theta$$

$$= 9\sin^2\theta\,\cos^2\theta(\cos^2\theta + \sin^2\theta)$$

$$= 9\sin^2\theta\,\cos^2\theta$$

$$S_x = 2\pi \int_0^{\pi/2} y\sqrt{(x')^2 + (y')^2}\, d\theta$$

$$= 2\pi \int_0^{\pi/2} \sin^3\theta \cdot 3 \sin\theta \cos\theta\, d\theta$$

$$= 2\pi \int_0^{\pi/2} 3 \sin^4\theta \cos\theta\, d\theta = \frac{6\pi}{5} \sin^5\theta \Big|_0^{\pi/2} = \frac{6\pi}{5}$$

$$M_{yz} = 2\pi \int_0^{\pi/2} xy\sqrt{(x')^2 + (y')^2}\, d\theta$$

$$= 2\pi \int_0^{\pi/2} \cos^3\theta \, \sin^3\theta \cdot 3 \sin\theta \cos\theta\, d\theta$$

$$= 6\pi \int_0^{\pi/2} \sin^4\theta \cos^4\theta\, d\theta = 6\pi \int_0^{\pi/2} \frac{\sin^4 2\theta}{16}\, d\theta$$

$$= \frac{3\pi}{8} \int_0^{\pi/2} \frac{(1 - \cos 4\theta)^2}{4}\, d\theta$$

$$= \frac{3\pi}{32} \int_0^{\pi/2} (1 - 2\cos 4\theta + \cos^2 4\theta)\, d\theta$$

$$= \frac{3\pi}{32} \int_0^{\pi/2} \left(1 - 2\cos 4\theta + \frac{1 + \cos 8\theta}{2}\right) d\theta$$

$$= \frac{3\pi}{64}\left(3\theta - \sin 4\theta + \frac{1}{8}\sin 8\theta\right)\Big|_0^{\pi/2} = \frac{3\pi}{64} \frac{3\pi}{2} = \frac{9\pi^2}{128}$$

$$\bar{x} = \frac{M_{yz}}{S_x} = \frac{9\pi^2/128}{6\pi/5} = \frac{15\pi}{256}$$

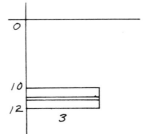

37. $F = \int_{10}^{12} (62.4x) 2\, dx$

$$= 62.4x^2 \Big|_{10}^{12} = 62.4(144 - 100)$$

$$= 2745.6 \text{ lb}$$

Chapter Seventeen
Limits and Continuity: The Epsilon-Delta Approach

<u>Section 17.1, p. 516</u>

1. Limit point

5. Not a limit point

9. Limit point

13. Suppose $0 < |x - 1| < \delta$.

$$|(x - 4) + 3| = |x - 1| < \delta$$

$$\delta = \epsilon$$

17. Suppose $0 < |x| < \delta$.

$$|x^2| < \delta^2 , \qquad \delta = \sqrt{\epsilon}$$

21. Suppose $0 < |x - 1| < \delta$ and $\delta \leqq 1$.

$|x - 1| < 1$ $|x| < 2$

$-1 < x - 1 < 1$ $|x^2 - x| = |x| \cdot |x - 1| < 2\delta$

$0 < x < 2$ $\delta = \min\{\epsilon/2, 1\}$

25. Suppose $0 < |x| < \delta$.

$$|x^3| < \delta^3 , \qquad \delta = \sqrt[3]{\epsilon}$$

29. Suppose $0 < |x| < \delta$ and $\delta \leqq \frac{1}{2}$.

$|x| < \frac{1}{2}$ $|x + 1| > \frac{1}{2}$

$-\frac{1}{2} < x < \frac{1}{2}$ $\frac{1}{|x + 1|} < 2$

$-\frac{3}{2} < x - 1 < -\frac{1}{2}$ $\left|\frac{1}{x - 1} + 1\right| = \left|\frac{x}{x - 1}\right| = \frac{1}{|x - 1|}|x| < 2\delta$

$$\delta = \min\{\epsilon/2, 1/2\}$$

1. Choose $\epsilon = 1$. If $\delta > 0$, let $1 - \delta < x < 1$.

   $-\delta < x - 1 < 0$
   $0 < |x - 1| < \delta$

   $x - 2 < -1$
   $|(x+1) - 3| > 1 = \epsilon$
   $|f(x) - 3| \not< \epsilon$

5. Choose $\epsilon = 1$. If $\delta > 0$, let $0 < x < \delta$.

   $0 < |x| < \delta$

   $f(x) = 1 = \epsilon$
   $|f(x)| \not< \epsilon$

9. Choose $\epsilon = 1$. If $\delta > 0$, let $0 < x < \min\{\delta, 1\}$.

   $0 < |x| < \delta$

   $|x^2| < 1$
   $\left|\dfrac{1}{x^2}\right| > 1 = \epsilon$
   $|f(x)| \not< \epsilon$

13. Suppose $0 < |x - 1| < \delta$.

    $\left|\dfrac{x^2 - 1}{x - 1} - 2\right| = |(x+1) - 2| = |x - 1| < \delta$

    $\delta = \epsilon$ The limit statement is true.

17. Choose $\epsilon = 1$. If $\delta > 0$, let $0 < x < \min\{1, \delta\}$.

    $0 < |x| < \delta$

    $0 < x < 1$
    $\dfrac{1}{x} > 1 = \epsilon$
    $|f(x)| \not< \epsilon$

    The limit statement is false.

21. Suppose $\lim_{x \to 0} f(x) = b$ and $b \geq 1$. Choose $\epsilon = \dfrac{1}{2}$.
    If $\delta > 0$, let $\max\{-\delta, -1/\sqrt{2}\} < x < 0$.

$$-\delta < x < 0 \qquad\qquad -\frac{1}{\sqrt{2}} < x < 0$$

$$0 < |x| < \delta \qquad\qquad 0 < f(x) < \frac{1}{2}$$

$$f(x) - b < -\frac{1}{2} = -\epsilon$$

$$|f(x) - b| > \epsilon$$

$$|f(x) - b| \not< \epsilon$$

Section 17.3, pp. 526-527

1. Suppose $0 < x < \delta$. Then $f(x) = x$ and $|f(x)| = x$.
   Hence $|f(x) - 0| < \delta$. Let $\delta = \epsilon$. Then $|f(x) - 0| < \epsilon$
   whenever $0 < x < \delta$.

5. Suppose $0 < |x| < \delta$.

   $$\left|\frac{1}{x}\right| > \frac{1}{\delta} \qquad\qquad \frac{1}{x^2} > \frac{1}{\delta^2}$$

   $$\frac{-1}{x^2} < \frac{-1}{\delta^2}$$

   Choose δ such that $-1/\delta^2 = N$ or $\delta = 1/\sqrt{-N}$ when N is
   negative; choose any δ when $N \geq 0$.

9. Suppose $x > N$ (N positive).

   $$\frac{1}{x^2} < \frac{1}{N^2}$$

   Choose N such that $1/N^2 = \epsilon$ or $N = 1/\sqrt{\epsilon}$.

13. Suppose $x < N$.

    $$e^x < e^N$$

    Choose N such that $e^N = \epsilon$ or $N = \ell n\ \epsilon$.

17. Choose $N = 1$. If $\delta > 0$, let $-\delta < x < 0$.

    $$-\delta^3 < x^3 \qquad\qquad \frac{1}{x^3} < \frac{-1}{\delta^3} < N$$

    $$\frac{1}{x^3} \not> N$$

21. Choose $\epsilon = 1$. If N is a number, let $x > \max\{N, 1\}$.

   $x > N$ $\qquad\qquad\qquad$ $x > 1$

$$(x + 1) - 1 > 1 = \epsilon$$
$$\left|\frac{x^2 - 1}{x - 1} - 1\right| \not< \epsilon$$

25. Suppose $x > M$.

   $x^2 > M^2$

   Choose M such that $N = M^2$ or $M = \sqrt{N}$ if N is positive; choose any M if N is negative.

Section 17.4, p. 531

1. Suppose $|x| < \delta$.

   Then $|f(x)| = |x| < \delta$

   $\delta = \epsilon$

5. Suppose $|x| < \delta$.

   Then $|f(x) - 1| = \left|\frac{x^2 - 1}{x - 1} - 1\right| = |(x + 1) - 1| = |x| < \delta$

   $\delta = \epsilon$

9. Suppose $|x - 1| < \delta$ and $\delta \leq 1$. Then

   $|x - 1| < 1$ $\qquad\qquad$ $|x| < 2$

   $-1 < x - 1 < 1$ $\qquad$ $|x^2 + x + 1| \leq |x|^2 + |x| + 1 < 4 + 2 + 1 = 7$

   $\qquad 0 < x < 2$ $\qquad\quad$ $|x^3 - 1| = |x^2 + x + 1| \cdot |x - 1| < 7\delta$

   $$\delta = \min\{\epsilon/7, 1\}$$

13. Suppose $|x - 1| < \delta$ and $\delta < 1$.

   Then $|f(x) - 1| = |1 - 1| = 0 < \epsilon$

   δ can be any number less than 1.

17. Choose $\epsilon = 1$. If $\delta > 0$, let $0 < x < \delta$.

   $|x| < \delta$ $\qquad\qquad\qquad$ $x + 1 > 1 = \epsilon$

   $\qquad\qquad\qquad$ $|f(x)| \not< \epsilon$

21. Choose $\epsilon = 1$. If $\delta > 0$, let $0 < x < \delta$, $x = \dfrac{(2n+1)\pi}{2}$

$|x| < \delta$ $\qquad\qquad f(x) = \pm 1$

$\qquad\qquad\qquad\qquad |f(x)| = 1 = \epsilon$

$\qquad\qquad\qquad\qquad |f(x)| \not< \epsilon$

25. $f(x) = x^2$, $-1 < x < 2$, has
a minimum but no maximum.
But Theorem 17.5 again tells
us nothing about $f(x)$.

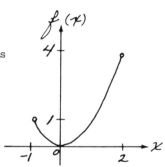

29. Suppose not. The velocity v cannot be less than
v_{av} for all $a \leqq t \leqq b$, nor can it be greater than v_{av}
for all $a \leqq t \leqq b$. Hence there must be times t_0
and t_1 such that $v(t_0) < v_{av} < v(t_1)$. Since the
velocity of a moving object is continuous, there is
a time t_2 between t_0 and t_1 such that $v(t_2) = v_{av}$.

Section 17.5, pp. 535-536

1. $a = 0$, $b = 4$

   f is continuous on $[0,4]$;
   $f'(x) = 2x - 4$ exists for
   all x in $(0,4)$; $f(0) = f(4)$
   $= 0$. Thus there is a
   number x_0 between 0 and 4
   such that $f'(x_0) = 0$.
   $f'(x_0) = 2x_0 - 4 = 0$ when $x_0 = 2$

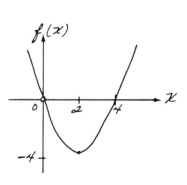

5. $a = 0$, $b = 1$

f is continuous on $[0,1]$;

$f'(x) = 3x^2 - 2x$ exists for all

x in $(0,1)$; $f(0) = f(1) = 0$.

Thus there is a number x_o

between 0 and 1 such that

$f'(x_o) = 0$.

$f'(x_o) = 3x_o^2 - 2x_o = 0$ when $x_o = \frac{2}{3}$

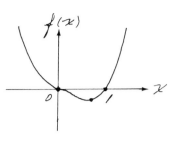

9. f is continuous on $[0,\pi]$;

$f'(x) = \cos x$ exists for all

x in $(0,\pi)$; $f(0) = f(\pi) = 0$.

Thus there is a number x_o

between 0 and π such that

$f'(x_o) = 0$.

$f'(x_o) = \cos x_o = 0$ when $x_o = \frac{\pi}{2}$

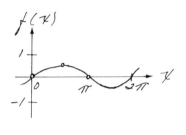

13. Rolle's theorem tells us nothing because $f'(0)$ does not exist ($f(x) = 0$ only at $x = -1$ and $x = 1$).

17. Rolle's theorem tells us that there is a number x_o between -1 and 1 such that $f'(x_o) = 0$.

21. Case II $f(x) < 0$ for some x between a and b:

By Theorem 17.3, there is a number x_o in $[a,b]$ such that $f(x_o) \leq f(x)$ for every x in $[a,b]$. Since $f(x) < 0$ for some x between a and b and $f(x_o) \leq f(x)$, it follows that $f(x) < 0$. Thus x_o is neither a nor b, since $f(a) = f(b) = 0$, and x_o is between a and b. Now define a function F by

$$F(x) = \frac{f(x_o) - f(x)}{x_o - x}$$

for all x except x_o in $[a,b]$. Since $f(x_o) \leq f(x)$ for all x in $[a,b]$,

$$f(x_o) - f(x) \leq 0$$

for all such values of x. If $x < x_o$, then $x_o - x > 0$

and $F(x) \leqq 0$. If $x > x_o$, then $x_o - x < 0$ and $F(x) \geqq 0$.
By definition,

$$\lim_{x \to x_o} F(x) = \lim_{x \to x_o} \frac{f(x_o) - f(x)}{x_o - x} = f'(x_o)$$

Since we are given that $f'(x)$ exists for all x in (a,b), we know that $\lim_{x \to x_o} F(x)$ is some number L. Let us assume L to be positive. If $\epsilon = L/2$, then, by the definition of limit, there is a positive number δ such that

$$\left| F(x) - L \right| < \frac{L}{2}$$

whenever

$$0 < \left| x - x_o \right| < \delta$$

Suppose $x > a$ and $x_o - \delta < x < x_o$. For this choice of x, $\left| x - x_o \right| < \delta$. But, since $x < x_o$, $F(x) \leqq 0$ and

$$F(x) - L \leqq -L$$

Thus

$$\left| F(x) - L \right| \geqq L > \frac{L}{2}$$

which is a contradiction. Thus the assumption that L is positive is wrong.

Assume now that L is negative. If $\epsilon = -\frac{L}{2}$, then by the definition of limit, there is a positive number δ such that

$$\left| F(x) - L \right| < -\frac{L}{2}$$

whenever

$$0 < \left| x - x_o \right| < \delta$$

Suppose $x < b$ and $x_o < x < x_o + \delta$. For this choice of x, $\left| x - x_o \right| < \delta$. But, since $x > x_o$, $F(x) \geqq 0$ and

$$F(x) - L \geqq -L > 0$$

Thus

$$\left| F(x) - L \right| \geqq -L > -\frac{L}{2}$$

which is a contradiction. The assumption that L is negative is also wrong. Since L is neither positive nor negative, L must be 0, which means $f'(x_o) = 0$.

25. Since f is continuous on $[a,b]$, $f'(x)$ exists for all x in (a,b), and $f(a) = f(b) = 0$, there is a number p between a and b such that $f'(p) = 0$. Similarly there is a number q between b and c such that $f'(q) = 0$.

Now we use Rolle's theorem for $f'(x)$ on (p,q). We have shown that $f'(p) = f'(q) = 0$ and $f''(x)$ exists for all x in (p,q). Since $f''(x)$ exists for all x in (a,c) and $a < p < q < c$, f' is continuous on $[p,q]$. Hence there is a number d between p and q (and therefore between a and c) such that $f''(d) = 0$.

1. f is continuous on $[-1,2]$; $f'(x) = 2x$ exists on $(-1,2)$. Thus there is a number x_o between -1 and 2 such that $f'(x_o) = \dfrac{f(2) - f(-1)}{2 - (-1)}$. $2x_o = \dfrac{4 - 1}{3} = 1$ when $x_o = \dfrac{1}{2}$

5. f is continuous on $[1,2]$; $f'(x) = -\dfrac{1}{x^2}$ exists on $(1,2)$. Thus there is a number x_o between 1 and 2 such that $f'(x_o) = \dfrac{f(2) - f(1)}{2 - 1}$. $-\dfrac{1}{x_o^2} = \dfrac{1/2 - 1}{1} = -\dfrac{1}{2}$ when $x_o = \sqrt{2}$

9. f is continuous on $[0,1]$; $f'(x) = e^x$ exists on $(0,1)$. Thus there is a number x_o between 0 and 1 such that $f'(x_o) = \dfrac{f(1) - f(0)}{1 - 0}$. $e^{x_o} = \dfrac{e - 1}{1}$ when x_o $\ell n(e - 1)$

13. The mean-value theorem tells us nothing because $f'(0)$ does not exist.

17. f and g are continuous on $[0,1]$; $f'(x) = 2$ and $g'(x) = 2x$ exist on $(0,1)$; $g'(x)$ is not 0 for any number in $(0,1)$. Thus there is a number x_o between 0 and 1 such that $\dfrac{f'(x_o)}{g'(x_o)} = \dfrac{f(1) - f(0)}{g(1) - g(0)}$.

$\dfrac{2}{2x_o} = \dfrac{3-1}{1-0}$ when $x_o = 1/2$

21. Assume $g(a) = g(b)$. By Problem 22 of the previous section, there is a number x_o between a and b such that $g'(x_o) = 0$. This contradicts the statement that $g'(x) \neq 0$ for all x in (a,b). Thus $g(a) \neq g(b)$.

Review 17, pp. 540-541

1. True. Suppose $0 < |x - 2| < \delta$.

$-\delta < x - 2 < \delta \qquad\qquad -2\delta < 2x - 4 < 2\delta$

$|f(x) - 3| = |(2x - 1) - 3| = |2x - 4| < 2\delta$

Choose $\delta = \epsilon/2$. Then $|f(x) - 3| < \epsilon$ whenever $0 < |x - 2| < \delta$.

5. True. Let $\epsilon > 0$. Suppose, for some $\delta > 0$,

$0 < |x + 2| < \delta \qquad\qquad -\delta < x + 2 < \delta$

$-4 - \delta < x - 2 < -4 + \delta$

Let us choose $\delta \leq 1$. Then

$-5 \leq -4 - \delta < x - 2 < -4 + \delta \leq -3, \qquad |x - 2| < 5$

$|f(x) - 4| = |x^2 - 4| = |(x+2)(x-2)| = |x + 2||x - 2| < 5\delta$

Choose $\delta = \min\{1, \epsilon/5\}$. Then $|f(x) - 4| < \epsilon$ whenever $0 < |x + 2| < \delta$.

9. True. Suppose $N \geq 1$. Let $0 < x - 2 < \delta$ for some positive δ. Then $\dfrac{1}{x-2} > \dfrac{1}{\delta}$. Choose $\delta = 1/N$. Then $\dfrac{1}{x-2} > N$ whenever $0 < x - 2 < \delta$. (If $N < 1$, choose $\delta = 1$.)

13. $\sin u = 1$ when $u = \pi/2 + 2\pi n = (4n+1)\pi/2$

$\sin \dfrac{1}{x} = 1$ when $x = 2/\pi(4n+1)$

$\sin u = -1$ when $u = -\pi/2 + 2\pi n = (4n-1)\pi/2$

$\sin \dfrac{1}{x} = 1$ when $x = 2/\pi(4n-1)$

Assume $\lim\limits_{x\to 0} \sin(1/x) = L \neq 1$. Let $\epsilon = |1 - L|/2$.
For any $\delta > 0$, let $|x| < \delta$ and $x = 2/\pi(4n+1)$ for some integer n. Then $\sin(1/x) = 1$ and

$$\left|\sin(1/x) - L\right| = \left|1 - L\right| \not< \epsilon .$$

Thus if the limit exists, it must be 1. Assume $L = 1$. Let $\epsilon = 1$. For any $\delta > 0$, let $|x| < \delta$ and $x = 2/(4n-1)\pi$ for some integer n. Then $\sin(1/x) = -1$ and

$$\left|\sin(1/x) - L\right| = \left|-1 - 1\right| = 2 \not< \epsilon .$$

Hence $\lim\limits_{x\to 0} \sin(1/x)$ does not exist.

17. Suppose $|x - 1| < \delta$ for some choice of δ.

$-\delta < x - 1 < \delta$

$2 - \delta < x + 1 < 2 + \delta$

$|x + 1| < 2 + \delta$

Choose $\delta \leq 1$. Then $|x + 1| < 3$.

$\left|f(x) - 2\right| = \left|(x^2+1)-2\right| = \left|x^2-1\right| = \left|x+1\right| \left|x-1\right| \leq 3\delta$

Choose $\delta = \min\{1, \epsilon/3\}$.

21. $f(-1) = f(1) = 0$. But $f(x)$ is not continuous on $[-1, 1]$ nor does f' exist on $(-1, 1)$ since f is not defined at $x = 0$. Therefore Rolle's theorem tells us nothing about this function.

25. This requires several applications of Rolle's theorem. First we apply it to f on $[a, b]$. Since the conditions of Rolle's theorem are satisfied, there is a number d between a and b such that $f'(c) = 0$. Now we apply it to f' on $[a, d]$ and $[d, b]$. This implies that there are numbers p and q such that $a < p < d < q < b$ and $f''(p) = f''(q) = 0$. Finally we apply it to f'' on $[p, q]$. Thus there is a number c between p and q (and hence between a and b) such that $f''(c) = 0$.

Section 18.1, pp. 547-548

1. $\lim\limits_{x\to 1} \dfrac{x^2 - 1}{x - 1} = \lim\limits_{x\to 1} \dfrac{2x}{1} = 2$

5. $\lim\limits_{x\to +\infty} \dfrac{x^2 + x}{x^2 - 1} = \lim\limits_{x\to +\infty} \dfrac{2x + 1}{2x} = \lim\limits_{x\to +\infty} \dfrac{2}{2} = 1$

9. $\lim\limits_{x\to 1} \dfrac{\ln x}{\sqrt{1 - x}} = \lim\limits_{x\to 1} \dfrac{1/x}{-1/2\sqrt{1 - x}} = \lim\limits_{x\to 1} -\dfrac{2\sqrt{1 - x}}{x} = 0$

13. $\lim\limits_{x\to +\infty} \dfrac{\ln x}{x} = \lim\limits_{x\to +\infty} \dfrac{1/x}{1} = 0$

17. $\lim\limits_{x\to 0} \dfrac{a^x - b^x}{x} = \lim\limits_{x\to 0} \dfrac{a^x \ln a - b^x \ln b}{1} = \ln a - \ln b$

21. $\lim\limits_{x\to 0} \dfrac{\ln \tan x}{\ln \tan 2x} = \lim\limits_{x\to 0} \dfrac{(\sec^2 x)/\tan x}{(2 \sec^2 2x)/\tan 2x}$

$= \lim\limits_{x\to 0} \dfrac{\sec^2 x \tan 2x}{2 \sec^2 2x \tan x}$

$= \lim\limits_{x\to 0} \dfrac{2 \sec^2 x \sec^2 2x + 2 \sec^2 x \tan x \tan 2x}{8 \sec^2 2x \tan 2x \tan x + 2 \sec^2 2x \sec^2 x} = \dfrac{2}{2} = 1$

25. $\lim\limits_{x\to 1} \dfrac{x^x - x}{1 - x + \ln x} = \lim\limits_{x\to 1} \dfrac{x^x(1 + \ln x) - 1}{-1 + 1/x}$

$= \lim\limits_{x\to 1} \dfrac{x^x/x + x^x(1 + \ln x)^2}{-1/x^2} = -2$

29. $\lim\limits_{x\to 0^+} \dfrac{e^{-1/x}}{x} = \lim\limits_{u\to +\infty} \dfrac{e^{-u}}{1/u} = \lim\limits_{u\to +\infty} \dfrac{u}{e^u} = \lim\limits_{u\to +\infty} \dfrac{1}{e^u} = 0$

33. $\displaystyle\int_1^{+\infty} \frac{\ell n\ x}{x^2}dx = \lim_{k\to+\infty} \int_1^k \frac{\ell n\ x}{x^2}dx$

$$u = \ell n\ x \qquad\qquad v' = 1/x^2$$
$$u' = 1/x \qquad\qquad v = -1/x$$

$$= \lim_{k\to+\infty} \left[-\frac{\ell n\ x}{x}\Big|_1^k + \int_1^k \frac{1}{x^2}\ dx \right]$$

$$= \lim_{k\to+\infty} -\frac{1+\ell n\ x}{x}\Big|_1^k = -\lim_{k\to+\infty} (1 - \frac{1+\ell n\ k}{k})$$

$$= 1 - \lim_{k\to+\infty} \frac{1/k}{1} = 1$$

Section 18.2, pp. 550-551

1. $\displaystyle\lim_{x\to 0} x^2 \ell n\ x = \lim_{x\to 0} \frac{\ell n\ x}{1/x^2} = \lim_{x\to 0} \frac{1/x}{-2/x^3} = \lim_{x\to 0} -\frac{x^2}{2} = 0$

5. $\displaystyle\lim_{x\to\pi/4^-} (1-\tan x)\sec 2x = \lim_{x\to\pi/4^-} \frac{1-\tan x}{\cos^2 x}$

$$= \lim_{x\to\pi/4^-} \frac{-\sec^2 x}{-2\sin 2x} = \frac{-2}{-2} = 1$$

9. $\displaystyle\lim_{x\to+\infty} e^{-x}\ell n\ x = \lim_{x\to+\infty} \frac{\ell n\ x}{e^x} = \lim_{x\to+\infty} \frac{1/x}{e^x} = 0$

13. $\displaystyle\lim_{x\to+\infty} x\ \ell n(1+\frac{a}{x}) = \lim_{x\to+\infty} \frac{\ell n(1+a/x)}{1/x} = \lim_{x\to+\infty} \frac{\frac{-a/x^2}{1+a/x}}{-1/x^2}$

$$= \lim_{x\to+\infty} \frac{a}{1+a/x} = a$$

17. $\displaystyle\lim_{x\to\pi/2^-} (2x\tan x - \pi\sec x) = \lim_{x\to\pi/2^-} \frac{2x\sin x - \pi}{\cos x}$

$$\lim_{x\to\pi/2^-} \frac{2x\cos x + 2\sin x}{-\sin x} = \frac{2}{-1} = -2$$

21. $\lim\limits_{x\to 0}\left(\dfrac{1}{x^2} - \dfrac{1}{x\tan x}\right) = \lim\limits_{x\to 0}\dfrac{\tan x - x}{x^2 \tan x}$

$\qquad = \lim\limits_{x\to 0}\dfrac{\sec^2 x - 1}{x^2\sec^2 x + 2x\tan x}$

$\qquad = \lim\limits_{x\to 0}\dfrac{2\sec^2 x\tan x}{2x^2\sec^2 x\tan x + 4x\sec^2 x + 2\tan x}$

$\qquad = \lim\limits_{x\to 0}\dfrac{\sec^4 x + 2\sec^2 x\tan^2 x}{x^2\sec^4 x + 2x^2\sec^2 x\tan^2 x + 6x\sec^2 x\tan x + 3\sec^2 x}$

$\qquad = \dfrac{1}{3}$

25. $\lim\limits_{x\to 1^+}\left(\underset{\downarrow}{\dfrac{1}{1-x}} - \underset{\downarrow}{\dfrac{1}{\ell n\, x}}\right) = -\infty$
$\qquad\qquad\;\; -\infty \qquad +\infty$

26. $\lim\limits_{x\to \pi^-}(\underset{\downarrow}{\cot x} - \underset{\downarrow}{\csc x}) = -\infty$
$\qquad\qquad\quad -\infty \qquad +\infty$

29. $\displaystyle\int_0^{+\infty}\dfrac{dx}{x(x+1)} = \lim\limits_{\varepsilon\to 0^+}\int_\varepsilon^1\dfrac{dx}{x(x+1)} + \lim\limits_{k\to +\infty}\int_1^k\dfrac{dx}{x(x+1)}$

$\qquad = \lim\limits_{\varepsilon\to 0^+}\int_\varepsilon^1\left(\dfrac{1}{x} - \dfrac{1}{x+1}\right)dx + \lim\limits_{k\to +\infty}\int_1^k\left(\dfrac{1}{x} - \dfrac{1}{x+1}\right)dx$

$\qquad = \lim\limits_{\varepsilon\to 0^+}\ell n\left|\dfrac{x}{x+1}\right|\Big|_\varepsilon^1 + \lim\limits_{k\to +\infty}\ell n\left|\dfrac{x}{x+1}\right|\Big|_1^k$

$\qquad = \lim\limits_{\varepsilon\to 0^+}\left(\ell n\,\dfrac{1}{2} - \ell n\,\dfrac{\varepsilon}{\varepsilon+1}\right) + \lim\limits_{k\to +\infty}\left(\ell n\,\dfrac{k}{k+1} - \ell n\,\dfrac{1}{2}\right)$

$\qquad = \ell n\,\dfrac{1}{2} + \infty - \ell n\,\dfrac{1}{2} = +\infty$

Section 18.3, p. 554

1. $\lim\limits_{x\to 0^+} x^2\ell n\, x = \lim\limits_{x\to 0^+}\dfrac{\ell n\, x}{1/x^2} = \lim\limits_{x\to 0^+}\dfrac{1/x}{-2/x^3} = \lim\limits_{x\to 0^+} -\dfrac{x^2}{2} = 0$

$\quad \lim\limits_{x\to 0^+} x^{x^2} = e^0 = 1$

5. $\lim\limits_{x\to+\infty} x \, \ell n(1+\frac{1}{x}) = \lim\limits_{x\to+\infty} \dfrac{\ell n(1+1/x)}{1/x} = \lim\limits_{x\to+\infty} \dfrac{(-1/x^2)/(1+1/x)}{-1/x^2}$

$= \lim\limits_{x\to+\infty} \dfrac{1}{1+1/x} = 1$

$\lim\limits_{x\to+\infty} (1+\frac{1}{x})^x = e^1 = e$

9. $\lim\limits_{x\to\pi/2^-} \cos x \, \ell n \, \tan x = \lim\limits_{x\to\pi/2^-} \dfrac{\ell n \, \tan x}{\sec x} = \lim\limits_{x\to\pi/2^-} \dfrac{\sec^2 x/\tan x}{\sec x \tan x}$

$= \lim\limits_{x\to\pi/2^-} \dfrac{\sec x}{\tan^2 x} = \lim\limits_{x\to\pi/2^-} \dfrac{1/\cos x}{\sin^2 x/\cos^2 x} = \lim\limits_{x\to\pi/2^-} \dfrac{\cos x}{\sin^2 x} = 0$

$\lim\limits_{x\to\pi/2^-} (\tan x)^{\cos x} = e^0 = 1$

13. $\lim\limits_{x\to 0} \dfrac{1}{x} \ell n(e^x + x) = \lim\limits_{x\to 0} \dfrac{\ell n(e^x + x)}{x} = \lim\limits_{x\to 0} \dfrac{(e^x + 1)/(e^x + x)}{1} = 2$

$\lim\limits_{x\to 0} (e^x + x)^{1/x} = e^2$

17. $\lim\limits_{x\to 1} \dfrac{1}{1-x} \ell n \, x = \lim\limits_{x\to 1} \dfrac{\ell n \, x}{1-x} = \lim\limits_{x\to 1} \dfrac{1/x}{-1} = -1$

$\lim\limits_{x\to 1} x^{1/(1-x)} = e^{-1} = \dfrac{1}{e}$

21. $\lim\limits_{x\to 0} \dfrac{b}{x} \ell n \cos ax = \lim\limits_{x\to 0} \dfrac{b \, \ell n \cos ax}{x} = \lim\limits_{x\to 0} \dfrac{-ab \sin ax/\cos ax}{1} = 0$

$\lim\limits_{x\to 0} (\cos ax)^{b/x} = e^0 = 1$

25. Suppose $\lim\limits_{x\to a} f(x) = 0$ and $\lim\limits_{x\to a} g(x) = -\infty$. Then

$\lim\limits_{x\to a} \ell n \, f(x) = -\infty$ and $\lim\limits_{x\to a} g(x) \, \ell n \, f(x) = +\infty$

Thus $\lim\limits_{x\to a} [f(x)]^{g(x)} = +\infty$

1. $\lim\limits_{x \to 0} \dfrac{\sqrt{2+x} - \sqrt{2-x}}{\sqrt{4+x} - \sqrt{4-x}} = \lim\limits_{x \to 0} \dfrac{\dfrac{1}{2\sqrt{2+x}} + \dfrac{1}{2\sqrt{2-x}}}{\dfrac{1}{2\sqrt{4+x}} + \dfrac{1}{2\sqrt{4-x}}}$

$= \dfrac{1/2\sqrt{2} + 1/2\sqrt{2}}{1/4 + 1/4} = \dfrac{1/\sqrt{2}}{1/2} = \sqrt{2}$

5. $\lim\limits_{x \to 0} (e^x - e^{-x})^x = y$

$\ln y = \lim\limits_{x \to 0} \ln(e^x - e^{-x})^x = \lim\limits_{x \to 0} x \, \ln(e^x - e^{-x})$

$= \lim\limits_{x \to 0} \dfrac{\ln(e^x - e^{-x})}{1/x} = \lim\limits_{x \to 0} \dfrac{\dfrac{e^x + e^{-x}}{e^x - e^{-x}}}{-1/x^2} = \lim\limits_{x \to 0} \dfrac{x^2(e^x + e^{-x})}{e^x - e^{-x}}$

$= \lim\limits_{x \to 0} \dfrac{x^2(e^x - e^{-x}) + 2x(e^x + e^{-x})}{-e^{-x} - e^x} = \dfrac{0}{-2} = 0$

$y = e^0 = 1$

9. $\lim\limits_{x \to +\infty} \dfrac{x^3}{e^x - x} = \lim\limits_{x \to +\infty} \dfrac{3x^2}{e^x - 1} = \lim\limits_{x \to +\infty} \dfrac{6x}{e^x} = \lim\limits_{x \to +\infty} \dfrac{6}{e^x} = 0$

13. $\lim\limits_{x \to 0} \dfrac{e^x - e^{-x}}{\cot 2x} = 0$

17. $\lim\limits_{x \to +\infty} \dfrac{\sin x}{x^2} = 0$

21. $\displaystyle\int_1^{+\infty} \dfrac{x \, dx}{(x+1)(x+2)} = \lim\limits_{k \to +\infty} \int_1^k \dfrac{x \, dx}{(x+1)(x+2)}$

$= \lim\limits_{k \to +\infty} \int_1^k \left(\dfrac{2}{x+2} - \dfrac{1}{x+1}\right) dx = \lim\limits_{k \to +\infty} \left[2\,\ln(x+2) - \ln(x+1) \right]\Big|_1^k$

$= \lim\limits_{k \to +\infty} \ln \dfrac{(x+2)^2}{x+1}\Big|_1^k = \lim\limits_{k \to +\infty} \left[\ln \dfrac{(k+2)^2}{k+1} - \ln \dfrac{9}{2} \right] = +\infty$

Chapter Nineteen
Infinite Series

1. $1/3,\ 1/9,\ 1/27,\ 1/81,\ \cdots$

5. $1,\ 1/2,\ 1/4,\ 1/8,\ \cdots$

9. $s_n = \dfrac{1}{3^n}$, $n = 0, 1, 2, \cdots$

13. $s_n = \dfrac{n + (-1)^n}{n}$, $\qquad n = 2, 3, 4, \cdots$

17. $\displaystyle\lim_{n\to+\infty} \dfrac{1}{2^n} = 0$; converges to 0

21. $\displaystyle\lim_{n\to+\infty} \dfrac{n}{n+1} = 1$; converges to 1

25. $\displaystyle\lim_{n\to+\infty} \dfrac{n}{3^n} = \lim_{n\to+\infty} \dfrac{1}{3^n \ell n\ 3} = 0$; converges to 0

29. Since $\{s_n\}$ is bounded, there is a number k such that
    $s_n \geqq k$ for every integer n. Let $S = \{x \mid x \leqq s_n$ for
    every integer n$\}$. This set is not empty since k is
    in it. Let u be the largest number in S.

    Let $\epsilon > 0$, and assume that

    $$s_n \geqq u + \epsilon$$

    for every positive integer n. Then $u + \epsilon$ is in S.
    But $u + \epsilon > u$, contradicting the statement that u is
    the largest number in S. Thus the assumption is
    wrong, and there is a number N such that $s_n < u + \epsilon$.
    If $n > N$, then $s_n < s_N < u + \epsilon$ and $s_n > u$. Thus there
    is a number N such that, if $n > N$, $|s_n - u| < \epsilon$; and
    $\{s_n\}$ converges to u.

1. $\dfrac{1}{3} + \dfrac{1}{3^2} + \dfrac{1}{3^3} + \dfrac{1}{3^4} + \cdots = \dfrac{1}{3} + \dfrac{1}{9} + \dfrac{1}{27} + \dfrac{1}{81} + \cdots$

5. $\dfrac{1}{1^2 + 1} + \dfrac{1}{2^2 + 1} - \dfrac{1}{3^2 + 1} + \dfrac{1}{4^2 + 1} - \cdots = -\dfrac{1}{2} + \dfrac{1}{5} - \dfrac{1}{10} + \dfrac{1}{17} - \cdots$

9. $\displaystyle\sum_{n=1}^{\infty} (2n - 1) \cdot 2n$

13. $0.1111\cdots = \dfrac{1}{10} + \dfrac{1}{100} + \dfrac{1}{1000} + \cdots = \dfrac{1/10}{1 - 1/10} = \dfrac{1}{9}$

17. $0.9999 \cdots = \dfrac{9}{10} + \dfrac{9}{100} + \dfrac{9}{1000} + \cdots = \dfrac{9/10}{1 - 1/10} = 1$

21. $a = 1/3, \quad r = 1/3;$ converges

$s = \dfrac{a}{1 - r} = \dfrac{1/3}{1 - 1/3} = \dfrac{1}{2}$

25. $\dfrac{n^2 + 3n + 1}{n^2(n + 1)^2} = \dfrac{An + B}{n^2} + \dfrac{Cn + D}{(n+1)^2}$

$n^2 + 3n + 1 = A(n^3 + 2n^2 + n) + B(n^2 + 2n + 1) + Cn^3 + Dn^2$

$A + C = 0 \qquad\qquad 2A + B + D = 1$

$A + 2B = 3 \qquad\qquad B = 1$

$A = 1, \quad B = 1, \quad C = -1, \quad D = -2$

$\dfrac{n^2 + 3n + 1}{n^2(n + 1)^2} = \dfrac{n + 1}{n^2} - \dfrac{n + 2}{(n + 1)^2} = \dfrac{n + 1}{n^2} - \dfrac{(n + 1) + 1}{(n + 1)^2}$

$s_n = 2 - \dfrac{n + 2}{(n + 1)^2} \to 2 \qquad$ Converges to 2

29. $\lim\limits_{n \to +\infty} a_n = \lim\limits_{n \to +\infty} (-1)^{n+1} n$ does not exist. Diverges

33. Assume it converges. Then $2 \displaystyle\sum_{n=1}^{\infty} \dfrac{1}{2n}$ converges. But

$2 \displaystyle\sum_{n=1}^{\infty} \dfrac{1}{2n} = \displaystyle\sum_{n=1}^{\infty} \dfrac{1}{n}$ diverges. Thus the original series

diverges.

37. Since $\sum\limits_{n=N}^{\infty} a_n$ converges, $\lim\limits_{n\to+\infty} t_n$ exists, where $t_n = a_N + a_{N+1} + \cdots + a_{N+n-1}$. The k-th partial sum of $\sum\limits_{n=1}^{\infty} a_n$ is $s_k = a_1 + a_2 + \cdots + a_k$. Letting $k = N + n - 1$ $(k \geqq N-1)$, $s_k = s_{N+n-1} = a_1 + a_2 + \cdots + a_{N-1} + t_n$. Since $\lim\limits_{n\to+\infty} t_n$ exists, $\lim\limits_{n\to+\infty} s_{N+n-1}$ exists and $\lim\limits_{k\to+\infty} s_k$ exists. Thus $\sum\limits_{n=1}^{\infty} a_n$ converges.

Section 19.3, pp. 579-580

1. $\lim\limits_{n\to+\infty} \dfrac{1/(n^2+1)}{1/n^2} = \lim\limits_{n\to+\infty} \dfrac{n^2}{n^2+1} = \lim\limits_{n\to+\infty} \dfrac{1}{1+1/n^2} = 1$

$\sum \dfrac{1}{n^2+1}$ and $\sum \dfrac{1}{n^2}$ are of the same order of magnitude.

Since $\sum \dfrac{1}{n^2}$ converges (p-series, $p = 2 > 1$), the given series converges.

5. $\lim\limits_{n\to+\infty} \dfrac{n}{n+1} = \lim\limits_{n\to+\infty} \dfrac{1}{1} = 1$. Diverges

9. $\lim\limits_{n\to+\infty} \dfrac{1/n!}{1/n^2} = \lim\limits_{n\to+\infty} \dfrac{n}{(n-1)!} = 0$

Thus $\sum \dfrac{1}{n!}$ is of a lesser order of magnitude than $\sum \dfrac{1}{n^2}$. Since $\sum \dfrac{1}{n^2}$ converges (p-series, $p = 2 > 1$), the given series converges.

13. $\lim\limits_{n\to+\infty} \dfrac{1/(n+1)(n+3)}{1/n^2} = \lim\limits_{n\to+\infty} \dfrac{1}{(1+1/n)(1+3/n)} = 1$

Hence $\sum \dfrac{1}{(n+1)(n+3)}$ and $\sum \dfrac{1}{n^2}$ are of the same order of magnitude. Since $\sum \dfrac{1}{n^2}$ converges (p-series, $p = 2 > 1$), the given series converges.

17. $\lim\limits_{n\to+\infty} \dfrac{n/(n+1)(n+2)(n+3)}{1/n^2} = \lim\limits_{n\to+\infty} \dfrac{1}{(1+1/n)(1+2/n)(1+3/n)} = 1$

Hence $\sum \dfrac{n}{(n+1)(n+2)(n+3)}$ and $\sum \dfrac{1}{n^2}$ are of the same

order of magnitude. Since $\sum \dfrac{1}{n^2}$ converges (p–series,

$p = 2 > 1$), the given series converges.

21. $\lim\limits_{n\to+\infty} \dfrac{1/\sqrt{n(n+1)}}{1/n} = \lim\limits_{n\to+\infty} \dfrac{1}{\sqrt{1 + 1/n}} = 1$

Hence $\sum \dfrac{1}{\sqrt{n(n+1)}}$ is of the same order of magnitude

as $\sum \dfrac{1}{n}$. Since $\sum \dfrac{1}{n}$ diverges (harmonic series),

$\sum \dfrac{1}{\sqrt{n(n+1)}}$ diverges.

25. $\lim\limits_{n\to+\infty} \dfrac{\ln(2n+1)/n(n+1)}{1/n^{3/2}} = \lim\limits_{n\to+\infty} \dfrac{\sqrt{n}\,\ln(2n+1)}{n+1}$

$= \lim\limits_{n\to+\infty} \dfrac{\ln(2n+1)}{\sqrt{n} + 1/\sqrt{n}} = \lim\limits_{n\to+\infty} \dfrac{2/(2n+1)}{1/2\sqrt{n} - 1/2n^{3/2}}$

$= \lim\limits_{n\to+\infty} \dfrac{4n^{3/2}}{(2n+1)(n-1)} = \lim\limits_{n\to+\infty} \dfrac{4n^{3/2}}{2n^2 - n - 1}$

$= \lim\limits_{n\to+\infty} \dfrac{6n^{1/2}}{4n-1} = \lim\limits_{n\to+\infty} \dfrac{3n^{-1/2}}{4} = 0$

Hence $\sum \dfrac{\ln(2n+1)}{n(n+1)}$ is of a lesser order of magnitude

than $\sum \dfrac{1}{n^{3/2}}$. Since $\sum \dfrac{1}{n^{3/2}}$ converges (p-series,

$p = 3/2 > 1$), the given series converges.

29. Let $\sum a_n = \sum(-n)$ and $\sum b_n = \sum \dfrac{1}{n^2}$. $\sum b_n$ converges

and $a_n < b_n$ for all n, but $\sum a_n$ diverges.

33. Suppose $\sum a_n$ is of a greater order of magnitude than

$\sum b_n$. Then

$$\lim\limits_{n\to+\infty} \dfrac{a_n}{b_n} = +\infty ,$$

or

$$\lim_{n \to +\infty} \frac{b_n}{a_n} = 0$$

If $\sum a_n$ converges, then the second part of this theorem assures us that $\sum b_n$ also converges. Since this is a contradiction, $\sum a_n$ must diverge.

Section 19.4, p. 586

1. $\lim_{n \to +\infty} \dfrac{n+2}{(n+1)!} \dfrac{n!}{n+1} = \lim_{n \to +\infty} \dfrac{n+2}{(n+1)^2} = 0$; converges

5. $\lim_{n \to +\infty} \dfrac{4^{n+1}}{(n+1)!} \dfrac{n!}{4^n} = \lim_{n \to +\infty} \dfrac{4}{n+1} = 0$; converges

9. $\lim_{n \to +\infty} \dfrac{2 \cdot 4 \cdot 6 \cdots (2n+2)}{1 \cdot 3 \cdot 5 \cdots (2n+1)} \dfrac{1 \cdot 3 \cdot 5 \cdots (2n-1)}{2 \cdot 4 \cdot 6 \cdots 2n}$

   $= \lim_{n \to +\infty} \dfrac{2n+2}{2n+1} = 1$;

   the ratio test fails

   $\dfrac{2}{1} \cdot \dfrac{4}{3} \cdot \dfrac{6}{5} \cdots \dfrac{2n}{2n-1} > 1 \cdot 1 \cdot 1 \cdots 1 = 1$, $\sum 1$ diverges;

   $\sum \dfrac{2 \cdot 4 \cdot 6 \cdots 2n}{1 \cdot 3 \cdot 5 \cdots (2n-1)}$ diverges

13. $\displaystyle\int_2^\infty \dfrac{dx}{x \, \ell n^2 x} = \lim_{k \to +\infty} \int_2^k \dfrac{dx}{x \, \ell n^2 x} = \lim_{k \to +\infty} \left. -\dfrac{1}{\ell n \, x} \right|_2^k$

    $= \lim_{k \to +\infty} \left(\dfrac{1}{\ell n \, 2} - \dfrac{1}{\ell n \, k} \right) = \dfrac{1}{\ell n \, 2}$; converges

17. $\displaystyle\int_1^\infty \dfrac{x}{e^x} \, dx = \lim_{n \to +\infty} \int_1^n x e^{-x} dx = \lim_{n \to +\infty} \left. -e^{-x}(x+1) \right|_1^n$

    $= \lim_{n \to +\infty} \left(-\dfrac{n+1}{e^n} + \dfrac{2}{e} \right) = \dfrac{2}{e}$; converges

21. $\dfrac{1}{2^n+1} < \dfrac{1}{2^n}$, $\sum \dfrac{1}{2^n}$ converges; $\sum \dfrac{1}{2^n+1}$ converges

25. $\lim\limits_{n\to+\infty} \dfrac{2n+1}{2^{n+1}} \dfrac{2^n}{2n-1} = \lim\limits_{n\to+\infty} \dfrac{2n+1}{2(2n-1)} = \dfrac{1}{2}$; converges

29. $\displaystyle\int_1^\infty \text{Arccsc } x \, dx = \lim\limits_{k\to+\infty} \int_1^k \text{Arccsc } x \, dx$

$\qquad u = \text{Arccsc } x \qquad\qquad v' = 1$

$\qquad u' = \dfrac{-1}{x\sqrt{x^2-1}} \qquad\qquad v = x$

$\qquad = \lim\limits_{k\to+\infty} \left\{ x \text{ Arccsc } x \Big|_1^k + \int_1^k \dfrac{dx}{\sqrt{x^2-1}} \right.$

$\qquad\qquad x = \csc\theta \, , \qquad\qquad dx = -\csc\theta \cot\theta d\theta$

$\qquad = \lim\limits_{k\to+\infty} \left\{ k \text{ Arccsc } k - \text{Arccsc } 1 + \int_{\pi/2}^{\text{Arccsc } k} \dfrac{-\csc\theta \cot\theta d\theta}{\sqrt{\csc^2\theta - 1}} \right.$

$\qquad = \lim\limits_{k\to+\infty} \left\{ k \text{ Arccsc } k - \pi/2 - (\ln|\csc\theta - \cot\theta|) \Big|_{\pi/2}^{\text{Arccsc } k} \right.$

$\qquad = \lim\limits_{k\to+\infty} \left\{ k \text{ Arccsc } k - \pi/2 - \ln(k - \sqrt{k^2-1}) \right\}$

$\qquad = \lim\limits_{k\to+\infty} k \text{ Arccsc } k = \lim\limits_{k\to+\infty} \dfrac{\text{Arccsc } k}{1/k} = \lim\limits_{k\to+\infty} \dfrac{-1/k\sqrt{k^2-1}}{-1/k^2}$

$\qquad = \lim\limits_{k\to+\infty} \dfrac{k}{\sqrt{k^2-1}} = 1$; diverges

Section 19.5, pp. 592–593 $\qquad \dfrac{(-1)^{n+1}}{2(n+1)} \cdot \dfrac{2n}{(-1)^n}$

1. $\lim\limits_{n\to+\infty} \dfrac{1}{2n} = 0$. $\quad \dfrac{1}{2(n+1)} < \dfrac{1}{2n}$; converges

$\quad \sum \dfrac{1}{2n}$ diverges

The original series converges conditionally.

5. $\lim\limits_{n\to+\infty} \dfrac{1}{\sqrt{n}} = 0$, $\quad \dfrac{1}{\sqrt{n+1}} < \dfrac{1}{\sqrt{n}}$; converges

$\quad \sum \dfrac{1}{\sqrt{n}}$ diverges

The original series converges conditionally.

9. $\sum \dfrac{1}{2^n}$ converges

The original series converges absolutely.

13. $\lim\limits_{n \to +\infty} \dfrac{n}{\ell n\, n} = \lim\limits_{n \to +\infty} \dfrac{1}{1/n} = \lim\limits_{n \to +\infty} n = +\infty\,;$ diverges

17. $\lim\limits_{n \to +\infty} \dfrac{n+1}{e^{n+1}} \dfrac{e^n}{n} = \lim\limits_{n \to +\infty} \dfrac{n+1}{e^n} = \dfrac{1}{e}\,;\ \sum \dfrac{n}{e^n}$ converges

The original series converges absolutely.

21. Consider the series

$$(-1)^{n+2} a_{n+1} + (-1)^{n+3} a_{n+2} + (-1)^{n+4} a_{n+3} + \cdots$$

$$+ (-1)^{n+m+1} a_{n+m} + \cdots$$

Let $t_m = (-1)^{n+2} a_{n+1} + (-1)^{n+3} a_{n+2} + \cdots + (-1)^{n+m+1} a_{n+m}$.

If s_n represents the n-th partial sum of the original series $\sum a_i$, then

$$s_{n+m} = s_n + t_m \quad \text{or} \quad t_m = s_{n+m} - s_n$$

and

$$\lim\limits_{m \to +\infty} t_m = s - s_n$$

Suppose $(-1)^{n+2} a_{n+1} > 0$; then, by the argument of p. 587, $0 < t_m < a_{n+1}$. Furthermore $t_m < a_{n+1} - a_{n+2}$ $+ a_{n+3} < a_{n+1}$ for all $m \geqq 3$. Thus

$$0 < \lim\limits_{m \to +\infty} t_m < a_{n+1}$$

and $\left| s - s_n \right| < \left| a_{n+1} \right|$

Now suppose $(-1)^{n+2} a_{n+1} < 0$. By a similar argument,

$$-a_{n+1} < \lim\limits_{m \to +\infty} t_m < 0$$

and $\left| s - s_n \right| < \left| a_{n+1} \right|$

25. $\sum_{n=1}^{\infty} \frac{(-1)^{n+1} n}{(n+1)^2} \approx \frac{1}{4} - \frac{2}{9} + \frac{3}{16} - \frac{4}{25} \approx 0.1$

29. Take positive terms of the given series (beginning with 1 and going in descending order) until the partial sum is greater than 1. Then take negative terms until the partial sum is less than 1. Then take positive terms until the partial sum exceeds 2 and negative terms until the partial sum is less than 2. Repeat using 3, 4, etc. The resulting series diverges to $+\infty$.

Section 19.6, pp. 597-598

1. $\lim_{n \to +\infty} \left| \frac{x^{n+1}}{n+2} \frac{n+1}{x^n} \right| = \lim_{n \to +\infty} \left| \frac{n+1}{n+2} x \right| = |x| < 1; \quad -1 < x < 1$

$x = 1: \quad \sum \frac{1}{n+1}$ diverges

$x = -1: \quad \sum \frac{(-1)^n}{n+1}$ converges

5. $\lim_{n \to +\infty} \left| \frac{(n+1) x^{n+1}}{2^{n+1}} \frac{2^n}{nx^n} \right| = \lim_{n \to +\infty} \left| \frac{(n+1)x}{2n} \right| = \left| \frac{x}{2} \right| < 1$

$-2 < x < 2$

$x = 2: \quad \sum n$ diverges. $\quad x = -2: \quad \sum (-1)^n$ diverges

Converges for $-2 < x < 2$

9. $\lim_{n \to +\infty} \left| \frac{(n+1)(n+2) x^{n+1}}{5^{n+1}} \frac{5^n}{n(n+1)x^n} \right| = \lim_{n \to +\infty} \left| \frac{(n+2)x}{5n} \right|$

$= \left| \frac{x}{5} \right| < 1; \quad -5 < x < 5$

$x = 5: \quad \sum n(n+1)$ diverges

$x = -5$: $\sum(-1)^n n(n+1)$ diverges

Converges for $-5 < x < 5$

13. $\displaystyle\lim_{n\to+\infty} \left| \frac{(x+4)^{n+1}}{2^{n+1}} \frac{2^n}{(x+4)^n} \right| = \lim_{n\to+\infty} \left| \frac{x+4}{2} \right| = \left| \frac{x+4}{2} \right| < 1$

$-1 < \dfrac{x+4}{2} < 1$, $\quad -2 < x+4 < 2$, $\quad -6 < x < -2$

$x = -2$: $\sum 1$ diverges. $\quad x = -6$: $\sum(-1)^n$ diverges

Converges for $-6 < x < -2$

17. $\displaystyle\lim_{n\to+\infty} \left| \frac{(n+1)\,2^{n+1}(x-1)^{n+1}}{n+2} \frac{n+1}{n\cdot 2^n (x-1)^n} \right|$

$\quad = \displaystyle\lim_{n\to+\infty} \left| \frac{2(n+1)^2 (x-1)}{n(n+2)} \right| = \left| 2(x-1) \right| < 1$

$-1 < 2x - 2 < 1$, $\quad 1 < 2x < 3$, $\quad \dfrac{1}{2} < x < \dfrac{3}{2}$

$x = \dfrac{3}{2}$: $\sum \dfrac{n}{n+1}$ diverges. $\quad x = \dfrac{1}{2}$: $\sum \dfrac{(-1)^n n}{n+1}$ diverges

Converges for $\dfrac{1}{2} < x < \dfrac{3}{2}$

21. $\displaystyle\lim_{n\to+\infty} \left| \frac{(x+4)^{n+1}}{(n+1)(n+2)} \frac{n(n+1)}{(x+4)^n} \right| = \lim_{n\to+\infty} \left| \frac{n(x+4)}{n+2} \right|$

$\quad = \left| x+4 \right| < 1$

$-1 < x+4 < 1$, $\quad -5 < x < -3$

$x = -3$: $\sum \dfrac{1}{n(n+1)}$ converges

$x = -5$: $\sum \dfrac{(-1)^n}{n(n+1)}$ converges

Converges for $-5 \leqq x \leqq -3$

25. $\displaystyle\lim_{n\to+\infty} \left| \frac{2^{n+1}\sin^{n+1}x}{2^n \sin^n x} \right| = \lim_{n\to+\infty} 2\left| \sin x \right| = 2\left| \sin x \right| < 1$

$\left| \sin x \right| < \dfrac{1}{2}$, $\quad -\dfrac{\pi}{6} + n\pi < x < \dfrac{\pi}{6} + n\pi$

$x = \dfrac{\pi}{6}$: $\displaystyle\sum_{n=0}^{\infty} 2^n \frac{1}{2^n} = \sum_{n=0}^{\infty} 1$ diverges

$$x = -\frac{\pi}{6}: \quad \sum_{n=0}^{\infty} 2^n \frac{(-1)^n}{2^n} = \sum_{n=1}^{\infty} (-1)^n \quad \text{diverges}$$

Converges for $-\frac{\pi}{6} + n\pi < x < \frac{\pi}{6} + n\pi$

Section 19.7, pp. 602-603

1. $f(x) = e^x$ $\qquad$ $f(1) = e$

$f'(x) = e^x$ $\qquad$ $f'(1) = e$

$\vdots$ $\qquad\qquad\qquad$ $\vdots$

$e^x = e + e(x-1) + \frac{e}{2!}(x-1)^2 + \frac{e}{3!}(x-1)^3 + \cdots$

$\quad = \sum_{n=0}^{\infty} \frac{e}{n!}(x-1)^n$

$\lim_{n \to +\infty} \left| \frac{e(x-1)^{n+1}}{(n+1)!} \frac{n!}{e(x-1)^n} \right| = \lim_{n \to +\infty} \left| \frac{x-1}{n+1} \right| = 0$

Converges for all x

5. $f(x) = \cos x$ $\qquad$ $f(\pi/6) = \sqrt{3}/2$

$f'(x) = -\sin x$ $\qquad$ $f'(\pi/6) = -1/2$

$f''(x) = -\cos x$ $\qquad$ $f''(\pi/6) = -\sqrt{3}/2$

$f'''(x) = \sin x$ $\qquad$ $f'''(\pi/6) = 1/2$

$f^{(4)}(x) = \cos x$ $\qquad$ $f^{(4)}(\pi/6) = \sqrt{3}/2$

$\cos x = \frac{\sqrt{3}}{2} - \frac{1}{2}(x - \pi/6) - \frac{\sqrt{3}}{2} \frac{(x - \pi/6)^2}{2!} + \frac{1}{2} \frac{(x - \pi/6)^3}{3!} + \cdots$

$\lim_{n \to +\infty} \left| \frac{c_{n+1}(x - \pi/6)^{n+1}}{(n+1)!} \frac{n!}{c_n(x - \pi/6)^n} \right| = \lim_{n \to +\infty} \left| \frac{c_{n+1}(x - \pi/6)}{c_n(n+1)} \right| = 0$

Converges for all x

9. $f(x) = \sinh x$ $\qquad$ $f(0) = 0$

$f'(x) = \cosh x$ $\qquad$ $f'(0) = 1$

$f''(x) = \sinh x$ $\qquad$ $f''(0) = 0$

$\vdots$ $\qquad\qquad\qquad$ $\vdots$

$$\sinh x = x + \frac{x^3}{3!} + \frac{x^5}{5!} + \cdots + \frac{x^{2n-1}}{(2n-1)!} + \cdots$$

$$\lim_{n \to +\infty} \left| \frac{x^{2n+1}}{(2n+1)!} \frac{(2n-1)!}{x^{2n-1}} \right| = \lim_{n \to +\infty} \left| \frac{x^2}{2n(2n+1)} \right| = 0$$

Converges for all x

13.
$$f(x) = x^3 + x^2 - 2x + 1 \qquad\qquad f(1) = 1$$
$$f'(x) = 3x^2 + 2x - 2 \qquad\qquad f'(1) = 3$$
$$f''(x) = 6x + 2 \qquad\qquad f''(1) = 8$$
$$f'''(x) = 6 \qquad\qquad f'''(1) = 6$$
$$f^{(4)}(x) = 0 \qquad\qquad f^{(4)}(1) = 0$$
$$\vdots \qquad\qquad \vdots$$

$$x^3 + x^2 - 2x + 1 = 1 + 3(x-1) + \frac{8}{2!}(x-1)^2 + \frac{6}{3!}(x-1)^3$$
$$= 1 + 3(x-1) + 4(x-1)^2 + (x-1)^3$$

Converges for all x

17.
$$f(x) = \frac{1}{x^2 + 1} \qquad\qquad f(1) = \frac{1}{2}$$

$$f'(x) = \frac{-2x}{(x^2+1)^2} \qquad\qquad f'(1) = -\frac{1}{2}$$

$$f''(x) = \frac{6x^2 - 2}{(x^2+1)^3} \qquad\qquad f''(1) = \frac{1}{2}$$

$$f'''(x) = \frac{24x - 24x^3}{(x^2+1)^4} \qquad\qquad f'''(1) = 0$$

$$f^{(4)}(x) = \frac{120x^4 - 240x^2 + 24}{(x^2+1)^5} \qquad\qquad f^{(4)}(1) = -3$$

$$f^{(5)}(x) = \frac{-720x^5 + 2400x^3 - 720x}{(x^2+1)^6} \qquad\qquad f^{(5)}(1) = 15$$

$$\vdots \qquad\qquad \vdots$$

$$\frac{1}{x^2+1} = \frac{1}{2} - \frac{1}{2}(x-1) + \frac{1}{2}\frac{(x-1)^2}{2!} - 3\frac{(x-1)^4}{4!} + \frac{245}{16}\frac{(x-1)^5}{5!} + \cdots$$

$$= \frac{1}{2} - \frac{1}{2}(x-1) + \frac{1}{4}(x-1)^2 - \frac{1}{8}(x-1)^4 + \frac{1}{8}(x-1)^5 + \cdots$$

21. $f(x) = \ln x$ does not have a Maclaurin's series expansion because there is no interval containing 0 in which f and all its derivatives exist.

25. $\dfrac{1}{1+x^2} = 1 - x^2 + x^4 - x^6 + \cdots$

$\displaystyle\int (1 - x^2 + x^4 - x^6 + \cdots)\,dx = C + x - \frac{x^3}{3} + \frac{x^5}{5} - \frac{x^7}{7} + \cdots$

$\text{Arctan } x = x - \dfrac{x^3}{3} + \dfrac{x^5}{5} - \dfrac{x^7}{7} + \cdots$

Section 19.8, p. 608

1. $e^x = 1 + x + \dfrac{x^2}{2!} + R_2$, $\quad e^{0,2} \doteq 1 + 0.2 + \dfrac{0.04}{2} = 1.22$

   $R_2 = \dfrac{x^3}{3!}e^c$, where $0 < c < 0.2 < \dfrac{(0.2)^3}{6} \cdot 2 \doteq 0.003$

5. $5° = 0.0873$, $\quad \cos x = 1 - \dfrac{x^2}{2!} + \dfrac{x^4}{4!} + R_4$

   $\cos 5° \doteq 1 - \dfrac{0.0873^2}{2!} + \dfrac{0.0873^4}{4!} = 0.9962$

   $R_4 = \dfrac{x^5}{5!}\sin c$, where $0 < c < 0.0873 < \dfrac{0.1^5}{5!} \doteq 0.000\,000\,1$

9. $5° = 0.0873$, $\quad\quad\quad \tan x = x + \dfrac{x^3}{3} + R_4$

   $\tan 5° \doteq 0.0873 + \dfrac{0.0873^3}{3} = 0.0875$

   $R_4 = \dfrac{x^5}{5!}(16\sec^6 c + 88\sec^4 c\tan^2 c + 16\sec^2 c\tan^4 c)$,

   where $0 < c < 0.0873 < \dfrac{0.0873^5}{5!} \cdot 136 = 0.000\,001$

13. $R_n = \dfrac{x^{n+1}}{(n+1)!}f^{(n+1)}(c)$, where $0 < c < 0.0873$

    $< \dfrac{(0.1)^{n+1}}{(n+1)!} < 0.00005$

    When $n = 4$, from Problem 3, $\sin 5° \doteq 0.0872$

17. $R_n = \dfrac{(x - \pi/6)^{n+1}}{(n + 1)!} f^{(n+1)}(c)$, where $\dfrac{\pi}{6} < c < 0.61087$

$\qquad < \dfrac{(0.1)^{n+1}}{(n + 1)!} < 0.00005$. When $n = 4$,

$\sin x = \dfrac{1}{2} + \dfrac{\sqrt{3}}{2}(x - \pi/6) - \dfrac{1}{2}\dfrac{(x - \pi/6)^2}{2!} - \dfrac{\sqrt{3}}{2}\dfrac{(x - \pi/6)^3}{3!}$

$\qquad + \dfrac{1}{2}\dfrac{(x - \pi/6)^4}{4!} + R_4$

$\sin 35° \doteq \dfrac{1}{2} + \dfrac{\sqrt{3}}{2}(0.0873) - \dfrac{1}{2}\dfrac{(0.0873)^2}{2!} - \dfrac{\sqrt{3}}{2}\dfrac{(0.0873)^3}{3!}$

$\qquad + \dfrac{1}{2}\dfrac{(0.0873)^4}{4!}$

$\qquad = 0.5736$

21. $R_n = \dfrac{(x - 1)^{n+1}}{(n + 1)!} f^{(n+1)}(c)$, where $1 < c < 2$

$\qquad < \dfrac{1}{(n + 1)!} n! = \dfrac{1}{n + 1} < 0.1$. When $n = 10$,

$\ln x = (x - 1) - \dfrac{(x - 1)^2}{2} + \dfrac{(x + 1)^3}{3} - \cdots - \dfrac{(x - 1)^{10}}{10} + R_{10}$

$\ln 2 \doteq 1 - \dfrac{1}{2} + \dfrac{1}{3} - \dfrac{1}{4} + \cdots - \dfrac{1}{10} \doteq 0.65$

Section 19.9, pp. 612-613

1. $f(x) = \dfrac{-1}{1 + x}$, $g(x) = \dfrac{1}{(1 + x)^2}$

$f(x) = -1 + x - x^2 + x^3 - \cdots$

$g(x) = f'(x) = 1 - 2x + 3x^2 - 4x^3 + \cdots$

Both series converge for $-1 < x < 1$.

5. $f(x) = 1 - x^2 + x^4 - x^6 + \cdots$ converges for $-1 < x < 1$

$g(x) = \displaystyle\int f(x)\,dx = C + x - \dfrac{x^3}{3} + \dfrac{x^5}{5} - \dfrac{x^7}{7} + \cdots$

Since $g(0) = 0$, $C = 0$

$g(x) = x - \dfrac{x^3}{3} + \dfrac{x^5}{5} - \dfrac{x^7}{7} + \cdots$ converges for $-1 \le x \le 1$

9. $f(x) = x + \dfrac{1}{3} x^3 + \dfrac{2}{15} x^5 + \dfrac{17}{315} x^7 + \cdots$

$g(x) = -\int f(x)\,dx = c - \dfrac{x^2}{2} - \dfrac{x^4}{12} - \dfrac{x^6}{45} - \dfrac{17x^8}{2520} - \cdots$

Since $g(0) = 0$, $\quad C = 0$

$g(x) = -\dfrac{x^2}{2} - \dfrac{x^4}{12} - \dfrac{x^6}{45} - \dfrac{17x^8}{2520} - \cdots$

Both series converge for $-\pi/2 < x < \pi/2$.

13. $\displaystyle\int \cos x^2\,dx = \int \left(1 - \dfrac{x^4}{2!} + \dfrac{x^8}{4!} - \dfrac{x^{12}}{6!} + \cdots \right) dx$

$= C + x - \dfrac{x^5}{5\cdot 2!} + \dfrac{x^9}{9\cdot 4!} - \dfrac{x^{13}}{13\cdot 6!} + \cdots$

17. $\displaystyle\int \sqrt{1+x^3}\,dx = \int \left(1 + \dfrac{1}{2}x^3 - \dfrac{1}{2^2\cdot 2!}x^6 + \dfrac{1\cdot 3}{2^3\cdot 3!}x^9 - \dfrac{1\cdot 3\cdot 5}{2^4\cdot 4!}x^{12}\right.$

$\left. + \cdots \right) dx = C + x + \dfrac{1}{2}\dfrac{x^4}{4} - \dfrac{1}{2^2\cdot 2!}\dfrac{x^7}{7} + \dfrac{1\cdot 3}{2^3\cdot 3!}\dfrac{x^{10}}{10}$

$- \dfrac{1\cdot 3\cdot 5}{2^4\cdot 4!}x^{12} + \cdots$

21. $\displaystyle\int_0^{1/2} \cos x^3\,dx = \int_0^{1/2} \left(1 - \dfrac{x^6}{2!} + \dfrac{x^{12}}{4!} - \cdots\right) dx$

$= \left. \left(x - \dfrac{x^7}{7\cdot 2!} + \dfrac{x^{13}}{13\cdot 4!} - \cdots\right)\right|_0^{1/2}$

$= \dfrac{1}{2} - \dfrac{1}{2^7\cdot 7\cdot 2!} + \dfrac{1}{2^{13}\cdot 13\cdot 4!} - \cdots$

$\approx 0.50000 - 0.00056 = 0.4994$

25. $\displaystyle\lim_{x\to 0} \dfrac{e^x - 1}{\sin x} = \lim_{x\to 0} \dfrac{x + \dfrac{x^2}{2!} + \dfrac{x^3}{3!} + \dfrac{x^4}{4!} + \cdots}{x - \dfrac{x^3}{3!} + \dfrac{x^5}{5!} - \dfrac{x^7}{7!} + \cdots}$

$= \displaystyle\lim_{x\to 0} \dfrac{1 + \dfrac{x}{2!} + \dfrac{x^2}{3!} + \dfrac{x^3}{4!} + \cdots}{1 - \dfrac{x^2}{3!} + \dfrac{x^4}{5!} - \dfrac{x^6}{7!} + \cdots} = 1$

1. $s_n = n \cdot 2^{n-1}$, $\quad n = 1, 2, 3, \cdots$

5. Since $\lim\limits_{n \to +\infty} \dfrac{2^n + 1}{3^n - 1} = \lim\limits_{n \to +\infty} \dfrac{(2/3)^n + (1/3)^n}{1 - (1/3)^n} = 0$,

   $\lim\limits_{n \to +\infty} s_n = 0$. The sequence converges to 0.

9. $\dfrac{2}{4n^2 - 1} = \dfrac{1}{2n - 1} - \dfrac{1}{2n + 1}$, $\quad s_n = 1 - \dfrac{1}{2n + 1} \to 1$

   The series converges to 1.

13. $\lim\limits_{n \to +\infty} \dfrac{n(n + 1)}{(2n + 1)(2n - 1)} = \lim\limits_{n \to +\infty} \dfrac{1 + 1/n}{(2 + 1/n)(2 - 1/n)} = \dfrac{1}{4}$

   Diverges

17. $\lim\limits_{n \to +\infty} \dfrac{(n + 2)/2^{n+1}}{(n + 1)/2^n} = \lim\limits_{n \to +\infty} \dfrac{n + 2}{2(n + 1)} = \lim\limits_{n \to +\infty} \dfrac{1}{2} = \dfrac{1}{2}$

   Converges

21. $\displaystyle\int_2^\infty \dfrac{dx}{x \, \ell n^4 x} = \lim\limits_{k \to +\infty} \int_2^k \dfrac{dx}{x \, \ell n^4 x} = \lim\limits_{k \to +\infty} \dfrac{-1}{3 \, \ell n^3 x} \Big|_2^k$

   $= \lim\limits_{k \to +\infty} \left(\dfrac{1}{3 \, \ell n^3 2} - \dfrac{1}{3 \, \ell n^3 k} \right) = \dfrac{1}{3 \, \ell n^3 2}$. Converges

25. Since $\lim\limits_{n \to +\infty} \dfrac{n}{n + 1} = 1$, the series diverges.

29. The series $\dfrac{1}{2} + \dfrac{1}{2^2} + \dfrac{1}{2^3} + \dfrac{1}{2^4} + \cdots$ of positive terms

   converges. The series $-1 - \dfrac{1}{2^2} - \dfrac{1}{3^2} - \cdots$ also converges

   (p-series, $p = 2$). Hence the given series converges

   absolutely.

33. $\lim\limits_{n\to+\infty}\left|\dfrac{(n+1)!\,x^{n+1}/(n+1)^4\,4^{n+1}}{n!\,x^n/n^4\,4^n}\right| = \lim\limits_{n\to+\infty}\left|\dfrac{n^4 x}{(n+1)^3\cdot 4}\right|$

$\qquad = +\infty \qquad (x \neq 0)$

Converges for $x = 0$

37. $\dfrac{1}{1+x^3} = 1 - x^3 + x^6 - \cdots = \sum\limits_{n=0}^{\infty}(-1)^n x^{3n}$

$\qquad \lim\limits_{n\to+\infty}\left|\dfrac{x^{3n+3}}{x^{3n}}\right| = |x^3| < 1$. Converges for $|x| < 1$

41. $e^x = 1 + x + \dfrac{x^2}{2!} + \dfrac{x^3}{3!} + R_3$

$\qquad \sqrt{e} = e^{1/2} \approx 1 + \dfrac{1}{2} + \dfrac{1}{4\cdot 2!} + \dfrac{1}{8\cdot 3!}$

$\qquad\qquad = 1 + 0.5 + 0.125 + 0.021 = 1.646$

$\qquad R_n = \dfrac{f^{(n+1)}(c)}{(n+1)!}(x-a)^{n+1} \qquad a < c < x$

$\qquad R_3 = \dfrac{e^c}{4!}\dfrac{1}{2^4} < \dfrac{e^{1/2}}{24\cdot 16} \approx \dfrac{1.6}{24\cdot 16} = 0.004$

45. $f(x) = \dfrac{1}{1+x^3} = 1 - x^3 + x^6 - x^9 + \cdots$

$\qquad f'(x) = \dfrac{-3x^2}{(1+x^3)^2} = -3x^2 + 6x^5 - 9x^8 + 12x'' - \cdots$

$\qquad g(x) = \dfrac{x^2}{(1+x^3)^2} = x^2 - 2x^5 + 3x^8 - 4x'' + \cdots$

49. $\dfrac{\sin x - x + x^3/6}{x^5} = \dfrac{x - x^3/6 + x^5/5! - x^7/7! + \cdots - x + x^3/6}{x^5}$

$\qquad = \dfrac{1}{5!} - \dfrac{x^2}{7!} + \dfrac{x^4}{9!} - \cdots$

$\qquad \lim\limits_{x\to 0}\dfrac{\sin x - x + x^3/6}{x^5} = \lim\limits_{x\to 0}\left(\dfrac{1}{5!} - \dfrac{x^2}{7!} + \dfrac{x^4}{9!} - \cdots\right) = \dfrac{1}{5!} = \dfrac{1}{120}$

Section 20.1, p. 619

1. $d = \sqrt{(2+3)^2 + (5-1)^2 + (0-3)^2} = \sqrt{50} = 5\sqrt{2}$

5. $d = \sqrt{(-5-4)^2 + (0-1)^2 + (2+5)^2} = \sqrt{131}$

9. $d = \sqrt{(4-3)^2 + (7+1)^2 + (-1-3)^2} = \sqrt{81} = 9$

13. $x = x_1 + r(x_2 - x_1) = -2 + \frac{1}{4}(10+2) = 1$

$y = y_1 + r(y_2 - y_1) = 0 + \frac{1}{4}(8-0) = 2$

$z = z_1 + r(z_2 - z_1) = 1 + \frac{1}{4}(5-1) = 2$

17. $x = \dfrac{x_1 + x_2}{2} = \dfrac{5-3}{2} = 1$, $y = \dfrac{y_1 + y_2}{2} = \dfrac{-2+4}{2} = 1$

$z = \dfrac{z_1 + z_2}{2} = \dfrac{3+7}{2} = 5$

21. $x = x_1 + r(x_2 - x_1)$, $6 = 5 + \frac{1}{3}(x_2 - 5)$, $x_2 = 8$

$y = y_1 + r(y_2 - y_1)$, $0 = -2 + \frac{1}{3}(y_2 + 2)$, $y_2 = 4$

$z = z_1 + r(z_2 - z_1)$, $0 = 3 + \frac{1}{3}(z_2 - 3)$, $z_2 = -6$

25. $6 = \sqrt{(5-1)^2 + (1-y)^2 + (0-2)^2}$

$36 = 16 + (1-y)^2 + 4$, $(1-y)^2 = 16$

$1 - y = \pm 4$, $y = 1 \pm 4 = 5, -3$

29. $6 = \sqrt{(\frac{x+1}{2} + 1)^2 + (1-5)^2 + (4-2)^2}$

$36 = \dfrac{(x+3)^2}{4} + 16 + 4$, $(x+3)^2 = 64$

$x + 3 = \pm 8$, $x = -3 \pm 8 = 5, -11$

33. If $r = 1/2$, then

$$x = x_1 + \frac{1}{2}(x_2 - x_1) = \frac{x_1 + x_2}{2}$$

$$y = y_1 + \frac{1}{2}(y_2 - y_1) = \frac{y_1 + y_2}{2}$$

$$z = z_1 + \frac{1}{2}(z_2 - z_1) = \frac{z_1 + z_2}{2}$$

Section 20.2, pp. 623-624

1. $v = (-4 - 2)i + (1 - 3)j + (2 + 5)k = -6i - 2j + 7k$

5. $u = \dfrac{4i + j - 2k}{\sqrt{16 + 1 + 4}} = \dfrac{4}{\sqrt{21}}i + \dfrac{1}{\sqrt{21}}j - \dfrac{2}{\sqrt{21}}k$

9.
$$x_2 - 2 = 2 \qquad\qquad y_2 - 1 = -1 \qquad\qquad z_2 - 5 = 3$$
$$x_2 = 4 \qquad\qquad y_2 = 0 \qquad\qquad z_2 = 8$$
$$B = (4, 0, 8)$$

13.
$$x_2 - x_1 = 4 \qquad\qquad y_2 - y_1 = -2 \qquad\qquad z_2 - z_1 = 1$$
$$\frac{x_2 + x_1}{2} = 2 \qquad\qquad \frac{y_2 + y_1}{2} = 5 \qquad\qquad \frac{z_2 + z_1}{2} = -1$$
$$x_2 + x_1 = 4 \qquad\qquad y_2 + y_1 = 10 \qquad\qquad z_2 + z_1 = -2$$
$$2x_2 = 8 \qquad\qquad 2y_2 = 8 \qquad\qquad 2z_2 = -1$$
$$x_2 = 4 \qquad\qquad y_2 = 4 \qquad\qquad z_2 = -1/2$$
$$x_1 = 0 \qquad\qquad y_1 = 6 \qquad\qquad z_1 = -3/2$$
$$A = (0, 6, -3/2), \qquad B = (4, 4, -1/2)$$

17. $\cos\theta = \dfrac{5 \cdot 4 - 1 \cdot 5 + 3(-2)}{\sqrt{25 + 1 + 9}\,\sqrt{16 + 25 + 4}} = \dfrac{9}{\sqrt{35}\,\sqrt{45}} = \dfrac{9}{15\sqrt{7}} = \dfrac{3\sqrt{7}}{35}$

$\approx 0.2268, \qquad \theta \approx 77°$

21. $u + v = (2 + 1)i + (-1 - 4)j + (6 - 1)k = 3i - 5j + 5k$

$u - v = (2 - 1)i + (-1 + 4)j + (6 + 1)k = i + 3j + 7k$

$u \cdot v = 2 \cdot 1 - 1 \, (-4) + 6(-1) = 0$.   Orthogonal

25. $w = \left(\dfrac{u \cdot v}{|v|}\right)\dfrac{v}{|v|} = \dfrac{u \cdot v}{|v|^2}v = \dfrac{4 \cdot 1 - 2(-2) - 1 \cdot 1}{1 + 4 + 1}(i - 2j + k)$

$= \dfrac{7}{6}(i - 2j + k) = \dfrac{7}{6}i - \dfrac{7}{3}j + \dfrac{7}{6}k$

29. $a_1 i + b_1 j + c_1 k$ is presented by $\overrightarrow{OP}$ from $(0, 0, 0)$ to

(a_1, b_1, c_1) and $a_2 i + b_2 j + c_2 k$ by $\overrightarrow{OP_2}$ from $(0, 0, 0)$

to (a_2, b_2, c_2), or by $\overrightarrow{P_1 P_3}$ from (a_1, b_1, c_1) to

$(a_1 + a_2, b_1 + b_2, c_1 + c_2)$. Hence the sum is

represented by $\overrightarrow{OP_3}$ or $(a_1 i + b_1 j + c_1 k) + (a_2 i + b_2 j + c_2 k)$

$= (a_1 + a_2)i + (b_1 + b_2)j + (c_1 + c_2)k$.

Let $(a_1 i + b_1 j + c_1 k) - (a_2 i + b_2 j + c_2 k) = ai + bj + ck$.

Then

$a_1 i + b_1 j + c_1 k = (a_2 i + b_2 j + c_2 k) + (ai + bj + ck)$

$\qquad\qquad = (a_2 + a)i + (b_2 + b)j + (c_2 + c)k$

$a_1 = a_2 + a \qquad\qquad b_1 = b_2 + b \qquad\qquad c_1 = c_2 + c$

$a = a_1 - a_2 \qquad\qquad b = b_1 - b_2 \qquad\qquad c = c_1 - c_2$

$ai + bj + ck$ is represented by $\overrightarrow{OP}$ from $(0, 0, 0)$ to

(a, b, c). Its length is

$\sqrt{(a - 0)^2 + (b - 0)^2 + (c - 0)^2} = \sqrt{a^2 + b^2 + c^2}$

$v = ai + bj + ck$ is represented by $\overrightarrow{OP}$ from $(0, 0, 0)$

to (a, b, c). $w = dai + dbj + dck$ is represented by

$\overrightarrow{OQ}$ from $(0, 0, 0)$ to (da, db, dc). Clearly these points

lie on the same line; so w is in the same direction

as or the opposite direction from v, depending upon

the sign of d. Furthermore

$$|w| = \sqrt{d^2 a^2 + d^2 b^2 + d^2 c^2} = \sqrt{d^2 (a^2 + b^2 + c^2)} = |d||v|$$

Hence $d(ai + bj + ck) = dai + dbj + dck$.

Section 20.3, pp. 628-629

1. $k^2 = a^2 + b^2 + c^2 = 1 + 16 + 64 = 81$, $k = 9$

   $\cos \alpha = \dfrac{1}{9} = 0.1111$, $\alpha = 84°$

   $\cos \beta = \dfrac{4}{9} = 0.4444$, $\beta = 64°$

   $\cos \gamma = \dfrac{8}{9} = 0.8889$, $\gamma = 27°$

5. $k^2 = a^2 + b^2 + c^2 = 1 + 1 + 1 = 3$, $k = -\sqrt{3}$

   $\cos \alpha = \cos \beta = \cos \gamma = -\dfrac{1}{\sqrt{3}} = -\dfrac{\sqrt{3}}{3} = -0.5773$,

   $\alpha = \beta = \gamma = 125°$

9. $\{2-0, \ 2-0, \ 1-3\} = \{2, 2, -2\}$

13. $(2+1, \ 3+5, \ -1+2) = (3, 8, 1)$

    $(2+2, \ 3+10, \ -1+4) = (4, 13, 3)$

17. $(1+4, \ 3+0, \ -1-1) = (5, 3, -2)$

    $(1+8, \ 3+0, \ -1-2) = (9, 3, -3)$

21. $(3, 4, 1), \ (4, 8, -1): \ \{-1, -4, 2\}$

    $(2, 3, -5), \ (0, -5, -1): \ \{2, 8, -4\}$

    Either parallel or coincident

    $(4, 8, -1), \ (2, 3, -5): \ \{2, 5, 4\}$

    Parallel

25. $(2, 1, 4), \ (4, -3, 12): \ \{-2, 4, -8\}$

    $(1, 3, 0), \ (6, -7, 20): \ \{-5, 10, -20\}$

Parallel or coincident

$(4,-3,12)$, $(1,3,0)$: $\{3,-6,12$

Coincident

29. $(2,1,3)$, $(5,-1,1)$: $\{-3,2,2\}$

$(3,4,-1)$, $(5,3,3)$: $\{-2,1,-4\}$

$(-3)(-2)+2\cdot 1+2(-4)=0$,   perpendicular

Section 20.4, pp. 633-634

1. $x=5+3t$, $y=1-2t$, $z=3+4t$

$$\frac{x-5}{3}=\frac{y-1}{-2}=\frac{z-3}{4}$$

5. $x=1+2t$, $y=1$, $z=1+t$

$$\frac{x-1}{2}=\frac{z-1}{1}, y=1$$

9. $\{4-2,\ 0-3,\ 5-1\}=\{2,-3,4\}$

$x=4+2t$, $y=-3t$, $z=5+4t$

13. $\{5-5,\ 2-1,\ 4-3\}=\{0,1,1\}$

$x=5$, $y=1+t$, $z=3+t$

$x=5$, $y-1=z-3$

17. $4+t=3+2s$ or $2s-t=1$

$-8-2t=-1+s$ or $s+2t=-7$

$12t=-3-3s$ or $s+4t=-1$

Solving the first two equations simultaneously, we get $s=-1$ and $t=-3$. Since this is not a solution of the third, the lines do not intersect.

21. $x = 2 + t,$ $\quad y = 3 - 2t,$ $\quad z = -1 + t$

$\quad x = 3 + 2s,$ $\quad y = 1 - 4s,$ $\quad z = 2s$

$\quad 2 + t = 3 + 2s \quad$ or $\quad 2s - t = -1$

$\quad 3 - 2t = 1 - 4s \quad$ or $\quad 2s - t = -1$

$\quad -1 + t = 2s \quad$ or $\quad 2s - t = -1$

Since there are infinitely many solutions for these three equations, the lines are identical.

25. $\{5, -2, 1\},$ $\{0, 2, 4\}$

$5 \cdot 0 - 2 \cdot 2 + 1 \cdot 4 = 0,$ perpendicular

29. $\{1, 3, -2\},$ $\{-3, -9, 6\}$

Parallel or coincident

$(-3, 4, -2)$ is on the first line but not the second.

Parallel

Section 20.5, pp. 639-641

1. $u \times v = (-1 \cdot 1 - 4 \cdot 1)i + (4 \cdot 2 - 3 \cdot 1)j + [3 \cdot 1 - (-1) \cdot 2]k$

$\quad = -5i + 5j + 5k$

5. $u \times v = (0 \cdot 0 - 1 \cdot 1)i + [1(-1) - 3 \cdot 0]j + [3 \cdot 1 - 0(-1)]k$

$\quad = -i - j + 3k$

9. $u = i - 2j - k,$ $\qquad v = -2i + j - k$

$\quad u \times v = 3i + 3j - 3k$

Direction numbers $\{1, 1, -1\}$

13. $u = -2i + j - k,$ $\qquad v = i + 2k - k$

$\quad u \times v = i - 3j - 5k$

Direction numbers $\{1,-3,-5\}$

$$x = 3 + t, \qquad y = 2 - 3t, \qquad z = 1 - 5t$$

17. $u = (4 + t - 2)i + (3 - 2t - 0)j + (1 + t - 5)k$

    $= (2 + t)i + (3 - 2t)j + (-4 + t)k$

    $v = i - 2j + k$

    $u \cdot v = (2 + t) \cdot 1 + (3 - 2t)(-2) + (-4 + t) \cdot 1 = 6t - 8 = 0$,

    $t = \dfrac{4}{3}$, $\quad (\dfrac{16}{3}, \dfrac{1}{3}, \dfrac{7}{3})$, $(2, 0, 5)$;

    direction numbers $\{\dfrac{10}{3}, \dfrac{1}{3}, -\dfrac{8}{3}\}$ or $\{10, 1, -8\}$

    $x = 2 + 10t, \qquad y = t, \qquad z = 5 - 8t$

21. $u = i - 5j + k, \qquad v = i + 2j + 4k$

    $u \times v = -22i - 3j + 7k$

    $w = (4 - 1)i + (5 - 1)j + (-3 - 2)k = 3i + 4j - 5k$

    $\left| \text{proj of } w \text{ on } u \times v \right| = \dfrac{|w \cdot (u \times v)|}{|u \times v|} = \dfrac{|-22 \cdot 3 - 3 \cdot 4 + 7(-5)|}{\sqrt{484 + 9 + 49}}$

    $\qquad\qquad = \dfrac{113}{\sqrt{542}}$

25. The area of $\triangle ABC$ is $A = \dfrac{1}{2} AB \cdot BC \cdot \sin \angle BAC$. If u is

    represented by $\overrightarrow{AB}$, v by $\overrightarrow{BC}$ and $\theta = \angle BAC$, then

    $\qquad A = \dfrac{1}{2}|u| \cdot |v| \sin \theta = \dfrac{1}{2}|u \times v|$

29. $u = (3 - 4)i + (-1 - 2)j = -i - 3j$

    $v = (-1 - 4)i + (0 - 2)j = -5i - 2j$

    $u \times v = -13j$, $\qquad A = \dfrac{1}{2}|u \times v| = \dfrac{13}{2}$

33. $u = 2i + j + 3k,$ $v = 5i + 3j + k,$ $w = 2i - j + 4k$

$v \times w = 13i - 18j - 11k$

$u \cdot (v \times w) = (2i + j + 3k) \cdot (13i - 18j - 11k)$

$$= 26 - 18 - 33 = -25$$

Section 20.6, p. 645

1.

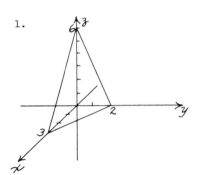

5.

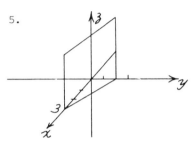

9. $3(x - 4) - (y - 1) - 2(z + 3) = 0$

$3x - y - 2z - 17 = 0$

13. u is a vector from $(1, 1, 0)$ to $(1, 3, 2)$, v from $(1, 3, 2)$ to $(2, -1, 1)$.

$u = 2j + 2k,$ $v = i - 4j - k$

17. The lines clearly intersect at $(4, 2, 1)$. If $\{a, b, c\}$ are direction numbers for a line perpendicular to both of the given lines, then

$a - b + 2c = 0,$ $-3a + 2b - c = 0$

$-5a + 3b = 0$

$a = 3,$ $b = 5,$ $c = 1$

$3(x - 4) + 5(y - 2) + (z - 1) = 0,$ $3x + 5y + z - 23 = 0$

21. The given lines are parallel. $(3, 4, 1)$ is on one line and $(-1, 3, 4)$ on the other. Direction numbers of the line through these two points are $\{4, 1, -3\}$. If $\{a, b, c\}$ are direction numbers of a line perpendicular to this line and the given lines, then

$$2a - b + c = 0, \qquad 4a + b - 3c = 0, \qquad 6a - 2c = 0$$

$$a = 1, \qquad c = 3, \qquad b = 5$$

$$(x - 3) + 5(y - 4) + 3(z - 1) = 0, \qquad x + 5y + 3z - 26 = 0$$

25. $(2, 5, -1), \quad \{2, -1, 3\}$

$$x = 2 + 2t, \qquad y = 5 - t, \qquad z = -1 + 3t$$

29. $\{3, 1, -5\}, \quad \{1, 2, 1\}$

$$3 \cdot 1 + 1 \cdot 2 - 5 \cdot 1 = 0, \qquad \text{perpendicular}$$

33. $\{2, -1, 3\}, \quad \{6, -3, 9\}$

Parallel or coincident

Since the coefficients of $x, y,$ and z of the second equation are 3 times those of the first, but the constant term of the second is not 3 times that of the first, they are parallel.

Section 20.7, pp. 649-650

1. $d = \dfrac{|2 \cdot 1 - 4 \cdot 3 + 4(-2) + 3|}{\sqrt{2^2 + (-4)^2 + 4^2}} = \dfrac{5}{2}$

5. $d = \dfrac{|1 \cdot 2 + 3 \cdot 1 + 1(-3) - 2|}{\sqrt{1^2 + 3^2 + 1^2}} = 0$

9. $1 = \dfrac{|8 \cdot 1 - 4 + 4z - 3|}{\sqrt{64 + 1 + 16}}, \qquad |1 + 4z| = 9$

$1 + 4z = \pm 9, \qquad z = \dfrac{-1 \pm 9}{4} = 2, \quad -\dfrac{5}{2}$

13. $u = i + 3j + k$

$(2, 4, -1)$, $(5, -2, 3)$: $\quad v = 3i - 6j + 4k$

$u \times v = 18i - j - 15k$

$d = |v| \sin \theta = \dfrac{|u \times v|}{|u|} = \dfrac{\sqrt{324 + 1 + 225}}{\sqrt{1 + 9 + 1}} = \sqrt{\dfrac{550}{11}} = 5\sqrt{2}$

17. $u = 3i + j - 2k$

$(1, 4, 2)$, $(1, -2, 4)$: $\quad v = 6j - 2k$

$u \times v = 10i + 6j + 18k$

$d = |v| \sin \theta = \dfrac{|u \times v|}{|u|} = \dfrac{2\sqrt{25 + 9 + 81}}{\sqrt{9 + 1 + 4}} = \sqrt{\dfrac{230}{7}}$

21. $\dfrac{3x + y - 4z - 3}{\sqrt{9 + 1 + 16}} = 0, \qquad \dfrac{3x + y - 4z - 3}{\sqrt{26}} = 0$

$\dfrac{6x + 2y - 8z + 5}{-\sqrt{36 + 4 + 64}} = 0, \qquad \dfrac{-6x - 2y + 8z - 5}{2\sqrt{26}} = 0$

$d = \dfrac{3}{\sqrt{26}} + \dfrac{5}{2\sqrt{26}} = \dfrac{11}{2\sqrt{26}}$

25. $\cos \theta = \dfrac{u \cdot v}{|u| \, |v|} = \dfrac{1 \cdot 3 + 3(-1) + 4 \cdot 4}{\sqrt{1 + 9 + 16}\, \sqrt{9 + 1 + 16}} = \dfrac{8}{13} = 0.6154$

$\theta = 52°$

29. $\cos \theta = \dfrac{-1(-1) + 2 \cdot 1 + 1 \cdot 2}{\sqrt{1 + 4 + 1}\, \sqrt{1 + 1 + 4}} = \dfrac{5}{6} = 0.8333$

$\theta = 34°$

33. A vector perpendicular to the plane is $v = Ai + Bj + Ck$. A, B, and C are not all 0. Suppose $A \neq 0$. Then a point in the plane is $(-D/A, 0, 0)$. The vector u from $(-D/A, 0, 0)$ to (x_1, y_1, z_1) is

$u = (x_1 + D/A)i + y_1 j + z_1 k$

$d = \dfrac{|u \cdot v|}{|v|} = \dfrac{|A(x_1 + D/A) + By_1 + Cz_1|}{\sqrt{A^2 + B^2 + C^2}} = \dfrac{|Ax_1 + By_1 + Cz_1 + D|}{\sqrt{A^2 + B^2 + C^2}}$

1.

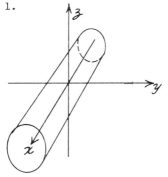

5.

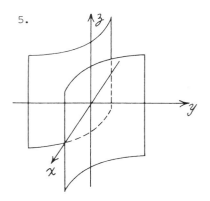

9.

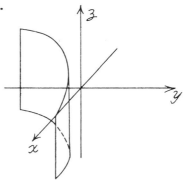

13. $x^2 + y^2 + z^2 - 8x + 4y - 10z + 46 = 0$

$x^2 - 8x + 16 + y^2 + 4y + 4 + z^2$

$- 10z + 25 = -46 + 16 + 4 + 25$

$(x - 4)^2 + (y + 2)^2 + (z - 5)^2 = -1$

No locus

17. $9x^2 + 9y^2 + 9z^2 - 6x + 6y + 12z - 2 = 0$

$x^2 + y^2 + z^2 - \frac{2}{3}x + \frac{2}{3}y + \frac{4}{3}z - \frac{2}{9} = 0$

$x^2 - \frac{2}{3}x + \frac{1}{9} + y^2 + \frac{2}{3}y + \frac{1}{9} + z^2 + \frac{4}{3}z + \frac{4}{9} = \frac{2}{9} + \frac{1}{9} + \frac{1}{9} + \frac{4}{9}$

$(x - \frac{1}{3})^2 + (y + \frac{1}{3})^2 + (z + \frac{2}{3})^2 = \frac{8}{9}$

Sphere: center $(\frac{1}{3}, -\frac{1}{3}, -\frac{2}{3})$, $r = \frac{2\sqrt{2}}{3}$

21. $(x - 4)^2 + (y - 1)^2 + (z + 2)^2 = 9$

$x^2 + y^2 + z^2 - 8x - 2y + 4z + 12 = 0$

25. The line through $(1, 4, -3)$ and perpendicular to the given plane is:

$$x = 1 + t, \qquad y = 4 - 3t, \qquad z = -3 + 4t$$

The sphere with center $(1, 4, -3)$ and $r = \sqrt{26}$ is

$$(x - 1)^2 + (y - 4)^2 + (z + 3)^2 = 26$$

$$t^2 + 9t^2 + 16t^2 = 26, \qquad 26t^2 = 26$$

$$t^2 = 1, \qquad t = \pm 1$$

The points of intersection of the line and sphere are $(2, 1, 1)$ and $(0, 7, -7)$. These are the centers of the desired spheres.

$$(x - 2)^2 + (y - 1)^2 + (z - 1)^2 = 26$$

$$x^2 + y^2 + z^2 - 4x - 2y - 2z - 20 = 0$$

$$x^2 + (y - 7)^2 + (z + 7)^2 = 26, \quad x^2 + y^2 + z^2 - 14y + 14z + 72 = 0$$

29. If (x, y, z) is on the sphere, its distance from the center is r.

$$\sqrt{(x - h)^2 + (y - k)^2 + (z - \ell)^2} = r$$

$$(x - h)^2 + (y - k)^2 + (z - \ell)^2 = r^2$$

The above argument is reversible since r is positive.

Section 20.9, p. 657

1. Ellipsoid

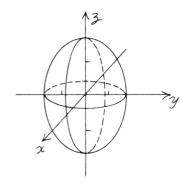

5. Circular cone

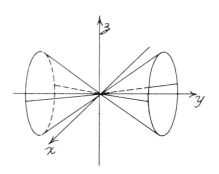

9. Hyperboloid of two sheets

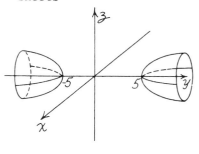

13. Hyperboloid of two sheets

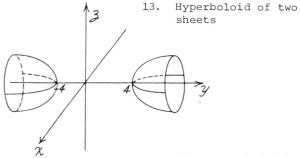

17. Hyperboloid of two sheets

21. Hyperbolic paraboloid

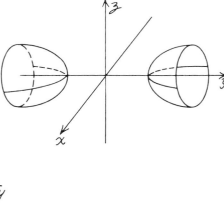

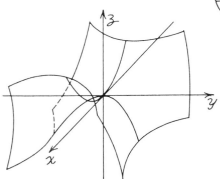

25. Hyperboloid of one sheet

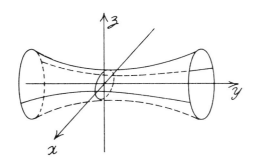

29. $x^2 + 4y^2 + 9z^2 + 2x + 16y - 18z - 10 = 0$

$(x^2 + 2x + 1) + 4(y^2 + 4y + 4) + 9(z^2 - 2z + 1) = 10 + 1 + 16 + 9$

$(x + 1)^2 + 4(y + 2)^2 + 9(z - 1)^2 = 36$

$\dfrac{(x + 1)^2}{36} + \dfrac{(y + 2)^2}{9} + \dfrac{(z - 1)^2}{4} = 1$

$\dfrac{x'^2}{36} + \dfrac{y'^2}{9} + \dfrac{z'^2}{4} = 1$

Section 20.10, p. 661

1. a) $x = r \cos \theta = 2 \cos 45° = \dfrac{2}{\sqrt{2}} = \sqrt{2}$

$y = r \sin \theta = 2 \sin 45° = \dfrac{2}{\sqrt{2}} = \sqrt{2}$

$(\sqrt{2}, \sqrt{2}, 1)$

b) $x = r \cos \theta = 3 \cos \dfrac{2\pi}{3} = 3(-\dfrac{1}{2}) = -\dfrac{3}{2}$

$y = 4 \sin \theta = 3 \sin \dfrac{2\pi}{3} = 3 \cdot \dfrac{\sqrt{3}}{2} = \dfrac{3\sqrt{3}}{2}$

$(-\dfrac{3}{2}, \dfrac{3\sqrt{3}}{2}, -2)$

3. a) $x = \rho \sin \varphi \cos \theta = 3 \sin 30° \cos 45°$

$\quad = 3 \cdot \dfrac{1}{2} \cdot \dfrac{1}{\sqrt{2}} = \dfrac{3}{2\sqrt{2}}$

$y = \rho \sin \varphi \sin \theta = 3 \sin 30° \sin 45°$

$\quad = 3 \cdot \dfrac{1}{2} \cdot \dfrac{1}{\sqrt{2}} = \dfrac{3}{2\sqrt{2}}$

$z = \rho \cos \varphi = 3 \cos 30° = 3 \cdot \dfrac{\sqrt{3}}{2} = \dfrac{3\sqrt{3}}{2}$

$(\dfrac{3}{2\sqrt{2}}, \dfrac{3}{2\sqrt{2}}, \dfrac{3\sqrt{3}}{2})$

b) $x = \rho \sin \varphi \cos \theta = 1 \sin 0 \cos \dfrac{\pi}{6} = 0$

$y = \rho \sin \varphi \sin \theta = 1 \sin 0 \sin \dfrac{\pi}{6} = 0$

$z = \rho \cos \varphi = 1 \cos 0 = 1 \cdot 1 = 1$

$(0, 0, 1)$

5. a) $x = r \cos \theta = 3 \cos 30° = 3 \cdot \dfrac{\sqrt{3}}{2} = \dfrac{3\sqrt{3}}{2}$

$y = r \sin \theta = 3 \sin 30° = 3 \cdot \dfrac{1}{2} = \dfrac{3}{2}$

$(\dfrac{3\sqrt{3}}{2}, \dfrac{3}{2}, 4)$

$\rho^2 = x^2 + y^2 + z^2 = \dfrac{27}{4} + \dfrac{9}{4} + 16 = 25, \qquad \rho = 5$

$z = \rho \cos \varphi, \quad 4 = 5 \cos \varphi, \quad \cos \varphi = \dfrac{4}{5},$

$\qquad \varphi = \text{Arccos } \dfrac{4}{5}$

$(5, 30°, \text{Arccos } \dfrac{4}{5})$

b) $x = r \cos \theta = 2 \cos \dfrac{\pi}{4} = 2 \cdot \dfrac{1}{\sqrt{2}} = \sqrt{2}$

$y = r \sin \theta = 2 \sin \dfrac{\pi}{4} = 2 \cdot \dfrac{1}{\sqrt{2}} = \sqrt{2}$

$(\sqrt{2}, \sqrt{2} - 2)$

$\rho^2 = x^2 + y^2 + z^2 = 2 + 2 + 4 = 8, \qquad \rho = 2\sqrt{2}$

$z = \rho \cos \varphi, \quad \pi - 2 = 2\sqrt{2} \cos \varphi,$

$\qquad \cos \varphi = -\dfrac{1}{\sqrt{2}}, \quad \varphi = \dfrac{3}{4}$

$(2\sqrt{2}, \dfrac{\pi}{4}, \dfrac{3\pi}{4})$

9. $x^2 + y^2 = z$

$\quad r^2 = z$

$x^2 + y^2 = z$

$\rho^2 \sin^2\varphi \cos^2\theta + \rho^2 \sin^2\varphi \sin^2\theta$

$\quad = \rho \cos \varphi$

$\rho^2 \sin^2\varphi = \rho \cos \varphi$

$\rho \sin^2\varphi = \cos \varphi$

$\rho = \dfrac{\cos \varphi}{\sin^2\varphi} = \csc \varphi \cot \varphi$

13. $x^2 + y^2 - z^2 = 1$

$\quad r^2 - z^2 = 1$

$\quad r^2 = 1 + z^2$

$x^2 + y^2 - z^2 = 1$

$\rho^2 \sin^2\varphi \cos^2\theta + \rho^2 \sin^2\varphi \sin^2\theta$

$\quad - \rho^2 \cos^2\varphi = 1$

$\rho^2 [\sin^2\varphi(\cos^2\theta + \sin^2\theta) - \cos^2\varphi]$

$\quad = 1$

$\rho^2 (\sin^2\varphi - \cos^2\varphi) = 1$

$-\rho^2 \cos 2\varphi = 1$

$\rho^2 = -\sec 2\varphi$

17. $z = r^2$

$\quad z = x^2 + y^2$

21. $\rho = \sin \varphi \cos \theta$

$\quad \rho^2 = \rho \sin \varphi \cos \theta$

$\quad x^2 + y^2 + z^2 = x$

1. $f'(t) = i + 3t^2 j$

    $f'(1) = i + 3j$

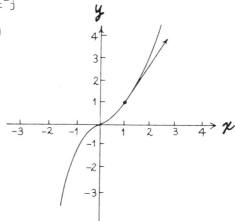

5. $f'(t) = i + 2tj + 2tk$

    $f'(1) = i + 2j + 2k$

9. $f'(t) = 2ti + 3t^2 j + 4t^3 k$,

    $f''(t) = 2i + 6tj + 12t^2 k$

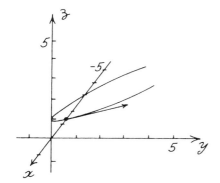

13. $v = i + 2tj$, $\quad a = 2j$

$\dfrac{ds}{dt} = |v| = \sqrt{1 + 4t^2}$

$\dfrac{d^2s}{dt^2} = \dfrac{8t}{2\sqrt{1 + 4t^2}} = \dfrac{4t}{\sqrt{1 + 4t^2}}$

At $t = 1$, $\quad v = i + 2j$

$\qquad a = 2j$

$\dfrac{ds}{dt} = \sqrt{5}$, $\quad \dfrac{d^2s}{dt^2} = \dfrac{4}{\sqrt{5}}$

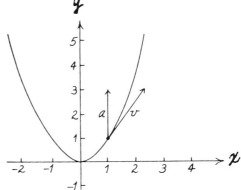

17. $v = i + 3t^2 j$, $\quad a = 6tj$

$\dfrac{ds}{dt} = |v| = \sqrt{1 + 9t^4}$

$\dfrac{d^2s}{dt^2} = \dfrac{36t^3}{2\sqrt{1 + 9t^4}}$

$\qquad = \dfrac{18t^3}{\sqrt{1 + 9t^4}}$

At $t = 1$, $v = i + 3j$

$\qquad a = 6j$

$\dfrac{ds}{dt} = \sqrt{10}$, $\quad \dfrac{d^2s}{dt^2} = \dfrac{18}{\sqrt{10}}$

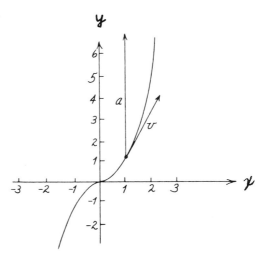

21. $v = i - \dfrac{1}{t^2}j$, $\quad a = \dfrac{2}{t^3}j$

$\dfrac{ds}{dt} = |v| = \sqrt{1 + 1/t^4}$

$\qquad = \dfrac{\sqrt{t^4 + 1}}{t^2}$

$\dfrac{d^2s}{dt^2} = \dfrac{t^2 \cdot 2t^3/\sqrt{t^4+1} - 2t\sqrt{t^4+1}}{t^4}$

$\qquad = \dfrac{-2}{t^3\sqrt{t^4 + 1}}$

At $t = 1$, $v = i - j$

$\qquad a = 2j$

$\dfrac{ds}{dt} = \sqrt{2}$, $\dfrac{d^2s}{dt^2} = \dfrac{-2}{\sqrt{2}} = -\sqrt{2}$

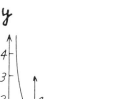

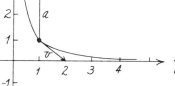

25. $v = i + e^t j$, $\quad a = e^t j$

$\dfrac{ds}{dt} = |v| = \sqrt{1 + e^{2t}}$

$\dfrac{d^2s}{dt^2} = \dfrac{2e^{2t}}{2\sqrt{1 + e^{2t}}} = \dfrac{e^{2t}}{\sqrt{1 + e^{2t}}}$

At $t = 1$, $v = i + ej$

$\qquad a = ej$

$\dfrac{ds}{dt} = \sqrt{1 + e^2}$, $\qquad \dfrac{d^2s}{dt^2} = \dfrac{e^2}{\sqrt{1 + e^2}}$

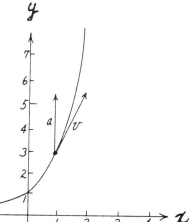

29. $v = 2ti - j + k$, $\quad a = 2i$

$$\frac{ds}{dt} = |v| = \sqrt{4t^2 + 1 + 1} = \sqrt{4t^2 + 2}$$

$$\frac{d^2s}{dt^2} = \frac{8t}{2\sqrt{4t^2 + 2}} = \frac{4t}{\sqrt{4t^2 + 2}}$$

33. $v = -\dfrac{1}{t^2}i + \sqrt{2}j + t^2 k$, $\quad a = \dfrac{2}{t^3}i + 2tk$

$$\frac{ds}{dt} = |v| = \sqrt{\frac{1}{t^4} + 2 + t^4} = \frac{1}{t^2} + t^2, \quad \frac{d^2s}{dt^2} = -\frac{2}{t^3} + 2t$$

Review 20, pp. 666-667

1. $(5,7,2)$, $(1,1,z)$: midpoint $(3,4,\dfrac{z+2}{2})$

$$3 = \sqrt{(5-3)^2 + (3-4)^2 + (\frac{z+2}{2} + 2)^2}$$

$$9 = 4 + 1 + \frac{(z+6)^2}{4}, \quad (z+6)^2 = 16$$

$z + 6 = \pm 4$, $\qquad z = -6 \pm 4 = -2, -10$

5. p = projection of u upon v

$$= \frac{u \cdot v}{|v|^2}v = \frac{2 \cdot 1 - 1 \cdot 4 + 3(-1)}{1 + 16 + 1}(i + 4j - k)$$

$$= -\frac{5}{18}(i + 4j - k) = -\frac{5}{18}i - \frac{10}{9}j + \frac{5}{18}k$$

$u = p + q$

$$2i - j + 3k = -\frac{5}{18}i - \frac{10}{9}j + \frac{5}{18}k + q$$

$$q = \frac{41}{18}i + \frac{1}{9}j + \frac{49}{18}k$$

9. $x = 2 + 5t$, $\quad y = -3 + 2t$, $\quad z = 5 - t$

$$\frac{x-2}{5} = \frac{y+3}{2} = \frac{z-5}{-1}$$

13. $(-1,3,1)$, $(4,2,3)$: $\quad u = 5i - j + 2k$

$(2,5,-3)$, $(4,2,3)$: $\quad v = 2i - 3j + 6k$

$u \times v = -26j - 13k$, $\quad$ Direction numbers $\{0, 2, 1\}$

$x = 4$, $\quad y = 2 + 2t$, $\quad z = 3 + t$

17. $\{-1, 2, 0\}$

$-1(x - 5) + 2(y - 0) + 0(z - 3) = 0, \quad -x + 2y + 5 = 0$

21. $d = \dfrac{|2 \cdot 4 - 2 \cdot 3 + 1 \cdot 1 - 4|}{\sqrt{4 + 4 + 1}} = \dfrac{1}{3}$

25. $u = 2i - 5j + k, \qquad v = i + j - 2k$

$\cos \theta = \dfrac{|u \cdot v|}{|u||v|} = \dfrac{|2 \cdot 1 - 5 \cdot 1 + 1(-2)|}{\sqrt{4 + 25 + 1}\,\sqrt{1 + 1 + 4}} = \dfrac{5}{\sqrt{30}\,\sqrt{6}} = \dfrac{\sqrt{5}}{6}$

$\qquad = 0.373$

$\theta = 68°$

29. $9x^2 + 9y^2 - 4z^2 = 36$

$\dfrac{x^2}{4} + \dfrac{y^2}{4} - \dfrac{z^2}{9} = 1$

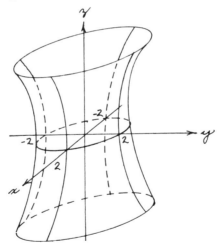

33. (a) $r^2 = x^2 + y^2 = 4 + 4 = 8, \qquad r = 2\sqrt{2}$

$\tan \theta = \dfrac{y}{x} = \dfrac{2}{2} = 1, \qquad\qquad \theta = \pi/4$

$(2\sqrt{2}, \pi/4, 1)$

$\rho^2 = x^2 + y^2 + z^2 = 4 + 4 + 1 = 9, \quad \rho = 3$

$\tan \theta = \dfrac{y}{x} = 1, \qquad\qquad \theta = \pi/4$

$$y = \rho \sin \varphi \sin \theta, \qquad 2 = 3 \sin \varphi \cdot 1/\sqrt{2}$$

$$\sin \varphi = 2\sqrt{2}/3, \qquad (3, \pi/4, \text{Arcsin } 2\sqrt{2}/3)$$

(b) $r^2 = x^2 + y^2 = 1 + 4 = 5, \qquad r = -\sqrt{5}$

$$\tan \theta = \frac{y}{x} = \frac{2}{-1} = -2, \qquad \theta = \text{Arctan } (-2)$$

$(-\sqrt{5}, \text{Arctan}(-2), -2)$

$$\rho^2 = x^2 + y^2 + z^2 = 1 + 4 + 4 = 9, \quad \rho = 3$$

$$\tan \theta = \frac{y}{x} = \frac{2}{-1} = -2, \quad \theta = \pi + \text{Arctan}(-2)$$

$$y = \rho \sin \varphi \sin \theta, \qquad 2 = 3 \sin \varphi \cdot 2/\sqrt{5}$$

$$\sin \varphi = \sqrt{5}/3,$$

$(3, \pi + \text{Arctan}(-2), \sqrt{5}/3)$

37.

t	x	y	z
-2	-3	3	-1
-1	-2	0	0
0	-1	-1	1
1	0	0	1
2	1	3	3
3	2	8	4

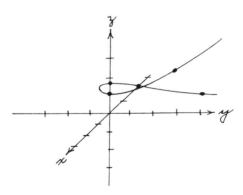

Chapter Twenty-One
Partial Derivatives

Section 21.2, pp. 673-674

1. $\dfrac{\partial z}{\partial x} = 2x$, $\dfrac{\partial z}{\partial y} = 2y$

5. $\dfrac{\partial z}{\partial x} = \dfrac{y}{2\sqrt{xy}}$, $\dfrac{\partial z}{\partial y} = \dfrac{x}{2\sqrt{xy}}$

9. $\dfrac{\partial z}{\partial x} = \dfrac{2x}{x^2 + y^2}$, $\dfrac{\partial z}{\partial y} = \dfrac{2y}{x^2 + y^2}$

13. $\dfrac{\partial z}{\partial x} = y = -1$, $\dfrac{\partial z}{\partial y} = x - 3y^2 = 2 - 3 \cdot 1 = -1$

17. $\dfrac{\partial w}{\partial x} = yz$, $\dfrac{\partial w}{\partial y} = xz$, $\dfrac{\partial w}{\partial z} = xy$

21. $\dfrac{\partial w}{\partial x} = 2x$, $\dfrac{\partial w}{\partial y} = 2y$, $\dfrac{\partial w}{\partial z} = 0$

25. $\dfrac{\partial z}{\partial x} = -2(x+y)^{-3}$, $\dfrac{\partial z}{\partial y} = -2(x+y)^{-3}$

$\dfrac{\partial^2 z}{\partial x^2} = 6(x+y)^{-4} = \dfrac{6}{(x+y)^4}$, $\dfrac{\partial^2 z}{\partial y \partial x} = 6(x+y)^{-4} = \dfrac{6}{(x+y)^4}$

$\dfrac{\partial^2 z}{\partial x \partial y} = 6(x+y)^{-4} = \dfrac{6}{(x+y)^4}$, $\dfrac{\partial^2 z}{\partial y^2} = 6(x+y)^{-4} = \dfrac{6}{(x+y)^4}$

29. Suppose $\left| x - 2 \right| < \delta$ and $\left| y - 3 \right| < \delta$ for some $\delta > 0$.

$\left| f(x) - 13 \right| = \left| 3x + 2y - 12 \right| = \left| 3x - 6 + 2y - 6 \right|$

$= \left| 3(x-2) + 2(y-3) \right| \leqq 3\left| x - 2 \right| + 2\left| y - 3 \right| < 5\delta$

Choose $\delta = \epsilon/5$.

33. Suppose $x = y$. Then $f(x,y) = f(y,y) = \dfrac{y^2}{2y^2} = \dfrac{1}{2}$.

$\lim_{\substack{x \to 0 \\ y \to 0}} f(x,y) = \lim_{y \to 0} f(y,y) = \lim_{y \to 0} \dfrac{1}{2} = \dfrac{1}{2}$

Suppose $x = 0$. Then $f(x,y) = f(0,y) = 0$.

$\lim_{\substack{x \to 0 \\ y \to 0}} f(x,y) = \lim_{y \to 0} f(0,y) = \lim_{y \to 0} 0 = 0$

1. $\dfrac{dz}{dt} = \dfrac{\partial z}{\partial x}\dfrac{dx}{dt} + \dfrac{\partial z}{\partial y}\dfrac{dy}{dt} = y \cdot 2 + x \cos t = 2y + x \cos t$

5. $\dfrac{dw}{dt} = \dfrac{\partial w}{\partial x}\dfrac{dx}{dt} + \dfrac{\partial w}{\partial y}\dfrac{dy}{dt} + \dfrac{\partial w}{\partial z}\dfrac{dz}{dt} = 2x \cdot 1 + 2y \cdot 1 + 2z \cdot 1$

$= 2x + 2y + 2z$

9. $\dfrac{\partial z}{\partial u} = \dfrac{\partial z}{\partial x}\dfrac{\partial x}{\partial u} + \dfrac{\partial z}{\partial y}\dfrac{\partial y}{\partial u} = 2x \cdot 2u + 2y \cdot 2u = 4(x+y)u$

$\dfrac{\partial z}{\partial v} = \dfrac{\partial z}{\partial x}\dfrac{\partial x}{\partial v} + \dfrac{\partial z}{\partial y}\dfrac{\partial y}{\partial v} = 2x \cdot 2v + 2y(-2v) = 4(x-y)v$

13. $\dfrac{\partial w}{\partial u} = \dfrac{\partial w}{\partial x}\dfrac{\partial x}{\partial u} + \dfrac{\partial w}{\partial y}\dfrac{\partial y}{\partial u} + \dfrac{\partial w}{\partial z}\dfrac{\partial z}{\partial u} = 2x \cdot 2u + 2y \cdot 2u + 2z \cdot v$

$= 4xu + 4yu + 2zv$

$\dfrac{\partial w}{\partial v} = \dfrac{\partial w}{\partial x}\dfrac{\partial x}{\partial v} + \dfrac{\partial w}{\partial y}\dfrac{\partial y}{\partial v} + \dfrac{\partial w}{\partial z}\dfrac{\partial z}{\partial v} = 2x \cdot 2v + 2y(-2v) + 2z \cdot u$

$= 4xv - 4yv + 2zu$

17. $\dfrac{\partial w}{\partial t} = \dfrac{\partial w}{\partial x}\dfrac{\partial x}{\partial t} + \dfrac{\partial w}{\partial y}\dfrac{\partial y}{\partial t} + \dfrac{\partial w}{\partial z}\dfrac{\partial z}{\partial t} = yz \cdot uv + xz\dfrac{1}{u} + xy \cdot 0$

$= yzuv + \dfrac{xz}{u}$

$\dfrac{\partial w}{\partial u} = \dfrac{\partial w}{\partial x}\dfrac{\partial x}{\partial u} + \dfrac{\partial w}{\partial y}\dfrac{\partial y}{\partial u} + \dfrac{\partial w}{\partial z}\dfrac{\partial z}{\partial u} = yz \cdot tv + xz(-\dfrac{t}{u^2}) + xy\dfrac{1}{v}$

$= yztv - \dfrac{xzt}{u^2} + \dfrac{xy}{v}$

$\dfrac{\partial w}{\partial v} = \dfrac{\partial w}{\partial x}\dfrac{\partial x}{\partial v} + \dfrac{\partial w}{\partial y}\dfrac{\partial y}{\partial v} + \dfrac{\partial w}{\partial z}\dfrac{\partial z}{\partial v} = yz \cdot tu + xz \cdot 0 + xy(-\dfrac{u}{v^2})$

$= yztu - \dfrac{xyu}{v^2}$

21. $\dfrac{dz}{dx} = \dfrac{\partial z}{\partial x} + \dfrac{\partial z}{\partial y}\dfrac{dy}{dx} = 2x + 2y(x \cos x + \sin x)$

25. $\dfrac{\partial z}{\partial r} = \dfrac{\partial z}{\partial x}\dfrac{\partial x}{\partial r} + \dfrac{\partial z}{\partial y}\dfrac{\partial y}{\partial r} = \dfrac{\partial z}{\partial x}\cos\theta + \dfrac{\partial z}{\partial y}\sin\theta$

$\dfrac{\partial z}{\partial \theta} = \dfrac{\partial z}{\partial x}\dfrac{\partial x}{\partial \theta} + \dfrac{\partial z}{\partial y}\dfrac{\partial y}{\partial \theta} = -\dfrac{\partial z}{\partial x}r\sin\theta + \dfrac{\partial z}{\partial y}r\cos\theta$

$(\dfrac{\partial z}{\partial r})^2 + \dfrac{1}{r^2}(\dfrac{\partial z}{\partial \theta})^2$

$= (\dfrac{\partial z}{\partial x}\cos\theta + \dfrac{\partial z}{\partial y}\sin\theta)^2 + \dfrac{1}{r^2}(-\dfrac{\partial z}{\partial x}r\sin\theta + \dfrac{\partial z}{\partial y}r\cos\theta)^2$

$= (\dfrac{\partial z}{\partial x})^2\cos^2\theta + 2\dfrac{\partial z}{\partial x}\dfrac{\partial z}{\partial y}\sin\theta\cos\theta + (\dfrac{\partial z}{\partial y})^2\sin^2\theta$

$$+ \left(\frac{\partial z}{\partial x}\right)^2 \sin^2\theta - 2 \frac{\partial z}{\partial x} \frac{\partial z}{\partial y} \sin\theta \cos\theta + \left(\frac{\partial z}{\partial y}\right)^2 \cos^2\theta$$

$$= \left(\frac{\partial z}{\partial x}\right)^2 (\cos^2\theta + \sin^2\theta) + \left(\frac{\partial z}{\partial y}\right)^2 (\sin^2\theta + \cos^2\theta)$$

$$= \left(\frac{\partial z}{\partial x}\right)^2 + \left(\frac{\partial z}{\partial y}\right)^2$$

29. $w = f(t, u, v)$, where $t = x - y$, $u = y - z$, $v = z - x$.

$$\frac{\partial w}{\partial x} = \frac{\partial f}{\partial t}\frac{\partial t}{\partial x} + \frac{\partial f}{\partial u}\frac{\partial u}{\partial x} + \frac{\partial f}{\partial v}\frac{\partial v}{\partial x} = \frac{\partial f}{\partial t} - \frac{\partial f}{\partial v}$$

$$\frac{\partial w}{\partial y} = \frac{\partial f}{\partial t}\frac{\partial t}{\partial y} + \frac{\partial f}{\partial u}\frac{\partial u}{\partial y} + \frac{\partial f}{\partial v}\frac{\partial v}{\partial y} = -\frac{\partial f}{\partial t} + \frac{\partial f}{\partial u}$$

$$\frac{\partial w}{\partial z} = \frac{\partial f}{\partial t}\frac{\partial t}{\partial z} + \frac{\partial f}{\partial u}\frac{\partial u}{\partial z} + \frac{\partial f}{\partial v}\frac{\partial v}{\partial z} = -\frac{\partial f}{\partial u} + \frac{\partial f}{\partial v}$$

$$\frac{\partial w}{\partial x} + \frac{\partial w}{\partial y} + \frac{\partial w}{\partial z} = \frac{\partial f}{\partial t} - \frac{\partial f}{\partial v} - \frac{\partial f}{\partial t} + \frac{\partial f}{\partial u} - \frac{\partial f}{\partial u} + \frac{\partial f}{\partial v}$$

$$= 0$$

33. $z_x = \dfrac{\partial z}{\partial x} = 3x^2 + 2y^2$

$$\frac{dz_x}{dt} = \frac{\partial z_x}{\partial x}\frac{dx}{dt} + \frac{\partial z_x}{\partial y}\frac{dy}{dt} = 6x\frac{dx}{dt} + 4y\frac{dy}{dt}$$

$$= 6 \cdot 2 \cdot 4 + 4 \cdot 5 \cdot 6 = 168 \text{ in}^2/\text{min}$$

Section 21.4, pp. 687-688

1. $\dfrac{dy}{dx} = -\dfrac{\partial f/\partial x}{\partial f/\partial y} = -\dfrac{3x^2 - 6xy}{-3x^2 + 3y^2} = \dfrac{x^2 - 2xy}{x^2 - y^2}$

5. $\dfrac{dy}{dx} = -\dfrac{\partial f/\partial x}{\partial f/\partial y} = -\dfrac{y \cos xy + 1}{x \cos xy}$

9. $\dfrac{\partial x}{\partial z} = -\dfrac{\partial f/\partial z}{\partial f/\partial x} = -\dfrac{x + y}{y + z}$ $\dfrac{\partial x}{\partial y} = -\dfrac{\partial f/\partial y}{\partial f/\partial x} = -\dfrac{x + z}{y + z}$

13. $x^2 - y^2 + 2z^2 - w = 0$ $\dfrac{\partial z}{\partial x} = -\dfrac{\partial f/\partial x}{\partial f/\partial z} = -\dfrac{2x}{4z} = -\dfrac{x}{2z}$

17. $\dfrac{dy}{dx} = -\dfrac{\dfrac{\partial(F,G)}{\partial(x,z)}}{\dfrac{\partial(F,G)}{\partial(y,z)}} = -\dfrac{\begin{vmatrix} \partial F/\partial x & \partial F/\partial z \\ \partial G/\partial x & \partial G/\partial z \end{vmatrix}}{\begin{vmatrix} \partial F/\partial y & \partial F/\partial z \\ \partial G/\partial y & \partial G/\partial z \end{vmatrix}} = -\dfrac{\begin{vmatrix} 2x & 2z \\ 1 & 1 \end{vmatrix}}{\begin{vmatrix} 2y & 2z \\ 1 & 1 \end{vmatrix}}$

$$= -\frac{2x - 2z}{2y - 2z} = \frac{x - z}{z - y}$$

21. $x^2 - y = 0$, $y^2 - z = 0$

$$\frac{dy}{dx} = -\frac{\dfrac{\partial(F,G)}{\partial(x,z)}}{\dfrac{\partial(F,G)}{\partial(y,z)}} = -\frac{\begin{vmatrix} \partial F/\partial x & \partial F/\partial z \\ \partial G/\partial x & \partial G/\partial z \\ \partial F/\partial y & \partial F/\partial z \\ \partial G/\partial y & \partial G/\partial z \end{vmatrix}}{} = -\frac{\begin{vmatrix} 2x & 0 \\ 0 & -1 \\ -1 & 0 \\ 2y & -1 \end{vmatrix}}{}$$

$$= -\frac{-2x}{1} = 2x$$

25. $\dfrac{dy}{dx} = -\dfrac{\dfrac{\partial(F,G,H)}{\partial(x,z,w)}}{\dfrac{\partial(F,G,H)}{\partial(y,z,w)}}$

$$= -\frac{\begin{vmatrix} \partial F/\partial x & \partial F/\partial z & \partial F/\partial w \\ \partial G/\partial x & \partial G/\partial z & \partial G/\partial w \\ \partial H/\partial x & \partial H/\partial z & \partial H/\partial w \\ \partial F/\partial y & \partial F/\partial z & \partial F/\partial w \\ \partial G/\partial y & \partial G/\partial z & \partial G/\partial w \\ \partial H/\partial y & \partial H/\partial z & \partial H/\partial w \end{vmatrix}}{} = \frac{\begin{vmatrix} 1 & 1 & 1 \\ 2x & 2z & 2w \\ wyz & wxy & xyz \\ 1 & 1 & 1 \\ 2y & 2z & 2w \\ wxz & wxy & xyz \end{vmatrix}}{}$$

$$= -\frac{2xyz^2 + 2w^2yz + 2wx^2y - 2wyz^2 - 2w^2xy - 2x^2yz}{2xyz^2 + 2w^2xz + 2wxy^2 - 2wxz^2 - 2w^2xy - 2xy^2z}$$

$$= -\frac{y(xz^2 + w^2z + wx^2 - wz^2 - w^2x - x^2z)}{x(yz^2 + w^2z + wy^2 - wz^2 - w^2y - y^2z)}$$

$$= -\frac{y(w-x)(x-z)(z-w)}{x(w-y)(y-z)(z-w)} = -\frac{y(w-x)(x-z)}{x(w-y)(y-z)}$$

29. $u - 2x - y = 0$, $v - x + 2y = 0$

$$\left(\frac{\partial x}{\partial v}\right)_u = -\frac{\dfrac{\partial(F,G)}{\partial(v,y)}}{\dfrac{\partial(F,G)}{\partial(x,y)}} = -\frac{\begin{vmatrix} \partial F/\partial v & \partial F/\partial y \\ \partial G/\partial v & \partial G/\partial y \\ \partial F/\partial x & \partial F/\partial y \\ \partial G/\partial x & \partial G/\partial y \end{vmatrix}}{} = -\frac{\begin{vmatrix} 0 & -1 \\ 1 & 2 \\ -2 & -1 \\ -1 & 2 \end{vmatrix}}{}$$

$$= -\frac{1}{-4-1} = \frac{1}{5}$$

$$\left(\frac{\partial y}{\partial v}\right)_u = -\frac{\dfrac{\partial(F,G)}{\partial(x,v)}}{\dfrac{\partial(F,G)}{\partial(x,y)}} = -\frac{\begin{vmatrix} \partial F/\partial x & \partial F/\partial v \\ \partial G/\partial x & \partial G/\partial v \\ \partial F/\partial x & \partial F/\partial y \\ \partial G/\partial x & \partial G/\partial y \end{vmatrix}}{} = -\frac{\begin{vmatrix} -2 & 0 \\ -1 & 1 \\ -2 & -1 \\ -1 & 2 \end{vmatrix}}{}$$

$$= -\frac{-2}{-4-1} = -\frac{2}{5}$$

$$\left(\frac{\partial z}{\partial v}\right)_u = \frac{y\left(\dfrac{\partial x}{\partial v}\right)_u - x\left(\dfrac{\partial y}{\partial v}\right)_u}{y^2} = \frac{y/5 + 2x/5}{y^2} = \frac{y + 2x}{5y^2} = \frac{1 + 2(x/y)}{5y}$$

$$= \frac{1 + 2z}{5y}$$

1. $df = \dfrac{\partial f}{\partial x}dx + \dfrac{\partial f}{\partial y}dy = 3x^2 dx - 3y^2 dy$

5. $df = \dfrac{\partial f}{\partial x}dx + \dfrac{\partial f}{\partial y}dy = 3(x + \tan xy)^2(1 + y\,\sec^2 xy)\,dx$

   $\qquad + 3(x + \tan xy)^2(x\,\sec^2 xy)\,dy$

   $\qquad = 3(x + \tan xy)^2[(1 + y\,\sec^2 xy)\,dx + x\,\sec^2 xy\;dy]$

9. $df = \dfrac{\partial f}{\partial x}dx + \dfrac{\partial f}{\partial y}dy + \dfrac{\partial f}{\partial z}dz = 2x\,dx + 2y\,dy + 2z\,dz$

13. $df = \dfrac{\partial f}{\partial x}dx + \dfrac{\partial f}{\partial y}dy + \dfrac{\partial f}{\partial z}dz$

   $\qquad = ye^{xy}dx + (xe^{xy} + z\cos yz)\,dy + y\cos yz\;dz$

17. $\dfrac{\partial P}{\partial y} = 2y\,,\quad \dfrac{\partial Q}{\partial x} = 2y\;;\quad$ exact

   $F = \displaystyle\int (y^2 + 3x^2)\,dx = xy^2 + x^3 + f(y)$

   $\dfrac{\partial F}{\partial y} = 2xy + f'(y) = 2xy$

   $f'(y) = 0\,,\quad f(y) = C$

   $F = xy^2 + x^3 + C$

21. $\dfrac{\partial P}{\partial y} = 2x\,,\quad \dfrac{\partial Q}{\partial x} = 2x\;;\quad$ exact

   $F = \displaystyle\int (1/x + 2\,xy)\,dx = \ln|x| + x^2 y + f(y)$

   $\dfrac{\partial F}{\partial y} = x^2 + f'(y) = 1/y + x^2 + 3y^2$

   $f'(y) = 1/y + 3y^2\,,\quad f(y) = \ln|y| + y^3 + C$

   $F = \ln|x| + x^2 y + \ln|y| + y^3 + C = \ln|xy| + y(x^2 + y^2) + C$

25. $\dfrac{\partial P}{\partial y} = 2x^2 y^3 e^{xy^2} + 4xy\,e^{xy^2} + 2$

   $\dfrac{\partial Q}{\partial y} = 2x^2 y^3 e^{xy^2} + 4\,xy\,e^{xy^2} + 2\;;\quad$ exact

   $F = \displaystyle\int (xy^2 e^{xy^2} + e^{xy^2} + 2y)\,dx = xe^{xy^2} + 2xy + f(y)$

   $\dfrac{\partial F}{\partial y} = 2x^2 ye^{xy^2} + 2x + f'(y) = 2x^2 ye^{xy^2} + 2x - 12y^2$

   $f'(y) = -12y^2\,,\quad f(y) = -4y^3$

   $xe^{xy^2} + 2xy - 4y^3 = C$

1. $z = xy^{1/2}$

$$dz = \frac{\partial z}{\partial x}dx + \frac{\partial z}{\partial y}dy = \sqrt{y}\ dx + \frac{x}{2\sqrt{y}}\ dy = 7(0.2) + \frac{20}{14}(1) = 2.83$$

$x = 20$ $\qquad$ $y = 49$ $\qquad$ $z = 140$

$dx = 0.2$ $\qquad$ $dy = 1$ $\qquad$ $dz = 2.83$

$20.2\sqrt{50} \approx 142.83$

5. $z = \frac{\sqrt[3]{x}}{y}$

$$dz = \frac{\partial z}{\partial x}dx + \frac{\partial z}{\partial y}dy = \frac{dx}{3y\ x^{2/3}} - \frac{\sqrt[3]{x}}{y^2}dy$$

$$= \frac{0.02}{3 \cdot 4 \cdot 4} - \frac{2}{16}(0.11) = -0.013$$

$x = 8$ $\qquad$ $y = 4$ $\qquad$ $z = 0.5$

$dx = 0.02$ $\qquad$ $dy = 0.11$ $\qquad$ $dz = 0.013$

$\frac{\sqrt[3]{8.02}}{4.11} \approx 0.487$

9. $V = xyz$, $\qquad dV = yz\ dx + xz\ dy + xy\ dz$

$$|dV| \leq yz|dx| + xz|dy| + xy|dz|$$

$$\leq 3 \cdot 6(0.01) + 4 \cdot 6(0.01) + 4 \cdot 3(0.01) = 0.54\ in^3$$

13. $\sin\theta = \frac{x}{y}$, $\qquad x = y\sin\theta$

$dx = \sin\theta\ dy + y\cos\theta d\theta$

$|dx| = |\sin\theta\ dy + y\cos\theta d\theta|$

$\qquad \leq \sin\theta|dy| + y\cos\theta|d\theta|$

$\qquad \leq (\sin 21°)(0.3) + 500(\cos 21°)(30")$

$\qquad = (0.36)(0.3) + 500(0.93)(.00015) = 0.81$

Section 21.7, pp. 700-701

1. $D_v z = \frac{\partial z}{\partial x}\cos\theta + \frac{\partial z}{\partial y}\sin\theta = (2x + y)\frac{1}{\sqrt{2}} + x \cdot \frac{1}{\sqrt{2}} = \frac{3x + y}{\sqrt{2}}$

5. $D_v z = \dfrac{\partial z}{\partial x}\cos\theta + \dfrac{\partial z}{\partial y}\sin\theta = 3x^2 \cdot \dfrac{\sqrt{2}}{\sqrt{3}} - 1 \cdot \dfrac{1}{\sqrt{3}}$

$\qquad = \dfrac{3\sqrt{2}x^2 - 1}{\sqrt{3}}$

9. $\dfrac{v}{|v|} = \dfrac{2}{\sqrt{10}}i - \dfrac{1}{\sqrt{10}}j - \dfrac{1}{\sqrt{2}}k$

$\quad D_v w = \dfrac{\partial w}{\partial x}\cos\alpha + \dfrac{\partial w}{\partial y}\cos\beta + \dfrac{\partial w}{\partial z}\cos\gamma$

$\qquad = 1(\dfrac{2}{\sqrt{10}}) - 2y(-\dfrac{1}{\sqrt{10}}) + 3z^2(-\dfrac{1}{\sqrt{2}}) = \dfrac{2 + 2y - 3\sqrt{5}z^2}{\sqrt{10}}$

13. $D_v z = \dfrac{\partial z}{\partial x}\cos\theta + \dfrac{\partial z}{\partial y}\sin\theta = (3x^2 + y^2)\dfrac{2}{\sqrt{13}} + 2xy \cdot \dfrac{3}{\sqrt{13}}$

$\qquad = \dfrac{6x^2 + 6xy + 2y^2}{\sqrt{13}}$

At $(1,1,2)$, $\quad D_v z = \dfrac{14}{\sqrt{13}}$

17. $\dfrac{v}{|v|} = \dfrac{1}{\sqrt{14}}i + \dfrac{3}{\sqrt{14}}j + \dfrac{2}{\sqrt{14}}k$

$\quad D_v w = \dfrac{\partial w}{\partial x}\cos\alpha + \dfrac{\partial w}{\partial y}\cos\beta + \dfrac{\partial w}{\partial z}\cos\gamma$

$\qquad = y \cdot \dfrac{1}{\sqrt{14}} + (x+z)\dfrac{3}{\sqrt{14}} + y \cdot \dfrac{2}{\sqrt{14}} = \dfrac{3(x+y+z)}{\sqrt{14}}$

At $(1,2,1,4)$, $\quad D_v w = \dfrac{3(1+2+1)}{\sqrt{14}} = \dfrac{12}{\sqrt{14}}$

21. $\operatorname{grad} z = \dfrac{\partial z}{\partial x}i + \dfrac{\partial z}{\partial y}j = 2xi - j$

$\quad \dfrac{df}{dn} = |\operatorname{grad} z| = \sqrt{4x^2 + 1}, \quad v = \dfrac{2xi - j}{\sqrt{4x^2 + 1}}$

25. $\operatorname{grad} w = \dfrac{\partial w}{\partial x}i + \dfrac{\partial w}{\partial y}j + \dfrac{\partial w}{\partial z}k = 2xi + 2yj + 2zk$

$\quad \dfrac{dw}{dn} = |\operatorname{grad} w| = \sqrt{4x^2 + 4y^2 + 4z^2} = 2\sqrt{x^2 + y^2 + z^2} = 2\sqrt{w}$

$\quad v = \dfrac{2xi + 2yj + 2zk}{2\sqrt{w}} = \dfrac{xi + yj + zk}{\sqrt{w}}$

29. $\operatorname{grad} w = \dfrac{\partial w}{\partial x}i + \dfrac{\partial w}{\partial y}j + \dfrac{\partial w}{\partial z}k = 2xyi + (x^2 + z)j + yk = 3j$

$\quad$ at $(1,0,2,0)$, $\quad \dfrac{dw}{dn} = |\operatorname{grad} w| = 3, \quad v = \dfrac{3j}{3} = j$

33. $P_0 = (2,-1)$, $\quad P_1 = (0,1)$; $\quad u_1 = -2i + 2j$

$$v_1 = \frac{u_1}{|u_1|} = \frac{-2i + 2j}{\sqrt{4 + 4}} = \frac{-i + j}{\sqrt{2}}$$

$P_0 = (2,-1)$, $\quad P_2 = (-1,1)$; $\quad u_2 = -3i + 2j$

$$v_2 = \frac{u_2}{|u_2|} = \frac{-3i + 2j}{\sqrt{9 + 4}} = \frac{-3i + 2j}{\sqrt{13}}$$

$$\frac{\partial f}{\partial x}\left(\frac{-1}{\sqrt{2}}\right) + \frac{\partial f}{\partial y}\left(\frac{1}{\sqrt{2}}\right) = 7 \ , \quad \frac{\partial f}{\partial x}\left(\frac{-3}{\sqrt{13}}\right) + \frac{\partial f}{\partial y}\left(\frac{2}{\sqrt{13}}\right) = 3$$

$$\frac{\partial f}{\partial x} = 14\sqrt{2} - 3\sqrt{13} \ , \qquad \frac{\partial f}{\partial y} = 21\sqrt{2} - 3\sqrt{13}$$

$P_0 = (2,-1)$, $\quad P_3 = (6,2)$; $\quad u_3 = 4i + 3j$

$$v_3 = \frac{u_3}{|u_3|} = \frac{4i + 3j}{5}$$

$$D_{v_3} f = \frac{\partial f}{\partial x} \cdot \frac{4}{5} + \frac{\partial f}{\partial y} \cdot \frac{3}{5} = \frac{4}{5}(14\sqrt{2} - 3\sqrt{13}) + \frac{3}{5}(21\sqrt{2} - 3\sqrt{13})$$

$$= \frac{119\sqrt{2} - 21\sqrt{13}}{5}$$

Section 21.8, p. 707

1. $\dfrac{\partial z}{\partial x} = 2x = 4$, $\qquad \dfrac{\partial z}{\partial y} = -2y = -2$

$4(x - 2) - 2(y - 1) - (z - 3) = 0$

$4x - 2y - z - 3 = 0$: $\quad$ tangent plane

$\dfrac{x - 2}{4} = \dfrac{y - 1}{-2} = \dfrac{z - 3}{-1}$: normal line

5. $\dfrac{\partial z}{\partial x} = -\dfrac{1}{x^2} = -1$, $\qquad \dfrac{\partial z}{\partial y} = -\dfrac{1}{y^2} = -1$

$-1(x - 1) - 1(y - 1) - (z - 2) = 0$

$x + y + z - 4 = 0$: $\quad$ tangent plane

$x - 1 = y - 1 = z - 2$: $\quad$ normal line

9. $\dfrac{\partial F}{\partial x} = 2x = 2$, $\qquad \dfrac{\partial F}{\partial y} = 2y = 4$, $\qquad \dfrac{\partial F}{\partial z} = 2z = 4$

$2(x - 1) + 4(y - 2) + 4(z - 2) = 0$

$(x - 1) + 2(y - 2) + 2(z - 2) = 0$

$x + 2y + 2z - 9 = 0$: tangent plane

$\dfrac{x - 1}{1} = \dfrac{y - 2}{2} = \dfrac{z - 2}{2}$: normal line

13. $\dfrac{\partial F}{\partial x} = y = 2$, $\qquad \dfrac{\partial F}{\partial y} = x + z = 2$, $\qquad \dfrac{\partial F}{\partial z} = y = 2$

$2(x - 1) + 2(y - 2) + 2(z - 1) = 0$

$(x - 1) + (y - 2) + (z - 1) = 0$

$x + y + z - 4 = 0$: tangent plane

$x - 1 = y - 2 = z - 1$: normal line

17. $\dfrac{dx}{dt} = -\sin t = 0$, $\quad \dfrac{dy}{dt} = \cos t = 1$, $\quad \dfrac{dz}{dt} = 1$, $(1, 0, 0)$

$x = 1$, $\quad y = t$, $\quad z = t$

21. $\dfrac{dx}{dt} = 3t^2 = 3$, $\dfrac{dy}{dt} = 2t = 2$, $\quad \dfrac{dz}{dt} = -\dfrac{1}{t^2} = 1$, $(1, 1, 1)$

$x = 1 + 3t$, $\quad y = 1 + 2t$, $\quad z = 1 - t$

25. $\left\{ \dfrac{\partial f}{\partial x}, \dfrac{\partial f}{\partial y}, \dfrac{\partial f}{\partial z} \right\}$ is a set of direction numbers for a line perpendicular to $f(x, y, z) = 0$; $\left\{ \dfrac{\partial g}{\partial x}, \dfrac{\partial g}{\partial y}, \dfrac{\partial g}{\partial z} \right\}$ is a set of direction numbers for a line perpendicular to $g(x, y, z) = 0$. Thus the two surfaces intersect orthogonally if and only if

$$\frac{\partial f}{\partial x} \cdot \frac{\partial g}{\partial x} + \frac{\partial f}{\partial y} \cdot \frac{\partial g}{\partial y} + \frac{\partial f}{\partial z} \cdot \frac{\partial g}{\partial z} = 0$$

Section 21.9, pp. 710–711

1. $\dfrac{\partial z}{\partial x} = 2x - 2$, $\quad \dfrac{\partial z}{\partial y} = 2y + 4$

$2x - 2 = 0$, $\quad 2y + 4 = 0$ gives $x = 1$, $\quad y = -2$, $\quad z = -7$

$\dfrac{\partial^2 z}{\partial x^2} = 2$, $\quad \dfrac{\partial^2 z}{\partial y^2} = 2$, $\quad \dfrac{\partial^2 z}{\partial y \partial x} = 0$, $\quad D = 2 \cdot 2 - 0^2 = 4$

$(1, -2, -7)$ is a relative minimum.

5. $\dfrac{\partial z}{\partial x} = 2x + 4y + 6$, $\qquad \dfrac{\partial z}{\partial y} = 4x + 2y$

$2x + 4y + 6 = 0$, $\ 4x + 2y = 0$ gives $x = 1$, $y = -2$, $z = 4$

$\dfrac{\partial^2 z}{\partial x^2} = 2$, $\quad \dfrac{\partial^2 z}{\partial y^2} = 2$, $\quad \dfrac{\partial^2 z}{\partial y \partial x} = 4$, $\quad D = 2 \cdot 2 - 4^2 = -12$

$(1, -2, 4)$ is neither a maximum nor a minimum.

9. $\dfrac{\partial z}{\partial x} = 3x^2 - 2x - 1$, $\qquad \dfrac{\partial z}{\partial y} = 2y + 2$

$3x^2 - 2x - 1 = 0$ gives $x = -\dfrac{1}{3}$, $x = 1$ $\qquad 6x - 2$

$2y + 2 = 0$ gives $y = -1$

Critical points: $\left(-\dfrac{1}{3}, -1, \dfrac{32}{27}\right)$, $(1, -1, 0)$

$\dfrac{\partial^2 z}{\partial x^2} = 6x - 2$, $\quad \dfrac{\partial^2 z}{\partial y^2} = 2$, $\quad \dfrac{\partial^2 z}{\partial y \partial x} = 0$, $\quad D = 12x - 4$

$\left(-\dfrac{1}{3}, -1, \dfrac{32}{27}\right)$: $\ D = -8$; neither maximum nor minimum.

$(1, -1, 0)$: $\ D = 8$, $\dfrac{\partial^2 z}{\partial x^2} = 4$, $\dfrac{\partial^2 z}{\partial y^2} = 2$; relative minimum.

13. $\dfrac{\partial z}{\partial x} = 2x + y - 1$, $\qquad \dfrac{\partial z}{\partial y} = x + 2y - 2$

$2x + y = 1$ and $x + 2y = 2$ gives $x = 0$, $y = 1$, $z = 0$

$\dfrac{\partial^2 z}{\partial x^2} = 2$, $\quad \dfrac{\partial^2 z}{\partial y^2} = 2$, $\quad \dfrac{\partial^2 z}{\partial y \partial x} = 1$, $\quad D = 2 \cdot 2 - 1^2 = 3$

$(0, 1, 0)$ is a relative minimum.

17. $x + y + z = 25$, $\qquad z = 25 - x - y$

$w = x^2 y^2 z = x^2 y^2 (25 - x - y) = 25 x^2 y^2 - x^3 y^2 - x^2 y^3$

$\dfrac{\partial w}{\partial x} = 50 x y^2 - 3 x^2 y^2 - 2 x y^3$, $\qquad \dfrac{\partial w}{\partial y} = 50 x^2 y - 2 x^3 y - 3 x^2 y^2$

$x y^2 (50 - 3x - 2y) = 0$ gives $3x + 2y = 50$ (x and y

must be positive); $x^2 y (50 - 2x - 3y) = 0$ gives

$2x + 3y = 50$; thus $x = 10$, $y = 10$, $z = 5$

21. $d = \sqrt{(x-1)^2 + (y-3)^2 + (z-4)^2}$, $z = 1 - 2x + y$

$D = (x-1)^2 + (y-3)^2 + (-3 - 2x + y)^2$

$\dfrac{\partial D}{\partial x} = 2(x-1) - 4(-3 - 2x + y) = 10x - 4y + 10$

$\dfrac{\partial D}{\partial y} = 2(y-3) + 2(-3 - 2x + y) = -4x + 4y - 12$

$5x - 2y = -5$, $-x + y = 3$

$x = \dfrac{1}{3}$, $y = \dfrac{10}{3}$, $z = \dfrac{11}{3}$

$d = \sqrt{(\dfrac{1}{3} - 1)^2 + (\dfrac{10}{3} - 3)^2 + (\dfrac{11}{3} - 4)^2} = \sqrt{\dfrac{6}{9}} = \dfrac{\sqrt{6}}{3}$

25. $P = (x - 65)(1200 - 40x + 20y) + (y - 80)(2650 + 10x - 30y)$

$= -40x^2 + 30xy - 30y^2 + 3000x + 3750y - 290,000$

$\dfrac{\partial P}{\partial x} = -80x + 30y + 3000 = 0$, $\dfrac{\partial P}{\partial y} = 30x - 60y + 3750 = 0$

$8x - 3y = 300$, $-x + 2y = 125$

$x = 75, \quad y = 100$

$\dfrac{\partial^2 P}{\partial x^2} = -80$, $\dfrac{\partial^2 P}{\partial y^2} = -60$, $\dfrac{\partial^2 P}{\partial y \partial x} = 30$

$D = \begin{vmatrix} -80 & 30 \\ 30 & -60 \end{vmatrix} = 4800 - 900 > 0$

Hence $x = 75$, $y = 100$ gives a maximum.

Section 21.10, p. 715

1. $u = x + y + z + \lambda(x^2 + y^2 + z^2 - 4)$

$\dfrac{\partial u}{\partial x} = 1 + 2\lambda x = 0$, $\dfrac{\partial u}{\partial y} = 1 + 2\lambda y = 0$, $\dfrac{\partial u}{\partial z} = 1 + 2\lambda z = 0$

$\dfrac{\partial u}{\partial \lambda} = x^2 + y^2 + z^2 - 4 = 0$

$2\lambda xyz = -yz, \quad 2\lambda xyz = -xz, \quad 2\lambda xyz = -xy, \quad yz = xz = xy$

Either two of the three are 0 and the third 2,

giving $w = 2$; or $x = y = z$, giving $3x^2 = 4$; or

$x = y = z = \dfrac{2}{\sqrt{3}}$ and $w = 2\sqrt{3}$

5. $u = 2x^2 + y^2 + z^2 + \lambda(x + 2y - 4z - 8)$

$\dfrac{\partial u}{\partial x} = 4x + \lambda = 0, \quad \dfrac{\partial u}{\partial y} = 2y + 2\lambda = 0, \quad \dfrac{\partial u}{\partial z} = 2z - 4\lambda = 0$

$-\lambda = 4x = y = -\dfrac{z}{2}, \quad 8x = 2y = -z$

$x + 8x + 32x = 8, \quad x = \dfrac{8}{41}, \quad y = \dfrac{32}{41}, \quad z = -\dfrac{64}{41}$

$w = \dfrac{128}{1681} + \dfrac{1024}{1681} + \dfrac{4096}{1681} = \dfrac{128}{41}$

9. $u = xyz + \lambda(x^2 + 2y^2 + z^2 - 2)$

$\dfrac{\partial u}{\partial x} = yz + 2\lambda x = 0, \quad \dfrac{\partial u}{\partial y} = xz + 4\lambda y = 0, \quad \dfrac{\partial u}{\partial z} = xy + 2\lambda z = 0$

$\dfrac{\partial u}{\partial \lambda} = x^2 + 2y^2 + z^2 - 2 = 0$

$-4\lambda xyz = 2y^2 z^2 = x^2 z^2 = 2x^2 y^2, \quad x^2 = z^2 = 2y^2$

$3x^2 = 2, \quad x = \pm\sqrt{\dfrac{2}{3}}, \quad y = \pm\sqrt{\dfrac{1}{3}}, \quad z = \pm\sqrt{\dfrac{2}{3}}$

We have a maximum when one or three are positive.

$w = xyz = \sqrt{\dfrac{2}{3} \cdot \dfrac{2}{3} \cdot \dfrac{1}{3}} = \dfrac{2}{3\sqrt{3}}$

13. $u = x^2 + y^2 + z^2 + \lambda(x + 2y - z - 2) + \mu(2x - y - z - 2)$

$\dfrac{\partial u}{\partial x} = 2x + \lambda + 2\mu = 0, \quad \dfrac{\partial u}{\partial y} = 2y + 2\lambda - \mu = 0$

$\dfrac{\partial u}{\partial z} = 2z - \lambda - \mu = 0, \quad \dfrac{\partial u}{\partial \lambda} = x + 2y - z - 2 = 0$

$\dfrac{\partial u}{\partial \mu} = 2x - y - z - 2 = 0$

$2x + \lambda + 2\mu = 0, \quad 2y + 2\lambda - \mu = 0, \text{ and } 2z - \lambda - \mu = 0 \quad \text{gives}$

$2x + 4y + 5\lambda = 0, \quad 2y - 2z + 3\lambda = 0, \text{ and } 3x + y + 5z = 0.$

Solving $3x + y + 5z = 0$ with $x + 2y - z = 2$ and

$2x - y - z = 2$, we have $x = \dfrac{6}{7}, \quad y = \dfrac{2}{7}, \quad z = -\dfrac{4}{7}.$

$w = x^2 + y^2 + z^2 = \dfrac{36}{49} + \dfrac{4}{49} + \dfrac{16}{49} = \dfrac{8}{7}$

17. $d = \sqrt{x^2 + y^2 + z^2}$ is to be a minimum. d is a minimum whenever $w = d^2$ is a minimum. Thus

$$w = x^2 + y^2 + z^2, \qquad -x^2 + y - z^2 = 2$$

$$u = x^2 + y^2 + z^2 + \lambda(-x^2 + y - z^2 - 2)$$

$$\frac{\partial u}{\partial x} = 2x - 2x\lambda = 0, \quad \frac{\partial u}{\partial y} = 2y + \lambda = 0, \quad \frac{\partial u}{\partial z} = 2z - 2z\lambda = 0$$

$$\frac{\partial u}{\partial \lambda} = -x^2 + y - z^2 - 2 = 0$$

$2x = 2x\lambda$ gives $x = 0$ or $\lambda = 1$. Assume $\lambda = 1$.

Then $y = -\dfrac{1}{2}$ and $x^2 + z^2 = -\dfrac{5}{2}$, which is impossible.

Thus $x = 0$. Similarly $2z = 2z\lambda$ gives $z = 0$.

Since $-x^2 + y - z^2 = 2$, $y = 2$. The nearest point is

$(0, 2, 0)$.

21. $V = xyz, \qquad u = xyz + \lambda(x + 2y + z - 2)$

$$\frac{\partial u}{\partial x} = yz + \lambda = 0, \quad \frac{\partial u}{\partial y} = xz + 2\lambda = 0, \quad \frac{\partial u}{\partial z} = xy + \lambda = 0$$

$$\frac{\partial u}{\partial \lambda} = x + 2y + z - 2 = 0$$

$-2\lambda = 2yz = xz = 2xy$. Thus $x = z = 2y$ (since none

of them can be zero at a maximum) and $x + 2y + z = 2$

becomes $3x = 2$ or $x = \dfrac{2}{3}$, $y = \dfrac{1}{3}$, $z = \dfrac{2}{3}$, and

$$v = xyz = \frac{2}{3} \cdot \frac{1}{3} \cdot \frac{2}{3} = \frac{4}{27}$$

Review 21, pp. 715-716

1. $\dfrac{\partial z}{\partial x} = 4x^3 + 12x^2 y \qquad\qquad \dfrac{\partial z}{\partial y} = 4x^3 - 8y^3$

$\dfrac{\partial^2 z}{\partial y \partial x} = 12x^2 \qquad\qquad \dfrac{\partial^2 z}{\partial x \partial y} = 12x^2$

5. $\dfrac{\partial w}{\partial u} = \dfrac{\partial w}{\partial x}\dfrac{\partial x}{\partial u} + \dfrac{\partial w}{\partial y}\dfrac{\partial y}{\partial u} + \dfrac{\partial w}{\partial z}\dfrac{\partial z}{\partial u}$

$\qquad = (y + 3z)2u + (x + 2z)2u + (2y + 3x)2$

$\qquad = 2u(x + y + 5z) + 6x + 4y$

$\dfrac{\partial w}{\partial v} = \dfrac{\partial w}{\partial x}\dfrac{\partial x}{\partial v} + \dfrac{\partial w}{\partial y}\dfrac{\partial y}{\partial v} + \dfrac{\partial w}{\partial z}\dfrac{\partial z}{\partial v}$

$$= (y + 3z)2v + (x + 2z)(-2v) + (2y + 3x)3$$

$$= 2v(-x + y + z) + 9x + 6y$$

9. $$\dfrac{dy}{dx} = -\dfrac{\dfrac{\partial(F, G)}{\partial(x, z)}}{\dfrac{\partial(F, G)}{\partial(y, z)}} = -\dfrac{\begin{vmatrix} \dfrac{\partial F}{\partial x} & \dfrac{\partial F}{\partial z} \\[2mm] \dfrac{\partial G}{\partial x} & \dfrac{\partial G}{\partial z} \end{vmatrix}}{\begin{vmatrix} \dfrac{\partial F}{\partial y} & \dfrac{\partial F}{\partial z} \\[2mm] \dfrac{\partial G}{\partial y} & \dfrac{\partial G}{\partial z} \end{vmatrix}} = -\dfrac{\begin{vmatrix} 4 & -2 \\[2mm] 3x^2 & -3z^2 \end{vmatrix}}{\begin{vmatrix} 3 & -2 \\[2mm] -3y^2 & -3z^2 \end{vmatrix}}$$

$$= -\dfrac{-12z^2 + 6x^2}{-9z^2 - 6y^2} = \dfrac{2x^2 - 4z^2}{2y^2 + 3z^2}$$

13. $\dfrac{\partial M}{\partial y} = -x \sin y, \quad \dfrac{\partial N}{\partial x} = 2y \sin x; \quad$ not exact

17. $\dfrac{v}{|v|} = \dfrac{4i + j - k}{3\sqrt{2}}$

$$D_v w = \dfrac{\partial w}{\partial x} \cdot \dfrac{4}{3\sqrt{2}} + \dfrac{\partial w}{\partial y} \cdot \dfrac{1}{3\sqrt{2}} + \dfrac{\partial w}{\partial z} \cdot \dfrac{-1}{3\sqrt{2}}$$

$$= (y + 4z)\dfrac{4}{3\sqrt{2}} + (x - 2z)\dfrac{1}{3\sqrt{2}} + (-2y + 4x)\dfrac{-1}{3\sqrt{2}}$$

$$= \dfrac{-3x + 6y + 14z}{3\sqrt{2}}$$

21. $\dfrac{\partial z}{\partial x} = 6x^2 = 6, \qquad\qquad \dfrac{\partial z}{\partial y} = -12y^2 = -12$

$$6(x - 1) - 12(y - 1) - (z + 2) = 0$$

$$6x - 12y - z + 4 = 0$$

$$x = 1 + 6t, \qquad y = 1 - 12t, \qquad z = -2 - t$$

25. $x = \sin t \qquad\qquad y = \cos 2t \qquad\qquad z = t$

$\dfrac{dx}{dt} = \cos t = \dfrac{\sqrt{3}}{2} \qquad \dfrac{dy}{dt} = -2 \sin 2t = -\sqrt{3} \qquad \dfrac{dz}{dt} = 1$

$x = \dfrac{1}{2} \qquad\qquad\qquad y = \dfrac{1}{2} \qquad\qquad\qquad z = \dfrac{\pi}{6}$

$x = \dfrac{1}{2} + \dfrac{\sqrt{3}}{2}t, \qquad\quad y = \dfrac{1}{2} - \sqrt{3}\, t \qquad\quad z = \dfrac{\pi}{6} + t$

29. $u = 2x + y + 4z + \lambda(x^2 + y^2 + z^2 - 9)$

$\dfrac{\partial u}{\partial x} = 2 + 2\lambda x = 0, \qquad \lambda = -1/x$

$\dfrac{\partial u}{\partial y} = 1 + 2\lambda y = 0, \qquad \lambda = -1/2y$

$\dfrac{\partial u}{\partial z} = 4 + 2\lambda z = 0, \qquad \lambda = -2/z$

$\dfrac{\partial u}{\partial \lambda} = x^2 + y^2 + z^2 - 9 = 0$

$-\dfrac{1}{x} = -\dfrac{1}{2y} = -\dfrac{2}{z}, \quad y = \dfrac{x}{2}, \quad z = 2x$

$x^2 + y^2 + z^2 = x^2 + \dfrac{x^2}{4} + 4x^2 = \dfrac{21x^2}{4} = 9$

$x^2 = \dfrac{36}{21}, \qquad x = \pm\dfrac{2}{7}\sqrt{21}$

$y = \pm\dfrac{\sqrt{21}}{7}, \qquad z = \pm\dfrac{4\sqrt{21}}{7}$

The maximum occurs when all three are positive.

$w = 2 \cdot \dfrac{2\sqrt{21}}{7} + \dfrac{\sqrt{21}}{7} + 4 \cdot \dfrac{4\sqrt{21}}{7} = 3\sqrt{21}$

Chapter Twenty-Two
Multiple Integrals

1. $\int_0^1 \int_0^x xy \, dy \, dx = \int_0^1 \frac{xy^2}{2}\Big|_0^x dx$

$= \int_0^1 \frac{x^3}{2} dx = \frac{x^4}{8}\Big|_0^1 = \frac{1}{8}$

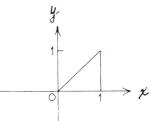

5. $\int_0^2 \int_0^1 (x^3 + y^3 - 3xy) \, dx \, dy$

$= \int_0^2 (\frac{x^4}{4} + xy^3 - \frac{3x^2y}{2})\Big|_0^1 dy$

$= \int_0^2 (\frac{1}{4} + y^3 - \frac{3y}{2}) dy$

$= (\frac{y}{4} + \frac{y^4}{4} - \frac{3y^2}{4})\Big|_0^2$

$= \frac{2}{4} + \frac{16}{4} - \frac{12}{4} = \frac{3}{2}$

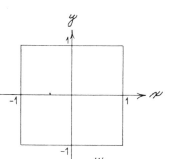

9. $\int_{-1}^1 \int_{-1}^1 dy \, dx = \int_{-1}^1 y\Big|_{-1}^1 dx$

$= \int_{-1}^1 2dx = 2x\Big|_{-1}^1 = 4$

$\int_{-1}^1 \int_{-1}^1 dx \, dy$

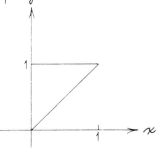

13. $\int_0^1 \int_0^y e^{x+y} dx \, dy = \int_0^1 e^{x+y}\Big|_0^y dy$

$= \int_0^1 (e^{2y} - e^y) dy = (\frac{1}{2}e^{2y} - e^y)\Big|_0^1$

$= \frac{1}{2}e^2 - e - \frac{1}{2} + 1 = \frac{1}{2}e^2 - e + \frac{1}{2}$

$\int_0^1 \int_x^1 e^{x+y} dy \, dx$

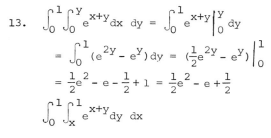

17. $\displaystyle\int_0^1\int_0^{3\sqrt{y}}(x^3-y^3)\,dx\,dy = \int_0^1(\frac{x^4}{4}-xy^3)\Big|_0^{y^{1/3}}\,dy$

$\displaystyle = \int_0^1(\frac{y^{4/3}}{4}-y^{10/3})\,dy = (\frac{3y^{7/3}}{28}-\frac{3y^{13/3}}{13})\Big|_0^1 = \frac{3}{28}-\frac{3}{13}$

$\displaystyle = -\frac{45}{364}$

21. $\displaystyle\int_0^1\int_{x^2}^x xy\,dy\,dx = \int_0^1\frac{xy^2}{2}\Big|_{x^2}^x dx = \int_0^1(\frac{x^3}{2}-\frac{x^5}{2})\,dx$

$\displaystyle = (\frac{x^4}{8}-\frac{x^6}{12})\Big|_0^1 = \frac{1}{8}-\frac{1}{12} = \frac{1}{24}$

Section 22.2, pp. 728-729

1. $\displaystyle V = \int_0^1\int_0^{1-x}(1-x-y)\,dy\,dx = \int_0^1(y-xy-\frac{y^2}{2})\Big|_0^{1-x}dx$

$\displaystyle = \int_0^1(1-x-x+x^2-\frac{1-2x+x^2}{2})\,dx = \int_0^1(\frac{1}{2}-x+\frac{x^2}{2})\,dx$

$\displaystyle = (\frac{1}{2}x-\frac{x^2}{2}+\frac{x^3}{6})\Big|_0^1 = \frac{1}{2}-\frac{1}{2}+\frac{1}{6} = \frac{1}{6}$

5. $\displaystyle V = \int_0^4\int_0^{1-x}xy\,dy\,dx = \int_0^4\frac{xy^2}{2}\Big|_0^{1-x}dx = \frac{1}{2}\int_0^4 x(1-x)^2\,dx$

$\displaystyle = \frac{1}{2}\int_0^4(x-2x^2+x^3)\,dx = \frac{1}{2}(\frac{x^2}{2}-\frac{2x^3}{3}+\frac{x^4}{4})\Big|_0^4$

$\displaystyle = \frac{1}{2}(8-\frac{128}{3}+64) = \frac{44}{3}$

9. $\displaystyle V = 4\int_1^3\int_0^{\sqrt{9-y^2}}\sqrt{9-y^2-z^2}\,dz\,dy$

$z = \sqrt{9-y^2}\sin\theta,\quad dz = \sqrt{9-y^2}\cos\theta\,d\theta$

$\displaystyle V = 4\int_1^3\int_0^{\pi/2}\sqrt{(9-y^2)(1-\sin^2\theta)}\,\sqrt{9-y^2}\cos\theta\,d\theta\,dy$

$\displaystyle = 4\int_1^3\int_0^{\pi/2}(9-y^2)\cos^2\theta\,d\theta\,dy$

$\displaystyle = 2\int_1^3(9-y^2)\int_0^{\pi/2}(1+\cos 2\theta)\,d\theta\,dy$

$\displaystyle = 2\int_1^3(9-y^2)(\theta+\sin\theta\cos\theta)\Big|_0^{\pi/2}dy$

$$= 2 \int_1^3 (9 - y^2) \frac{\pi}{2} dy = \pi(9y - \frac{y^3}{3}) \Big|_1^3$$

$$= \pi[(27 - 9) - (9 - 1/3)] = \frac{28\pi}{3}$$

13. $\quad V = 4 \int_0^r \int_0^{\sqrt{r^2-x^2}} \frac{h}{r}(r - \sqrt{x^2 + y^2}) \, dy \, dx$

$\quad\quad y = x \tan \theta, \quad dy = x \sec^2\theta \, d\theta$

$$= 4 \int_0^r \int_0^{\text{Arctan} \frac{\sqrt{r^2-x^2}}{x}} \frac{h}{r}(r - \sqrt{x^2 + x^2\tan^2\theta}) x \sec^2\theta \, d\theta \, dx$$

$$= \frac{4h}{r} \int_0^r \int_0^{\text{Arctan} \frac{\sqrt{r^2-x^2}}{x}} (rx \sec^2\theta - x^2\sec^3\theta) \, d\theta \, dx$$

$$= \frac{4h}{r} \int_0^r [rx \tan \theta$$
$$\quad - \frac{x^2}{2}(\sec \theta \tan \theta + \ln|\sec \theta + \tan \theta|)] \Big|_0^{\text{Arctan} \frac{\sqrt{r^2-x^2}}{x}} dx$$

$$= \frac{4h}{r} \int_0^r [r\sqrt{r^2 - x^2} - \frac{x^2}{2}(\frac{r}{x}\frac{\sqrt{r^2 - x^2}}{x} + \ln \frac{r + \sqrt{r^2 - x^2}}{x})] dx$$

$$= \frac{4h}{r} \int_0^r (\frac{r}{2}\sqrt{r^2 - x^2} - \frac{x^2}{2}\ln \frac{r + \sqrt{r^2 - x^2}}{x}) dx$$

$\quad\quad x = r \sin \theta, \quad dx = r \cos \theta \, d\theta$

$$= \frac{4h}{r} \int_0^{\pi/2} (\frac{r}{2}\sqrt{r^2 - r^2\sin^2\theta} - \frac{r^2\sin^2\theta}{2}\ln \frac{r + \sqrt{r^2 - r^2\sin^2\theta}}{r \sin \theta}) r \cos \theta \, d\theta$$

$$= 2hr \int_0^{\pi/2} (r \cos^2\theta - r \sin^2\theta \cos \theta \, \ln \frac{1 + \cos \theta}{\sin \theta}) \, d\theta$$

$\quad\quad u = \ln \frac{1 + \cos \theta}{\sin \theta} \quad\quad v' = -\sin^2\theta \, \cos \theta$

$\quad\quad u' = \frac{-\sin \theta}{1 + \cos \theta} - \frac{\cos \theta}{\sin \theta} \quad v = -\frac{\sin^3\theta}{3}$

$\quad\quad\quad = -\frac{1 - \cos \theta}{\sin \theta} - \frac{\cos \theta}{\sin \theta}$

$\quad\quad\quad = -\frac{1}{\sin \theta}$

$$= 2hr^2 [\int_0^{\pi/2} \frac{1 + \cos 2\theta}{2} d\theta - \frac{\sin^3\theta}{3}\ln \frac{1 + \cos \theta}{\sin \theta} \Big|_0^{\pi/2}$$
$$\quad - \int_0^{\pi/2} \frac{\sin^2\theta}{3} d\theta]$$

$$= 2hr^2 [(\frac{\theta}{2} + \frac{1}{4} \sin 2\theta - \frac{\sin^3\theta}{3}\ell n \frac{1 + \cos \theta}{\sin \theta})\Big|_0^{\pi/2}$$

$$- \frac{1}{6}\int_0^{\pi/2} (1 - \cos 2\theta) d\theta$$

$$= 2hr^2 (\frac{\theta}{2} + \frac{1}{4} \sin 2\theta - \frac{\sin^3\theta}{3}\ell n \frac{1 + \cos \theta}{\sin \theta} - \frac{\theta}{6} + \frac{1}{12} \sin 2\theta)\Big|_0^{\pi/2}$$

$$= 2hr^2 (\frac{\pi}{4} - \frac{\pi}{12}) = \frac{1}{3}\pi hr^2$$

17. $A = \int_0^1 \int_{x^2}^x dydx = \int_0^1 (x - x^2) dx = (\frac{x^2}{2} - \frac{x^3}{3})\Big|_0^1$

$$= \frac{1}{2} - \frac{1}{3} = \frac{1}{6}$$

21. $m = \int_0^1 \int_0^{x^3} xy\ dy\ dx = \int_0^1 \frac{xy^2}{2}\Big|_0^{x^3} dx = \int_0^1 \frac{x^7}{2}dx$

$$= \frac{x^8}{16}\Big|_0^1 = \frac{1}{16}$$

25. $m = 4\int_0^2 \int_0^{\sqrt{4-x^2}} (x + y) dydx = 4\int_0^2 (xy + \frac{y^2}{2})\Big|_0^{\sqrt{4-x^2}} dx$

$$= 4\int_0^2 (x\sqrt{4 - x^2} + \frac{4 - x^2}{2}) dx = 4(- \frac{1}{3}(4 - x^2)^{3/2} + 2x - \frac{1}{6}x^3)\Big|_0^2$$

$$= 4[(4 - \frac{8}{6}) + \frac{1}{3}\cdot 4^{3/2}] = \frac{64}{3}$$

Section 22.3, pp. 732-733

1. $V = \int_0^\pi \int_0^{\sin \theta} 3r\ drd\theta = \int_0^\pi \frac{3}{2}r^2\Big|_0^{\sin \theta} d\theta = \frac{3}{2}\int_0^\pi \sin^2\theta d\theta$

$$= \frac{3}{4}\int_0^\pi (1 - \cos 2\theta) d\theta = \frac{3}{4}(\theta - \frac{1}{2} \sin 2\theta)\Big|_0^\pi = \frac{3\pi}{4}$$

5. $V = 4\int_0^{\pi/2} \int_0^{\sin 2\theta} (1 + r^2) rdrd\theta = 4\int_0^{\pi/2} \int_0^{\sin 2\theta} (r + r^3) drd\theta$

$$= 4\int_0^{\pi/2} (\frac{r^2}{2} + \frac{r^4}{4})\Big|_0^{\sin 2\theta} d\theta = 4\int_0^{\pi/2} (\frac{\sin^2 2\theta}{2} + \frac{\sin^4 2\theta}{4}) d\theta$$

$$= 4\int_0^{\pi/2} (\frac{1 - \cos 4\theta}{4} + \frac{(1 - \cos 4\theta)^2}{16}) d\theta$$

$$= \frac{1}{4}\int_0^{\pi/2} (4 - 4 \cos 4\theta + 1 - 2 \cos 4\theta + \cos^2 4\theta)\, d\theta$$

$$= \frac{1}{4}\int_0^{\pi/2} (5 - 6 \cos 4\theta + \frac{1 + \cos 8\theta}{2})\, d\theta$$

$$= \frac{1}{4}(\frac{11\theta}{2} - \frac{3}{2} \sin 4\theta + \frac{1}{16} \sin 8\theta)\Big|_0^{\pi/2} = \frac{11\pi}{16}$$

9. $V = \int_0^{2\pi} \int_0^{1+\sin\theta} r^2 dr d\theta = \int_0^{2\pi} \frac{r^3}{3}\Big|_0^{1+\sin\theta} d\theta$

$$= \frac{1}{3}\int_0^{2\pi} (1 + 3 \sin \theta + 3 \sin^2\theta + \sin^3\theta)\, d\theta$$

$$= \frac{1}{3}\int_0^{2\pi} [1 + 3 \sin \theta + \frac{3}{2}(1 - \cos 2\theta) + (1 - \cos^2\theta)\sin \theta]\, d\theta$$

$$= \frac{1}{3}\int_0^{2\pi} [\frac{5}{2} + 4 \sin \theta - \frac{3}{2} \cos 2\theta + \cos^2\theta(-\sin \theta)]\, d\theta$$

$$= \frac{1}{3}(\frac{5\theta}{2} - 4 \cos \theta - \frac{3}{4} \sin 2\theta + \frac{\cos^3\theta}{3})\Big|_0^{2\pi}$$

$$= \frac{1}{3}(5\pi - 4 + \frac{1}{3} + 4 - \frac{1}{3}) = \frac{5\pi}{3}$$

13. $m = 4\int_0^{\pi/2} \int_0^{\sin 2\theta} r \sin 2\theta\, dr d\theta = 2\int_0^{\pi/2} r^2 \sin 2\theta\Big|_0^{\sin 2\theta} d\theta$

$$= 2\int_0^{\pi/2} \sin^3 2\theta\, d\theta = \int_0^{\pi/2} (1 - \cos^2 2\theta)2 \sin 2\theta\, d\theta$$

$$= (-\cos 2\theta + \frac{\cos^3 2\theta}{3})\Big|_0^{\pi/2} = 1 - \frac{1}{3} + 1 - \frac{1}{3} = \frac{4}{3}$$

17. $V = 4\int_0^{\pi/2} \int_0^1 4r\, dr d\theta = 8\int_0^{\pi/2} r^2\Big|_0^1 d\theta = 8\int_0^{\pi/2} d\theta$

$$= 8\theta\Big|_0^{\pi/2} = 4\pi$$

21. $V = 4\int_0^{\pi/2} \int_0^{1/\sqrt{2}} (\sqrt{1 - r^2} - r)r\, dr d\theta$

$$= 4\int_0^{\pi/2} [-\frac{1}{3}(1 - r^2)^{3/2} - \frac{r^3}{3}]\Big|_0^{1/\sqrt{2}} d\theta$$

$$= -\frac{4}{3}\int_0^{\pi/2} (\frac{1}{2\sqrt{2}} + \frac{1}{2\sqrt{2}} - 1)\, d\theta = -\frac{4}{3}(\frac{1}{\sqrt{2}} - 1)\theta\Big|_0^{\pi/2}$$

$$= -\frac{4}{3}(\frac{1 - \sqrt{2}}{\sqrt{2}})\frac{\pi}{2} = \frac{(2 - \sqrt{2})\pi}{3}$$

25.

$$\frac{\partial(x,y)}{\partial(r,\theta)} = \begin{vmatrix} \dfrac{\partial x}{\partial r} & \dfrac{\partial x}{\partial \theta} \\[2mm] \dfrac{\partial y}{\partial r} & \dfrac{\partial y}{\partial \theta} \end{vmatrix} = \begin{vmatrix} \cos\theta & -r\sin\theta \\[2mm] \sin\theta & r\cos\theta \end{vmatrix}$$

$$= r\cos^2\theta + r\sin^2\theta = r$$

Section 22.4, pp. 738-739

1. $m = \displaystyle\int_0^1 \int_0^{x^2} xy\,dy\,dx = \int_0^1 \frac{xy^2}{2}\Big|_0^{x^2} dx = \int_0^1 \frac{x^5}{2}dx = \frac{x^6}{12}\Big|_0^1 = \frac{1}{12}$

$M_x = \displaystyle\int_0^1 \int_0^{x^2} xy^2\,dy\,dx = \int_0^1 \frac{xy^3}{3}\Big|_0^{x^2} dx = \int_0^1 \frac{x^7}{3}dx = \frac{x^8}{24}\Big|_0^1 = \frac{1}{24}$

$M_y = \displaystyle\int_0^1 \int_0^{x^2} x^2 y\,dy\,dx = \int_0^1 \frac{x^2 y^2}{2}\Big|_0^{x^2} dx = \int_0^1 \frac{x^6}{2}dx = \frac{x^7}{14}\Big|_0^1 = \frac{1}{14}$

$\bar{x} = \dfrac{M_y}{m} = \dfrac{1/14}{1/12} = \dfrac{6}{7}$, $\bar{y} = \dfrac{M_x}{m} = \dfrac{1/24}{1/12} = \dfrac{1}{2}$

5. $m = \displaystyle\int_0^1 \int_0^{x^3} (2 - x - y)\,dy\,dx = \int_0^1 (2y - xy - \frac{y^2}{2})\Big|_0^{x^3} dx$

$\quad = \displaystyle\int_0^1 (2x^3 - x^4 - \frac{x^6}{2})\,dx = (\frac{x^4}{2} - \frac{x^5}{5} - \frac{x^7}{14})\Big|_0^1 = \frac{1}{2} - \frac{1}{5} - \frac{1}{14} = \frac{8}{35}$

$M_x = \displaystyle\int_0^1 \int_0^{x^3} y(2 - x - y)\,dy\,dx = \int_0^1 (y^2 - \frac{xy^2}{2} - \frac{y^3}{3})\Big|_0^{x^3} dx$

$\quad = \displaystyle\int_0^1 (x^6 - \frac{x^7}{2} - \frac{x^9}{3})\,dx = (\frac{x^7}{7} - \frac{x^8}{16} - \frac{x^{10}}{30})\Big|_0^1$

$\quad = \dfrac{1}{7} - \dfrac{1}{16} - \dfrac{1}{30} = \dfrac{79}{1680}$

$M_y = \displaystyle\int_0^1 \int_0^{x^3} x(2 - x - y)\,dy\,dx = \int_0^1 (2xy - x^2 y - \frac{xy^2}{2})\Big|_0^{x^3} dx$

$\quad = \displaystyle\int_0^1 (2x^4 - x^5 - \frac{x^7}{2})\,dx = (\frac{2x^5}{5} - \frac{x^6}{6} - \frac{x^8}{16})\Big|_0^1$

$\quad = \dfrac{2}{5} - \dfrac{1}{6} - \dfrac{1}{16} = \dfrac{41}{240}$

$\bar{x} = \dfrac{M_y}{m} = \dfrac{41/240}{8/35} = \dfrac{287}{384}$, $\bar{y} = \dfrac{M_x}{m} = \dfrac{79/1680}{8/35} = \dfrac{79}{384}$

9. $m = \int_0^1 \int_0^{x^2} xy\,dy\,dx = \int_0^1 \frac{xy^2}{2}\Big|_0^{x^2} dx = \int_0^1 \frac{x^5}{2}dx$

$\quad = \frac{x^6}{12}\Big|_0^1 = \frac{1}{12}$

$I_y = \int_0^1 \int_0^{x^2} x^3 y\,dy\,dx = \int_0^1 \frac{x^3 y^2}{2}\Big|_0^{x^2} dx$

$\quad = \int_0^1 \frac{x^7}{2}dx = \frac{x^8}{16}\Big|_0^1 = \frac{1}{16}$

$R = \sqrt{\frac{I}{m}} = \sqrt{\frac{1/16}{1/12}} = \frac{\sqrt{3}}{2}$

13. $m = \int_0^1 \int_0^{x^3} (x+y)\,dy\,dx = \int_0^1 (xy + \frac{y^2}{2})\Big|_0^{x^3} dx$

$\quad = \int_0^1 (x^4 + \frac{x^6}{2})\,dx = (\frac{x^5}{5} + \frac{x^7}{14})\Big|_0^1 = \frac{1}{5} + \frac{1}{14} = \frac{19}{70}$

$I_y = \int_0^1 \int_0^{x^3} x^2(x+y)\,dy\,dx = \int_0^1 (x^3 y + \frac{x^2 y^2}{2})\Big|_0^{x^3} dx$

$\quad = \int_0^1 (x^6 + \frac{x^8}{2})\,dx = (\frac{x^7}{7} + \frac{x^9}{18})\Big|_0^1 = \frac{1}{7} + \frac{1}{18} = \frac{25}{126}$

$R = \sqrt{\frac{I}{m}} = \sqrt{\frac{25/126}{19/70}} = \frac{5\sqrt{5}}{3\sqrt{19}}$

17. $V = \int_{-1}^1 \int_{-\sqrt{1-x^2}}^{\sqrt{1-x^2}} (2-x-y)\,dy\,dx$

$\quad = \int_{-1}^1 [(2-x)y - \frac{y^2}{2}]\Big|_{-\sqrt{1-x^2}}^{\sqrt{1-x^2}} dx = \int_{-1}^1 2(2-x)\sqrt{1-x^2}\,dx$

$\quad = \int_{-1}^1 4\sqrt{1-x^2}\,dx + \int_{-1}^1 -2x\sqrt{1-x^2}\,dx$

$\qquad x = \sin\theta, \quad dx = \cos\theta\,d\theta$

$\quad = \int_{-\pi/2}^{\pi/2} 4\cos^2\theta\,d\theta + \frac{2}{3}(1-x^2)^{3/2}\Big|_{-1}^1$

$\quad = 2\int_{-\pi/2}^{\pi/2} (1 + \cos 2\theta)\,d\theta = (2\theta + \sin 2\theta)\Big|_{-\pi/2}^{\pi/2} = 2\pi$

$M_{yz} = \int_{-1}^1 \int_{-\sqrt{1-x^2}}^{\sqrt{1-x^2}} x(2-x-y)\,dy\,dx$

$\quad = \int_{-1}^1 2x(2-x)\sqrt{1-x^2}\,dx$

$\quad = \int_{-1}^1 4x\sqrt{1-x^2}\,dx - \int_{-1}^1 2x^2\sqrt{1-x^2}\,dx$

$$x = \sin \theta , \quad dx = \cos \theta \, d\theta$$

$$= - \frac{4}{3}(1 - x^2)^{2/3} \Big|_{-1}^{1} - \int_{-\pi/2}^{\pi/2} 2 \sin^2\theta \cos^2\theta \, d\theta$$

$$= - \int_{-\pi/2}^{\pi/2} 2 \frac{1 - \cos 2\theta}{2} \frac{1 + \cos 2\theta}{2} d\theta$$

$$= - \frac{1}{2} \int_{-\pi/2}^{\pi/2} (1 - \cos^2 2\theta) \, d\theta$$

$$= - \frac{1}{2} \int_{-\pi/2}^{\pi/2} (1 - \frac{1 + \cos 4\theta}{2}) \, d\theta$$

$$= - \frac{1}{4} \int_{-\pi/2}^{\pi/2} (1 - \cos 4\theta) \, d\theta = - \frac{1}{4}(\theta - \frac{1}{4} \sin 4\theta) \Big|_{-\pi/2}^{\pi/2} = - \frac{\pi}{4}$$

$$M_{xz} = \int_{-1}^{1} \int_{-\sqrt{1-x^2}}^{\sqrt{1-x^2}} y(2 - x - y) \, dy \, dx$$

$$= \int_{-1}^{1} \int_{-\sqrt{1-x^2}}^{\sqrt{1-x^2}} [(2 - x)y - y^2] \, dy \, dx$$

$$= \int_{-1}^{1} [(2 - x)\frac{y^2}{2} - \frac{y^3}{3}] \Big|_{-\sqrt{1-x^2}}^{\sqrt{1-x^2}} \, dx = \int_{-1}^{1} - \frac{2}{3}(1 - x^2)^{3/2} dx$$

$$x = \sin \theta , \quad dx = \cos \theta \, d\theta$$

$$= \int_{-\pi/2}^{\pi/2} - \frac{2}{3} \cos^4\theta \, d\theta = \int_{-\pi/2}^{\pi/2} - \frac{1}{6}(1 + \cos 2\theta)^2 d\theta$$

$$= - \frac{1}{6} \int_{-\pi/2}^{\pi/2} (1 + 2 \cos 2\theta + \cos^2 2\theta) \, d\theta$$

$$= - \frac{1}{6} \int_{-\pi/2}^{\pi/2} (1 + 2 \cos 2\theta + \frac{1 + \cos 4\theta}{2}) \, d\theta$$

$$= - \frac{1}{12} \int_{-\pi/2}^{\pi/2} (3 + 4 \cos 2\theta + \cos 4\theta) \, d\theta$$

$$= - \frac{1}{12}(3\theta + 2 \sin 2\theta + \frac{1}{4} \sin 4\theta) \Big|_{-\pi/2}^{\pi/2}$$

$$= - \frac{1}{12} \cdot 3\pi = - \frac{\pi}{4}$$

$$M_{xy} = \int_{-1}^{1} \int_{-\sqrt{1-x^2}}^{\sqrt{1-x^2}} \frac{1}{2}(2 - x - y)^2 dy \, dx$$

$$= \frac{1}{2} \int_{-1}^{1} \int_{-\sqrt{1-x^2}}^{\sqrt{1-x^2}} (4 + x^2 + y^2 - 4x - 4y + 2xy) \, dy \, dx$$

$$= \frac{1}{2} \int_{-1}^{1} (4y + x^2 y + \frac{y^3}{3} - 4xy - 2y^2 + xy^2) \Big|_{-\sqrt{1-x^2}}^{\sqrt{1-x^2}} \, dx$$

$$= \frac{1}{2} \int_{-1}^{1} [2(x^2 - 4x + 4)\sqrt{1 - x^2} + \frac{(1 + x^2)^{3/2}}{3}] dx$$

$$x = \sin \theta, \quad dx = \cos \theta \, d\theta$$

$$= \frac{1}{2} \int_{-\pi/2}^{\pi/2} [2(\sin^2\theta - 4 \sin \theta + 4)\cos \theta + \frac{1}{3} \cos^3\theta]\cos \theta \, d\theta$$

$$= \frac{1}{2} \int_{-\pi/2}^{\pi/2} (2 \sin^2\theta \cos^2\theta - 8 \sin \theta \cos^2\theta + 8 \cos^2\theta$$

$$+ \frac{1}{3} \cos^4\theta) \, d\theta$$

$$= \frac{1}{2} \int_{-\pi/2}^{\pi/2} [\frac{1 - \cos^2 2\theta}{2} - 8 \sin \theta \cos^2\theta + 4(1 + \cos 2\theta)$$

$$+ \frac{1}{12}(1 + \cos 2\theta)^2] \, d\theta$$

$$= \frac{1}{2} \int_{-\pi/2}^{\pi/2} [\frac{55}{12} - \frac{1}{4}(1 + \cos 4\theta) - 8 \sin \theta \cos^2\theta + 4 \cos 2\theta$$

$$+ \frac{1}{6} \cos 2\theta + \frac{1}{24}(1 + \cos 4\theta)] \, d\theta$$

$$= \frac{1}{2}(\frac{55\theta}{12} - \frac{\theta}{4} - \frac{1}{16} \sin 4\theta + \frac{8}{3} \cos^3\theta + 2 \sin 2\theta + \frac{1}{12} \sin 2\theta$$

$$+ \frac{\theta}{24} + \frac{1}{96} \sin 4\theta) \Big|_{-\pi/2}^{\pi/2}$$

$$= \frac{55\pi}{24} - \frac{\pi}{8} + \frac{\pi}{48} = \frac{35\pi}{16}$$

$$\bar{x} = \frac{M_{yz}}{V} = \frac{-\pi/4}{2\pi} = -\frac{1}{8}, \quad \bar{y} = \frac{M_{xz}}{V} = \frac{-\pi/4}{2\pi} = -\frac{1}{8}$$

$$\bar{z} = \frac{M_{xy}}{V} = \frac{35\pi/16}{2\pi} = \frac{35}{32}$$

21.
$$V = 2\int_0^1 \int_0^{1-y^2} (1 - y^2) \, dz \, dy = 2\int_0^1 (1 - y^2)^2 \, dy$$

$$= 2\int_0^1 (1 - 2y^2 + y^4) \, dy = 2(y - \frac{2y^3}{3} + \frac{y^5}{5}) \Big|_0^1$$

$$= 2(1 - \frac{2}{3} + \frac{1}{5}) = \frac{16}{15}$$

$$I_x = 2\int_0^1 \int_0^{1-y^2} (y^2 + z^2)(1 - y^2) \, dz \, dy$$

$$= 2\int_0^1 \int_0^{1-y^2} (y^2 + z^2 - y^2 - y^2 z^2) \, dz \, dy$$

$$= 2\int_0^1 (y^2 z + \frac{z^3}{3} - y^4 z - \frac{y^2 z^3}{3}) \Big|_0^{1-y^2} \, dy$$

$$= 2\int_0^1 [y^2(1-y^2) + \frac{(1-y^2)^3}{3} - y^4(1-y^2) - \frac{y^2(1-y^2)^3}{3}] \, dy$$

308　　　_Section 22.4_

$$= 2\int_0^1 (\frac{1}{3} - \frac{y^2}{3} - \frac{y^6}{3} + \frac{y^8}{3})\,dy = \frac{2}{3}(y - \frac{y^3}{3} - \frac{y^7}{7} + \frac{y^9}{9})\Big|_0^1$$

$$= \frac{2}{3}(1 - \frac{1}{3} - \frac{1}{7} + \frac{1}{9}) = \frac{80}{189}$$

$$R = \sqrt{\frac{I}{V}} = \sqrt{\frac{80/189}{16/15}} = \frac{5}{3\sqrt{7}}$$

Section 22.5, pp. 744-745

1. $\displaystyle\int_0^2 \int_0^x \int_0^{x+y} x\,dz\,dy\,dx = \int_0^2 \int_0^x xz\Big|_0^{x+y}\,dy\ dx$

$$= \int_0^2 \int_0^x (x^2 + xy)\,dy\ dx = \int_0^2 (x^2 y + \frac{xy^2}{2})\Big|_0^x\,dx$$

$$= \int_0^2 (x^3 + \frac{x^3}{2})\,dx = \int_0^2 \frac{3x^3}{2}\,dx = \frac{3x^4}{8}\Big|_0^2 = 6$$

5. $\displaystyle\int_0^1 \int_{-x}^x \int_{-x}^x (x + z)\,dy\ dz\ dx = \int_0^1 \int_{-x}^x y(x+z)\Big|_{-x}^x dz\ dx$

$$= \int_0^1 \int_{-x}^x 2x(x+z)\,dz\ dx = \int_0^1 2x(xz + \frac{z^2}{2})\Big|_{-x}^x dx$$

$$= \int_0^1 2x(x^2 + \frac{x^2}{2} + x^2 - \frac{x^2}{2})\,dx = \int_0^1 4x^3\,dx = x^4\Big|_0^1 = 1$$

9. $\displaystyle V = \int_0^3 \int_{2x-6}^0 \int_0^{(6-2x+y)/3} dz\,dy\,dx = \int_0^3 \int_{2x-6}^0 \frac{6-2x+y}{3}\,dy\,dx$

$$= \frac{1}{3}\int_0^3 [(6-2x)y + \frac{y^2}{2}]\Big|_{2x-6}^0\,dx = \frac{1}{3}\int_0^3 [(2x-6)^2 - \frac{(2x-6)^2}{2}]\,dx$$

$$= \frac{1}{6}\int_0^3 (2x-6)^2\,dx = \frac{1}{12}\frac{(2x-6)^3}{3}\Big|_0^3 = 6$$

13. $\displaystyle V = 8\int_0^2 \int_0^{\sqrt{4-x^2}} \int_0^{\sqrt{4-x^2-y^2}} dz\,dy\,dx$

$$= 8\int_0^2 \int_0^{\sqrt{4-x^2}} \sqrt{4-x^2-y^2}\,dy\ dx$$

$$y = \sqrt{4 - x^2} \sin \theta, \quad dy = \sqrt{4 - x^2} \cos \theta \, d\theta$$

$$= 8 \int_0^2 \int_0^{\pi/2} \sqrt{(4 - x^2) - (4 - x^2)\sin^2\theta} \, \sqrt{4 - x^2} \, \cos \theta \, d\theta \, dx$$

$$= 8 \int_0^2 \int_0^{\pi/2} (4 - x^2) \cos^2\theta \, d\theta \, dx$$

$$= 4 \int_0^2 (4 - x^2) \int_0^{\pi/2} (1 + \cos 2\theta) \, d\theta \, dx$$

$$= 4 \int_0^2 (4 - x^2)(\theta + \tfrac{1}{2} \sin 2\theta) \Big|_0^{\pi/2} dx = 4 \int_0^2 (4 - x^2) \tfrac{\pi}{2} dx$$

$$= 2\pi (4x - \tfrac{x^3}{3}) \Big|_0^2 = 2\pi (8 - \tfrac{8}{3}) = \tfrac{32\pi}{3}$$

17. $\quad m = \int_0^1 \int_0^1 \int_0^1 xyz \, dz \, dy \, dx = \int_0^1 \int_0^1 \frac{xyz^2}{2} \Big|_0^1 dy \, dx$

$$= \int_0^1 \int_0^1 \frac{xy}{2} dy \, dx = \int_0^1 \frac{xy^2}{4} \Big|_0^1 dx = \int_0^1 \frac{x}{4} dx = \frac{x^2}{8} \Big|_0^1 = \frac{1}{8}$$

21. $\quad m = 8 \int_0^1 \int_0^{\sqrt{1-x^2}} \int_0^{\sqrt{1-y^2}} y \, dz \, dy \, dx = 8 \int_0^1 \int_0^{\sqrt{1-x^2}} y\sqrt{1 - y^2} \, dy \, dx$

$$= -4 \int_0^1 \frac{2}{3}(1 - y^2)^{3/2} \Big|_0^{\sqrt{1-x^2}} dx = -\frac{8}{3} \int_0^1 (x^3 - 1) \, dx$$

$$= -\frac{8}{3}(\frac{x^4}{4} - x) \Big|_0^1 = -\frac{8}{3}(\frac{1}{4} - 1) = 2$$

Section 22.6, p. 748

1. $\quad m = \frac{1}{8}$ (by Problem 17, Section 22.5)

$$M_{yz} = \int_0^1 \int_0^1 \int_0^1 x^2 yz \, dz \, dy \, dx = \int_0^1 \int_0^1 \frac{x^2 yz^2}{2} \Big|_0^1 dy \, dx$$

$$= \int_0^1 \int_0^1 \frac{x^2 y}{2} dy \, dx = \int_0^1 \frac{x^2 y^2}{4} \Big|_0^1 dx$$

$$= \int_0^1 \frac{x^2}{4} dx = \frac{x^3}{12} \Big|_0^1 = \frac{1}{12}$$

$$M_{xz} = \int_0^1 \int_0^1 \int_0^1 xy^2 z \, dz \, dy \, dx = \int_0^1 \int_0^1 \frac{xy^2 z^2}{2} \Big|_0^1 dy \, dx$$

$$= \int_0^1 \int_0^1 \frac{xy^2}{2} dy \, dx = \int_0^1 \frac{xy^3}{6} \Big|_0^1 dx$$

$$= \int_0^1 \frac{x}{6}dx = \frac{x^2}{12}\Big|_0^1 = \frac{1}{12}$$

$$M_{xy} = \int_0^1 \int_0^1 \int_0^1 xyz^2 dz\, dy\, dx = \int_0^1 \int_0^1 \frac{xyz^3}{3}\Big|_0^1 dy\, dx$$

$$= \int_0^1 \int_0^1 \frac{xy}{3}dy\, dx = \int_0^1 \frac{xy^2}{6}\Big|_0^1 dx = \int_0^1 \frac{x}{6}dx$$

$$= \frac{x^2}{12}\Big|_0^1 = \frac{1}{12}$$

$$\bar{x} = \frac{M_{yz}}{m} = \frac{1/12}{1/8} = \frac{2}{3}\ , \quad \bar{y} = \frac{M_{xz}}{W} = \frac{1/12}{1/8} = \frac{2}{3}\ ,$$

$$\bar{z} = \frac{M_{xz}}{m} = \frac{1/12}{1/8} = \frac{2}{3}$$

5. $m = 4\int_0^1 \int_0^{\sqrt{1-x^2}} \int_0^1 xyz\, dz\, dy\, dx = 4\int_0^1 \int_0^{\sqrt{1-x^2}} \frac{xyz^2}{2}\Big|_0^1 dy\, dx$

$$= 2\int_0^1 \int_0^{\sqrt{1-x^2}} xy\, dy\, dx = \int_0^1 xy^2\Big|_0^{\sqrt{1-x^2}} dx = \int_0^1 x(1-x^2)\, dx$$

$$= (\frac{x^2}{2} - \frac{x^4}{4})\Big|_0^1 = \frac{1}{2} - \frac{1}{4} = \frac{1}{4}$$

$$M_{xy} = 4\int_0^1 \int_0^{\sqrt{1-x^2}} \int_0^1 xyz^2 dz\, dx = 4\int_0^1 \int_0^{\sqrt{1-x^2}} \frac{xyz^3}{3}\Big|_0^1 dy\, dx$$

$$= \frac{4}{3}\int_0^1 \int_0^{\sqrt{1-x^2}} xy\, dy\, dx = \frac{2}{3}\int_0^1 xy^2\Big|_0^{\sqrt{1-x^2}} dx$$

$$= \frac{2}{3}\int_0^1 x(1-x^2)\, dx = \frac{2}{3}(\frac{x^2}{2} - \frac{x^4}{4})\Big|_0^1 = \frac{2}{3}(\frac{1}{2} - \frac{1}{4}) = \frac{1}{6}$$

$$\bar{x} = \bar{y} = 0 \ \text{(by symmetry)}, \quad \bar{z} = \frac{M_{xy}}{m} = \frac{1/6}{1/4} = \frac{2}{3}$$

9. $m = \frac{1}{8}$ (from Problem 1)

$$I_z = \int_0^1 \int_0^1 \int_0^1 xyz(x^2 + y^2)\, dz\, dy\, dx$$

$$= \frac{1}{2}\int_0^1 \int_0^1 xyz^2(x^2 + y^2)\Big|_0^1 dy\, dx$$

$$= \frac{1}{2}\int_0^1 \int_0^1 xy(x^2 + y^2)\, dy\, dx = \frac{1}{2}\int_0^1 (\frac{x^3 y^2}{2} + \frac{xy^4}{4})\Big|_0^1 dx$$

$$= \frac{1}{8}\int_0^1 (2x^3 + x)\, dx = \frac{1}{8}(\frac{x^4}{2} + \frac{x^2}{2})\Big|_0^1 = \frac{1}{8}$$

$$R = \sqrt{\frac{I}{m}} = \sqrt{\frac{1/8}{1/8}} = 1$$

13. $m = \int_0^1 \int_0^{x^2} \int_0^x 4 \, dz \, dy \, dx = \int_0^1 \int_0^{x^2} 4z \Big|_0^x dy \, dx$

$= \int_0^1 \int_0^{x^2} 4x \, dy \, dx = \int_0^1 4xy \Big|_0^{x^2} dx = \int_0^1 4x^3 dx = x^4 \Big|_0^1 = 1$

$I_x = \int_0^1 \int_0^{x^2} \int_0^x 4(y^2 + z^2) \, dz \, dy \, dx$

$= \int_0^1 \int_0^{x^2} 4(y^2 z + \frac{z^3}{3}) \Big|_0^x dy \, dx = \int_0^1 \int_0^{x^2} 4(xy^2 + \frac{x^3}{3}) \, dy \, dx$

$= \int_0^1 4(\frac{xy^3}{3} + \frac{x^3 y}{3}) \Big|_0^{x^2} dx = \frac{4}{3} \int_0^1 (x^7 + x^5) \, dx$

$= \frac{4}{3}(\frac{x^8}{8} + \frac{x^6}{6}) \Big|_0^1 = \frac{4}{3}(\frac{1}{8} + \frac{1}{6}) = \frac{7}{18}$

$R = \sqrt{\frac{I}{m}} = \sqrt{\frac{7/18}{1}} = \frac{\sqrt{7}}{3\sqrt{2}} = \frac{\sqrt{14}}{6}$

Section 22.7, pp. 754-755

1. $m = 4k\int_0^{\pi/2} \int_0^1 \int_0^4 r^2 dz \, dr \, d\theta$

$= 4k\int_0^{\pi/2} \int_0^1 4r^2 dr \, d\theta$

$= 4k\int_0^{\pi/2} \frac{4r^3}{3} \Big|_0^1 d\theta$

$= 4k\int_0^{\pi/2} \frac{4}{3} d\theta$

$= \frac{16k}{3} \cdot \frac{\pi}{2} = \frac{8\pi k}{3}$

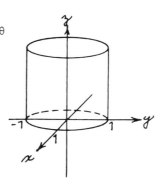

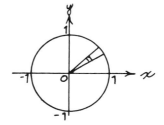

5. Cylinder: $r = 1$

Cone: $z = r$

$m = 4\int_0^{\pi/2} \int_0^1 \int_0^r zr \, dz \, dr \, d\theta = 4\int_0^{\pi/2} \int_0^1 \frac{z^2 r}{2} \Big|_0^r dr \, d\theta$

$= 2\int_0^{\pi/2} \int_0^1 r^3 dr \, d\theta = 2\int_0^{\pi/r} \frac{r^4}{4} \Big|_0^1 d\theta = \frac{1}{2}\int_0^{\pi/2} d\theta = \frac{\pi}{4}$

$$M_{xy} = 4\int_0^{\pi/2} \int_0^1 \int_0^r z^2 r \, dz \, dr \, d\theta$$

$$= 4\int_0^{\pi/2} \int_0^1 \left.\frac{z^3 r}{3}\right|_0^r dr \, d\theta$$

$$= \frac{4}{3}\int_0^{\pi/2} \int_0^1 r^4 dr \, d\theta$$

$$= \frac{4}{3}\int_0^{\pi/2} \frac{1}{5} d\theta$$

$$= \frac{4}{15} \cdot \frac{\pi}{2} = \frac{2\pi}{15}$$

$\bar{x} = \bar{y} = 0$ (by symmetry),

$$\bar{z} = \frac{M_{xy}}{m} = \frac{2\pi/15}{\pi/4} = \frac{8}{15}$$

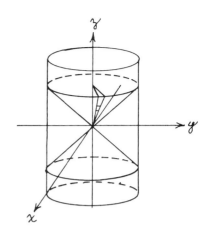

9. Cylinder: $r = 1$

Cone: $z = 2r$

$$I_z = 8\int_0^{\pi/2} \int_0^1 \int_0^{2r} r^2 z \cdot r \, dz \, dr \, d\theta$$

$$= 8\int_0^{\pi/2} \int_0^1 \left.\frac{r^3 z^2}{2}\right|_0^{2r} dr \, d\theta$$

$$= 8\int_0^{\pi/2} \int_0^1 2r^5 dr \, d\theta = 8\int_0^{\pi/2} \left.\frac{r^6}{3}\right|_0^1 d\theta$$

$$= \frac{8}{3}\int_0^{\pi/2} d\theta = \frac{8}{3} \cdot \frac{\pi}{2} = \frac{4\pi}{3}$$

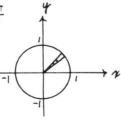

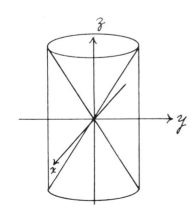

13. $m = 8 \int_0^{\pi/2} \int_0^{\pi/2} \int_r^R \rho^3 \sin\varphi \, d\rho \, d\theta \, d\varphi$

$\displaystyle = 8 \int_0^{\pi/2} \int_0^{\pi/2} \frac{\rho^4}{4} \sin\varphi \Big|_r^R \, d\theta \, d\varphi$

$\displaystyle = 2 \int_0^{\pi/2} \int_0^{\pi/2} (R^4 - r^4) \sin\varphi \, d\theta \, d\varphi$

$\displaystyle = 2 \int_0^{\pi/2} \frac{\pi}{2}(R^4 - r^4) \sin\varphi \, d\varphi$

$\displaystyle = -\pi(R^4 - r^4)\cos\varphi \Big|_0^{\pi/2} = \pi(R^4 - r^4)$

17. $m = 4 \int_0^{\pi/2} \int_0^{\pi/2} \int_1^2 \rho \cos\varphi \, \rho^2 \sin\varphi \, d\rho \, d\theta \, d\varphi$

$\displaystyle = 4 \int_0^{\pi/2} \int_0^{\pi/2} \frac{\rho^4}{4} \sin\varphi \cos\varphi \Big|_1^2 \, d\theta \, d\varphi$

$\displaystyle = \int_0^{\pi/2} \int_0^{\pi/2} 15 \sin\varphi \cos\varphi \, d\theta \, d\varphi$

$\displaystyle = \int_0^{\pi/2} \frac{15\pi}{2} \sin\varphi \cos\varphi \, d\varphi = \frac{15\pi}{4} \sin^2\varphi \Big|_0^{\pi/2} = \frac{15\pi}{4}$

$\displaystyle M_{xy} = 4 \int_0^{\pi/2} \int_0^{\pi/2} \int_1^2 \rho \cos\varphi \, \rho \cos\varphi \, \rho^2 \sin\varphi \, d\rho \, d\theta \, d\varphi$

$\displaystyle = 4 \int_0^{\pi/2} \int_0^{\pi/2} \frac{\rho^5}{5} \cos^2\varphi \sin\varphi \Big|_1^2 \, d\theta \, d\varphi$

$\displaystyle = \frac{4}{5} \int_0^{\pi/2} \int_0^{\pi/2} 31 \cos^2\varphi \sin\varphi \, d\theta \, d\varphi$

$\displaystyle = \frac{124}{5} \int_0^{\pi/2} \frac{\pi}{2} \cos^2\varphi \sin\varphi \, d\varphi$

$\displaystyle = -\frac{62\pi}{5} \frac{\cos^3\varphi}{3} \Big|_0^{\pi/2} = \frac{62\pi}{15}$

$\displaystyle \bar{x} = 0, \quad \bar{y} = 0, \quad \bar{z} = \frac{M_{xy}}{m} = \frac{62\pi/15}{15\pi/4} = \frac{248}{225}$

Review 22, pp. 755-756

1. $\displaystyle \int_0^3 \int_0^{4x-x^2} (x+y) \, dy \, dx$

$\displaystyle = \int_0^3 (xy + \frac{y^2}{2}) \Big|_0^{4x-x^2} \, dx$

$\displaystyle = \int_0^3 [x(4x - x^2) + \frac{1}{2}(4x - x^2)^2] \, dx$

$$= \int_0^3 [x(4x - x^2) + \frac{1}{2}(4x - x^2)^2]dx$$

$$= \int_0^3 (12x^2 - 5x^3 + \frac{x^4}{2})dx$$

$$= (4x^3 - \frac{5x^4}{4} + \frac{x^5}{10})\Big|_0^3$$

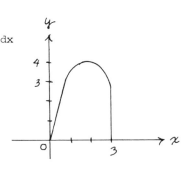

5. $$\int\int_R (x^2 - y)\,dA = \int_0^2 \int_y^{4-y} (x^2 - y)\,dx\,dy$$

$$= \int_0^2 (\frac{x^3}{3} - xy)\Big|_y^{4-y} dy$$

$$= \int_0^2 [\frac{(4-y)^3}{3} - (4-y)y - \frac{y^3}{3} + y^2]dy$$

$$= (-\frac{(4-y)^4}{12} - 2y^2 + \frac{y^3}{3} - \frac{y^4}{12} + \frac{y^3}{3})\Big|_0^2$$

$$= -\frac{16}{12} - 8 + \frac{8}{3} - \frac{4}{3} + \frac{8}{3} + \frac{256}{12}$$

$$= 16$$

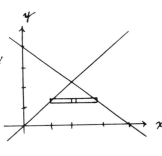

9. $$m = \int_0^{4/3} \int_y^{4-2y} (x + y)\,dx\,dy$$

$$= \int_0^{4/3} (\frac{x^2}{2} + xy)\Big|_y^{4-2y} dy$$

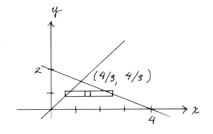

$$= \int_0^{4/3} [\frac{4(2-y)^2}{2} + (4-2y)y - \frac{3y^2}{2}]dy$$

$$= \int_0^{4/3} (8 - 4y - \frac{3y^2}{2})\,dy = (8y - 2y^2 - \frac{y^3}{2})\Big|_0^{4/3}$$

$$= \frac{32}{3} - \frac{32}{9} - \frac{32}{27} = \frac{160}{27}$$

13. $\quad m = 4\int_0^{\pi/2} \int_a^b r^2 dr \, d\theta = 4\int_0^{\pi/2} \frac{r^3}{3}\Big|_a^b d\theta$

$\qquad = \frac{4}{3}\int_0^{\pi/2}(b^3 - a^3)\,d\theta = \frac{4}{3}(b^3 - a^3)\frac{\pi}{2} = \frac{2\pi}{3}(b^3 - a^3)$

17. $\quad V = 2\int_0^2 \int_0^{4-x^2}(4 - z - x^2)\,dz\,dx$

$\qquad = 2\int_0^2 \Big[(4 - x^2)z - \frac{z^2}{2}\Big]\Big|_0^{4-x^2} dx$

$\qquad = 2\int_0^2 \frac{(4 - x^2)^2}{2}dx$

$\qquad = \int_0^2 (16 - 8x^2 + x^4)\,dx$

$\qquad = (16x - \frac{8x^3}{3} + \frac{x^5}{5})\Big|_0^2$

$\qquad = 32 - \frac{64}{3} + \frac{32}{5} = \frac{256}{15}$

$I_y = 2\int_0^2 \int_0^{4-x^2}(x^2 + z^2)(4 - z - x^2)\,dz\,dx$

$\qquad = 2\int_0^2 \int_0^{4-x^2}[(4 - x^2)x^2 - zx^2 + (4 - x^2)z^2 - z^3]\,dz\,dx$

$\qquad = 2\int_0^2 \Big[(4 - x^2)x^2 z - \frac{x^2 z^2}{2} + \frac{(4 - x^2)z^3}{3} - \frac{z^4}{4}\Big]\Big|_0^{4-x^2} dx$

$\qquad = 2\int_0^2 \Big[x^2(4 - x^2)^2 - \frac{x^2}{2}(4 - x^2)^2 + \frac{(4 - x^2)^4}{3} - \frac{(4-x^2)^4}{4}\Big]dx$

$\qquad = \int_0^2 \Big[x^2(4 - x^2)^2 + \frac{(4 - x^2)^4}{6}\Big]dx$

$\qquad = \frac{1}{6}\int_0^2 (x^8 - 10x^6 + 48x^4 - 160x^2 + 256)\,dx$

$\qquad = \frac{1}{6}(\frac{x^9}{9} - \frac{10x^7}{7} + \frac{48x^5}{5} - \frac{160x^3}{3} + 256x)\Big|_0^2$

$\qquad = \frac{1}{6}(\frac{512}{9} - \frac{1280}{7} + \frac{1536}{5} - \frac{1280}{3} + 512)$

$\qquad = \frac{41,984}{945}$

$R = \sqrt{\frac{I}{V}} = \sqrt{\frac{41,984/945}{256/15}} = \frac{2}{3}\sqrt{\frac{41}{7}}$

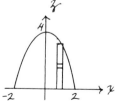

21. $m = 2\int_0^2 \int_{x^2}^4 \int_{\frac{y-4}{4}}^{\frac{4-y}{2}} (1 + y + z^2)\, dz\ dy\ dx$

$= 2\int_0^2 \int_{x^2}^4 \left[(1+y)z + \frac{z^3}{3}\right]\Big|_{\frac{y-4}{4}}^{\frac{4-y}{2}} dy\ dx$

$= 2\int_0^2 \int_{x^2}^4 \left[\frac{(1+y)(4-y)}{2} + \frac{(4-y)^3}{24} - \frac{(1+y)(y-4)}{4} - \frac{(y-4)^3}{192}\right] dy\ dx$

$= 2\int_0^2 \int_{x^2}^4 \left[\frac{3(4+3y-y^2)}{4} + \frac{3(4-y)^3}{64}\right] dy\ dx$

$= 2\int_0^2 \left(3y + \frac{9y^2}{8} - \frac{y^3}{4} - \frac{3(4-y)^4}{256}\right)\Big|_{x^2}^4 dx$

$= 2\int_0^2 \left(12 + 18 - 16 - 3x^2 - \frac{9x^4}{8} + \frac{x^6}{4} + \frac{3(4-x^2)^4}{256}\right) dx$

$= 2\int_0^2 \left(17 - 6x^2 + \frac{x^6}{16} + \frac{3x^8}{256}\right) dx$

$= \left(17x - 2x^3 + \frac{x^7}{112} + \frac{x^9}{768}\right)\Big|_0^2 = 2\left(34 - 16 + \frac{8}{7} + \frac{2}{3}\right) = \frac{832}{21}$

25. $m = \int_0^4 \int_0^{4-x} \int_0^{4-x-y} (1 + y)\, dz\ dy\ dx$

$= \int_0^4 \int_0^{4-x} (1 + y)(4 - x - y)\, dy\ dx$

$= \int_0^4 \int_0^{4-x} (4 - x + 3y - xy - y^2)\, dy\ dx$

$= \int_0^4 \left[(4 - x)y + (3 - x)\frac{y^2}{2} - \frac{y^3}{3}\right]\Big|_0^{4-x} dx$

$= \int_0^4 \left[(4 - x)^2 + \frac{(3 - x)(4 - x)^2}{2} - \frac{(4-x)^3}{3}\right] dx$

$= \int_0^4 \frac{1}{6}(112 - 72x + 15x^2 - x^3)\, dx$

$$= \frac{1}{6}(112x - 36x^2 + 5x^3 - \frac{x^4}{4})\Big|_0^4 = \frac{1}{6}(448 - 576 + 320 - 64) = \frac{64}{3}$$

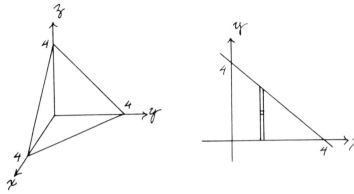

$$I_y = \int_0^4 \int_0^{4-x} \int_0^{4-x-y} (x^2 + z^2)(1 + y)\,dz\,dy\,dx$$

$$= \int_0^4 \int_0^{4-x} [x^2(1 + y)z + (1 + y)\frac{z^3}{3}]\Big|_0^{4-x-y}\,dy\,dx$$

$$= \int_0^4 \int_0^{4-x} [x^2(1 + y)(4 - x - y) + \frac{1}{3}(1 + y)(4 - x - y)^3]\,dy\,dx$$

$$= \frac{1}{3}\int_0^4 \int_0^{4-x} \{3x^2(4 - x) + (4 - x)^3 + [3x^2(4 - x) - 3x^2$$
$$- 3(4 - x)^2 + (4 - x)^3]y + [-3x^2 + 3(4 - x) - 3(4 - x)^2]y^2$$
$$+ [3(4 - x) - 1]y^3 - y^4\}\,dy\,dx$$

$$= \frac{1}{3}\int_0^4 \{[3x^2(4 - x) + (4 - x)^3]y + [3x^2(4 - x) - 3x^2$$
$$- 3(4 - x)^2 + (4 - x)^3]\frac{y^2}{2} + [-3x^2 + 3(4 - x)$$
$$- 3(4 - x)^2]\frac{y^3}{3} + [3(4 - x) - 1]\frac{y^4}{4} - \frac{y^5}{5}\}\Big|_0^{4-x}\,dx$$

$$= \frac{1}{3}\int_0^4 [\frac{1}{4}(4 - x)^4 + \frac{1}{20}(4 - x)^5 + 56x^2 - 36x^3 + \frac{15}{2}x^4 - \frac{1}{2}x^5]\,dx$$

$$= \frac{1}{3}[-\frac{(4 - x)^5}{20} - \frac{(4 - x)^6}{120} + \frac{56x^3}{3} - 9x^4 + \frac{3x^5}{2} - \frac{x^6}{12}]\Big|_0^4$$

$$= \frac{1}{3}[\frac{56 \cdot 64}{3} - 9 \cdot 256 + 3 \cdot 512 - \frac{1024}{3} + \frac{256}{5} + \frac{512}{15}] = \frac{512}{9}$$

$$R = \sqrt{\frac{I}{m}} = \sqrt{\frac{512/9}{64/3}} = \frac{2\sqrt{6}}{3}$$

318   *Review 22*

29. Sphere: $\rho = 2$, Cone: $\varphi = \pi/4$, Plane: $\varphi = \pi/2$

$$m = 4\int_{\pi/4}^{\pi/2} \int_0^{\pi/2} \int_0^2 (1+\rho \cos \varphi)\rho^2 \sin \varphi \, d\rho \, d\theta \, d\varphi$$

$$= 4\int_{\pi/4}^{\pi/2} \int_0^{\pi/2} \int_0^2 (\rho^2 \sin \varphi + \rho^3 \sin \varphi \cos \varphi) d\rho \, d\theta \, d\varphi$$

$$= 4\int_{\pi/4}^{\pi/2} \int_0^{\pi/2} (\frac{\rho^3 \sin \varphi}{3} + \frac{\rho^4 \sin\varphi \cos \varphi}{4}) \Big|_0^2 \, d\theta \, d\varphi$$

$$= 4\int_{\pi/4}^{\pi/2} \int_0^{\pi/2} (\frac{8}{3} \sin \varphi + 4 \sin \varphi \cos \varphi) d\theta \, d\varphi$$

$$= 2\pi \int_{\pi/4}^{\pi/2} (\frac{8}{3} \sin \varphi + 4 \sin \varphi \cos \varphi) d\varphi$$

$$= 2\pi (-\frac{8}{3} \cos \varphi + 2 \sin^2\varphi) \Big|_{\pi/4}^{\pi/2}$$

$$= 2\pi (0 + 2 + \frac{8}{3\sqrt{2}} - 1) = \frac{2\pi (3 + 4\sqrt{2})}{3}$$

Appendix

1. $2x + 5 < 3$

     $2x < -2$

     $x < -1$

5. $2x + 2 \leqq x - 4$

     $x \leqq -6$

9. $3x + 1 \geqq 2x + 2$

     $x \geqq 1$

13. $1 \leqq 2x + 1 \leqq 4$

     $0 \leqq 2x \leqq 3$

     $0 \leqq x \leqq \dfrac{3}{2}$

17. $2x - 1 \leqq x + 4$

     $x \leqq 5$

     $\dfrac{3}{2} \leqq x \leqq 5$

$x + 4 \leqq 3x + 1$

$-2x \leqq -3$

$x \geqq \dfrac{3}{2}$

21.   Case I: $x - 4 < 0$

     $2x + 1 \geqq x - 4$

     $x \geqq -5$ and $x < 4$

     $-5 \leqq x < 4$

Case II: $x - 4 > 0$

$2x + 1 \leqq x - 4$

$x \leqq -5$ and $x > 4$

No solution in this case

25. $\dfrac{a}{ab} < \dfrac{b}{ab}$, $\dfrac{1}{b} < \dfrac{1}{a}$, $\dfrac{1}{a} > \dfrac{1}{b}$

1. $|x + 2| = 5$

     $x + 2 = \pm 5$

     $x = -2 \pm 5 = 3, -7$

5. $|x - 2| = 0$

     $x - 2 = 0$

     $x = 2$

9. $\left|\dfrac{x-2}{x+1}\right| = 3$

   $\dfrac{x-2}{x+1} = \pm 3$

   $x - 2 = 3x + 3 \quad \text{or} \quad x - 2 = -3x - 3$

   $-2x = 5 \qquad\qquad\qquad 4x = -1$

   $x = -\dfrac{5}{2} \qquad\qquad\quad x = -\dfrac{1}{4}$

13. $\left|x + 3\right| < 1$

    $-1 < x + 3 < 1$

    $-4 < x < -2$

17. $\left|2x - 5\right| \leqq 4$

    $-4 \leqq 2x - 5 \leqq 4$

    $1 \leqq 2x \leqq 9$

    $\dfrac{1}{2} \leqq x \leqq \dfrac{9}{2}$

21. $\left|3x + 1\right| \geqq 4$

    $3x + 1 \leqq -4 \quad \text{or} \quad 3x + 1 \geqq 4$

    $3x \leqq -5 \qquad\qquad\qquad 3x \geqq 3$

    $x \leqq -\dfrac{5}{3} \qquad\qquad\qquad x \geqq 1$

25. $\left|f(x) - 3\right| = \left|x + 1 - 3\right| = \left|x - 2\right| < 1$

Section A.3, pp. 765-766

1. $\left|x^2 + 1\right| \leqq \left|x\right|^2 + 1 < 4 + 1 = 5$

5. $\left|x^4 + 4\right| \leqq \left|x\right|^4 + 4 < 5^4 + 4 = 629$

9. If $-5 < x < 1$, then $\left|x\right| < 5$

   $\left|x^2 - 3\right| \leqq \left|x\right|^2 + 3 < 5^2 + 3 = 28$

13. If $\left|x - 1\right| < \delta$ and $\delta \leqq 1$, then $\left|x - 1\right| < 1$

    $-1 < x - 1 < 1$, $\quad 2 < x + 2 < 4$, $\quad \left|x + 2\right| < 4$

    $\left|x^2 + x - 2\right| = \left|x + 2\right| \cdot \left|x - 1\right| < 4\delta$

17. If $\left|x - 1\right| < \delta$ and $\delta \leqq \dfrac{3}{4}$, then $\left|x - 1\right| < \dfrac{3}{4}$

    $-\dfrac{3}{4} < x - 1 < \dfrac{3}{4}$, $\quad \dfrac{1}{4} < x < \dfrac{7}{4}$

    $\left|x\right| > \dfrac{1}{4}$ or $\dfrac{1}{|x|} < 4$

    $\left|\dfrac{1}{x} - 1\right| = \left|\dfrac{1 - x}{x}\right| = \dfrac{1}{\left|x\right|} \cdot \left|x - 1\right| < 4\delta$